D0214899

Red Wine Color

ACS SYMPOSIUM SERIES **886**

Red Wine Color

Exploring the Mysteries

Andrew L. Waterhouse, Editor
University of California at Davis

James A. Kennedy, Editor
Oregon State University

Sponsored by the
ACS Division of Agricultural and Food Chemistry

American Chemical Society, Washington, DC

Library of Congress Cataloging-in-Publication Data

Red wine color : revealing the mysteries / Andrew L. Waterhouse, editor, James A. Kennedy, editor ; sponsored by the ACS Division of Agricultural and Food Chemistry.

p. cm.—(ACS symposium series ; 886)

Includes bibliographical references and index.

ISBN 0–8412–3900–2

1. Wine—Color—Congresses. 2. Wine and wine making—Analysis—Congresses.

I. Waterhouse, Andrew Leo. II. Kennedy, James A., 1962-. III. American Chemical Society. Division of Agricultural and Food Chemistry. IV. American Chemical Society. Meeting (225th : 2003 : New Orleans, La.) V. Series.

TP548.5.C64R43 2004
663′.2—dc22 2004043742

The paper used in this publication meets the minimum requirements of American National Standard for Information Sciences—Permanence of Paper for Printed Library Materials, ANSI Z39.48–1984.

Distributed by Oxford University Press

PRINTED IN THE UNITED STATES OF AMERICA

Foreword

The ACS Symposium Series was first published in 1974 to provide a mechanism for publishing symposia quickly in book form. The purpose of the series is to publish timely, comprehensive books developed from ACS sponsored symposia based on current scientific research. Occasionally, books are developed from symposia sponsored by other organizations when the topic is of keen interest to the chemistry audience.

Before agreeing to publish a book, the proposed table of contents is reviewed for appropriate and comprehensive coverage and for interest to the audience. Some papers may be excluded to better focus the book; others may be added to provide comprehensiveness. When appropriate, overview or introductory chapters are added. Drafts of chapters are peer-reviewed prior to final acceptance or rejection, and manuscripts are prepared in camera-ready format.

As a rule, only original research papers and original review papers are included in the volumes. Verbatim reproductions of previously published papers are not accepted.

ACS Books Department

Contents

Indexes

Preface

Symposia for the American Chemical Society are usually planned years in advance, so luck played a bit of a part in the timing of this symposium. Luck, because during the past two years, many advances in approaching the problem of the chemical composition of red wine color have occurred. As mentioned in Chapter 1, it was 70 years ago, after much progress in determining the chemical identity of grape pigments, that Professor Ribereau-Gayon proclaimed his frustration with making progress with wine pigments. Until just a few years ago the known chemistry of red wine color was just about at the same stage, although a number of theories had been advanced based on some known reactions of the red grape anthocyanin pigments. But recent advances in analytical instrumentation have advanced what can now be analyzed.

The use of high-performance liquid chromatography has been widely applied to analysis of wine composition since its early development in the 1970s. This tradition has now been expanded by the use of the efficient and capable electrospray ionization with mass spectral detectors to facilitate detection of high-mass compounds, and to measure their mass with some precision. This has truly opened the door to a new view of wine chemistry. When this capacity is combined with sample preparation techniques that use a different adsorption principle, the resulting chromatogram can have enough resolution to yield mass spectra of single peaks. Of course, the ability to obtain single mass traces gives even higher discrimination. Many of the reports herein take advantage of liquid chromotography/mass spectrometry (LC/MS) to report new substances as well as to advance new theories of how red wine pigments are formed.

However, the study of red wine color benefits from many other different approaches, including the use of radiolabeled tracers, time-of-flight matrix-assisted laser desorption-ionization mass spectrometry, spectral analysis that reveal copigmentation, predictive analysis using

Fourier transform infrared modeling, and protein precipitation assays. The approaches include not only direct analysis and monitoring during typical production but also intervention with possible reactants. The chapters not only include reports of current results but also considerable analysis of the extant literature. And we cannot forget a unique chapter on cranberry pigments, one that suggests the types of pigments observed in wine may be found in fresh plants—that idea sparked much debate indeed at the symposium.

It is important not only to identify the pigments but also to deduce their origins, because that information can be used to facilitate wine-making techniques that enhance red wine color. Because the depth of red wine color is generally accepted as a marker for red wine quality, it is likely that the conditions that can improve wine color may also lead to other chemical reactions that will enhance quality attributes in taste or aroma. Thus the knowledge gained by the pursuit of red wine color chemistry may well provide leads to improving wine quality in many ways. Many of the new pigments seem to be derived from reaction intermediates at the aldehyde or ketone oxidation state and those reactive substances may well modify many other grape components and fermentation products.

Our symposium began just as the war in Iraq started, so several speakers were unable to travel. We missed their participation. However, we managed to have a good attendance and the open time slots were used to great advantage to allow for much discussion of a number of complex issues. The symposium was attended by speakers from many wine-producing countries including France, Australia, Spain, Germany, and Italy in addition to the United States, and this monograph also includes authors from Morocco and Portugal. I hope future symposia will include speakers from South America, which is becoming an international wine production continent as well.

We gratefully thank the Division officers in the ACS Division of Food and Agricultural Chemistry, especially Carl Frey, the sessions organizer, who kept us on task and informed. We also thank our many speakers and contributors who managed to submit their manuscripts according to our deadlines, and we also thank the reviewers for their extra efforts. Peter Winterhalter deserves special mention for always being the first to submit. And finally we thank the staff and owner at the

Upperline Restaurant for the wonderful dinner they provided for our speakers and sponsors in uptown New Orleans.

In closing, we acknowledge the tremendous support we received from several organizations. Wine chemistry is a global effort with research teams in most of the major wine-producing countries and a representative survey of the field necessitates an international symposium. The supporting organizations include ETS Laboratories in St. Helena, California, one of the best wine analysis labs, E&J Gallo Wines, one of the largest wineries and one with a very large research enterprise, Canandaigua Wine Company, another of the largest and rapidly growing wineries, and Franciscan Estates, a wine company focusing on small, high-quality production in California and elsewhere.

Andrew Waterhouse
Professor of Enology
Department of Viticulture and Enology
University of California
Davis, CA 95616
530–752–4777 (telephone)
530–752–0382 (fax)
alwaterhouse@ucdavis.edu
http://waterhouse.ucdavis.edu

James A. Kennedy
Assistant Professor and Wine Chemist
Department of Food Science and Technology
Oregon State University
Corvallis, OR 97331
541–737–9150 (telephone)
541–737–1877 (fax)
james.kennedy@oregonstate.edu

Red Wine Color

Chapter 1

Short History of Red Wine Color

Alejandro Zimman[1,3], Andrew L. Waterhouse[1], and James A. Kennedy[2]

[1]Department of Viticulture and Enology, University of California, Davis, CA 95616–8749
[2]Department of Food Science and Technology, Oregon State University, Corvallis, OR 97331–6602
[3]Current address: Pathology and Laboratory Medicine, University of California at Los Angeles, Los Angeles, CA 90095–1732

Plant biology, breeding, genetics, horticulture, microbiology, engineering, chemistry and sensory science are disciplines that address the science needed to fully understand vines and wine. In addition, the real significance of grapes and wine can only be truly understood in a social and cultural context if their role in culture and tradition is considered as well. Consequently, there are many factors that influence the final enjoyment and/or judgement of red wine. One particularly important factor is color. Unlike many other variables, there is an insufficient understanding of the basic chemistry to allow full control or even to predict the final wine color. This is because the chemical composition of red wine has proven to be exceptionally complex, with pigmented molecules continually changing from the moment grapes are brought to the winery, during the crush, fermentation, pressing and then afterwards in the barrel and during bottle aging. The purpose of this chapter is to provide an overview of the key discoveries that have taken place over the past 140 years. To fully understand the significance of the surge of discoveries over the past 10 years, it is important to provide this background as a pretext to this book.

Most certainly, changes in red wine color, had to be noticed by anyone fermenting grapes and storing wine, but we will not attempt to refer to the earliest such account, probably cited in Greek, Roman, Egyptian or Sumerian literature. The first notable scientific observations were probably those of Louis Pasteur, where he conducted a series of experiments on the effects of oxygen on wine. The most obvious effect observed by Pasteur was a dramatic change in wine color (from red to brown) when exposed to oxygen. (*1*). He attributed the changes to oxygen and he proposed that a wine deprived of oxygen would maintain its color and in fact would not age. In the 1960's, Ribereau-Gayon and Peynaud undertook a comprehensive analysis of Pasteur's conclusions, hypothesizing that wine would age, even in the absence of oxygen. Their contentions were that Pasteur's conclusions were based on a limited number of wines and they felt that the source of wine he used would have led Pasteur to conclude that only oxygen exposure affected aging (*2*).

Additional early observations were made by Hilgard and Trillat (*3*)(*4*). Hilgard noticed in 1887 that the extraction of color and tannins did not follow the same extraction pattern, with color reaching a maximum of extraction prior to the end of fermentation, and tannins increasing throughout. He also noted that in the maximum in color extraction occured earlier at higher temperatures, and that wines fermented at higher temperatures did not result in wines with more color.. In 1906 Trillat observed that coloring material was precipitated with acetaldehyde, a key early observation on the chemical reactivity of colored components in red wine .

Wine color research, as expected,is linked to anthocyanin research. Wilstätter and Everest (*5*) were the first to characterize the aglycone portion of an anthocyanin, cyanidin. Contributions by Robinson and collaborators in the area of structure determination were also very significant. Among other anthocyanins, they malvidin-3-glucoside chloride (oenin chloride) and validated the structure with pigment extracted from "Fogarina" grapes (*6*). Wilstätter's anthocyanin research also included grapes and he was able to crystallize the anthocyanins found in grapes. But when he attempted to purify pigments in wine he did not obtain the crystals found in grapes (*7*). This result, together with the change in the appearance of color, led Wilstätter to the first observation that the pigments chemically change during the transformation from grapes into wine.

Ribereau-Gayon, not convinced that purified pigments in crystal form could not be islolated from red wine (as assurted by Willstätter), managed to obtain crystalline anthocyanin picrate from a young wine (although he was unable to so from an old wine) (*7*). He then concluded that:

"This impossibility, which we have found, of separating the pure pigment of old wines makes a chemical comparison of their pigments with the initial grape pigments very difficult, a problem which would be of great interest" (*8*).

Seventy years later, the mystery within this statement still resonates, and as such serves as an appropriate justification for this symposium.

The initial characterization of pigments in wine was elegantly studied by Ribereau-Gayon and Peynaud in two experiments. First, a colored precipitate appeared after cooling wine to 0°C. This resulted in a reduction in color. Secondly, an experiment consisting of a cellophane membrane separating a large volume of wine from a smaller volume of water was conducted. After equilibration, the inside of the membrane contained wine with slightly less color than the outside. Moreover, the wine inside the membrane did not produce a precipitate after chilling the wine, while the wine outside still produced a precipitate at low temperatures. The authors concluded that there were two types of coloring matter in red wine, small molecules and pigmented polymer, the latter being unable to cross the membrane (*9*). Even more interesting was the observation that after removing all precipitable pigments, wine regained the ability to form precipitates after three months at 15°C or one month at 25°C, showing once more the dynamic nature of wine pigments.

Wine researchers have taken advantage of new analytical techniques as they have appeared, and chromatography has been of particular value when addressing the complexity of wine. In 1948 Bate-Smith applied paper chromatography in the separation anthocyanins (*10*). Following this technique, Ribereau-Gayon characterized anthocyanins in *Vitis* sp. (*11*). The same anthocyanins were observed in young wine but as the wine aged, the spots corresponding to individual anthocyanins disappeared. The resulting colored material appeared as a diffuse band that moved more slowly than individual anthocyanins.

Different analytical techniques have been used to study the components responsible for color. Berg, for example, used nylon to separate anthocyanins and leucoanthocyanins (*12*). Mareca and Del Amo (*13*) used an alumina column in an attempt to characterize the evolution of color in Rioja wines. Somers divided the pigments into two groups using gel-filtration (*14*). In addition to the well-known set of anthocyanins, he provided additional information regarding the nature of the larger pigments. These pigments contained anthocyanins linked to tannins. Five years later, a follow up paper entitled "The polymeric nature of wine pigments" by the same author became a key reference in the field of pigmented polymers (*15*). Among other results, it showed that the contribution of pigmented polymers to overall red wine color grew as the wine aged. Two important characteristics of the pigmented polymers were: 1) the spectral response was different from grape anthocyanins and 2) color was not lost by SO_2 "bleaching".

Since it appears that anthocyanins and tannins condense in wine to form secondary wine pigments, the nature of the linkage between these compounds has been the focus of many investigations. There are a large number of possibilities but this chapter will discuss only two for historical purposes. Others will be discussed by other authors in this book. Based upon the structure of a natural compound, Jurd synthesized a dimer from catechin and a synthetic anthocyanidin (Figure 1). The flavylium form of the anthocyanidin has an electrophilic carbon at C-4 that will condense with the phloroglucinol ring of catechin. The new product, a flavene, will be oxidized by a second flavylium (*16*). The second mechanism is based on observations by Haslam regarding breakdown of interflavonoid bonds under acidic conditions. As a result of this, a carbocationic proanthocyanidin intermediate could react with an anthocyanin at the phloroglucinol ring (Figure 2).

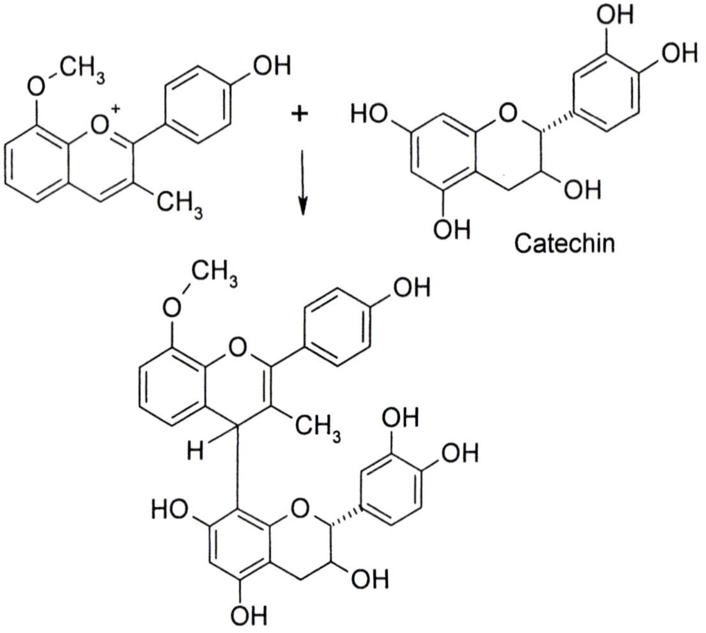

Figure 1.

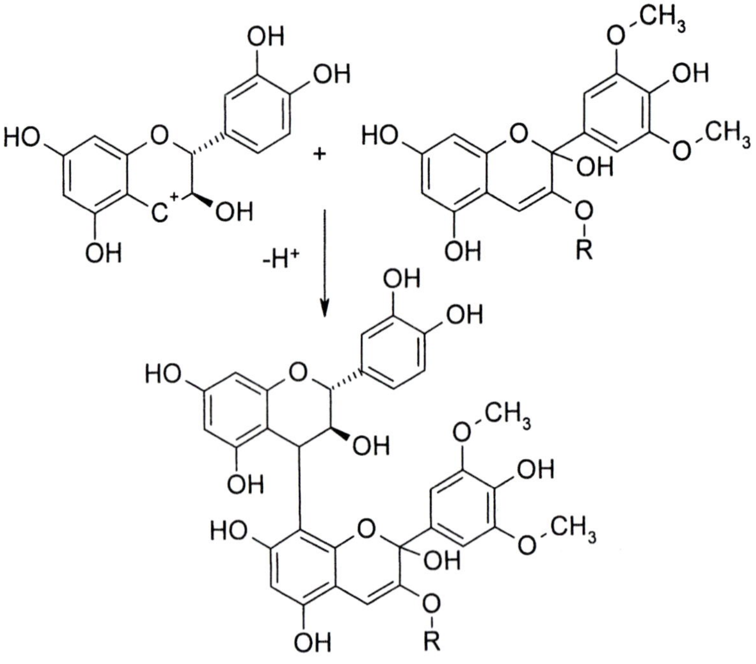

Figure 2.

During the late 70's there were two major contributions in the field of wine color studies. Brouillard and collaborators published a series of papers that fully characterized the proton transfer, hydration and ring-opening of anthocyanins (*17,18*). Understanding the chemistry of free anthocyanins is an essential step toward understanding the chemistry of condensation and the behavior of anthocyanin containing pigments in wine solution. The second contribution was the development and use of HPLC to separate anthocyanins in *Vitis vinifera* (*19*). This allowed for an easier separation and determination of concentration.

There are obvious reasons for the slow progress in the characterization of colored matter in red wine. If one takes into account that the number of different products depends upon the different reagents, then it is clear that tannins are responsible for the complexity. Even if an HPLC method were able to separate each compound into discrete peaks, characterization would be a major problem. Another issue is that wine does not have a fixed composition. Differences are obvious from wine to wine, but also in the same wine as it ages. Another difficulty is the pH of the analysis. Most HPLC methods used for pigment analysis need to maintain a mobile phase pH below 2. Therefore, large peaks may have little or no contribution at wine pH (*20*).

Additional historical work will be noted in other presentations in this same volume. Despite the difficulties, significant progress is taking place. It is notable that the complexity of the problem may be an indication of wine's unique properties and attributes.

References

1. Pasteur, L. Oeuvres de Pasteur réunies par Pasteur Vallery-Radot, 1924; Vol. 3. Études sur le vinaigre et le vin.
2. Ribereau-Gayon, J.; Peynaud, E. Traité d'oenologie. II. Compostion, transformations et traitements des vins; Librarie Polytechnique Ch. Béranger: Paris, 1961.
3. Hilgard, E. W. The extraction of color and tannin during red-wine fermentation. Calif. Agric. Exper. Stat. Bull. 1887, 77, 1-3.
4. de Almeida, H. Interprétation des phénomenes d'oxydo-réduction au cours du vieillissement du vin de "Porto" en bouteilles. Bull. O.I.V. 1962, 371, 60-84.
5. Willslatter, r.; Everest, A. E. Untersuchungen uber die anthocyane, 1913; Vol. 401.
6. Levy, L. F.; Posternack, T.; Robinson, R. Experiments on the Synthesis of the Anthocyanins. Part VIII. A synthesis of OEnin Chloride. J. Chem. Soc. 1931, 2701-2715.

7. Ribereau Gayon, J. Substances oxydables du vin. In Contribution B létude des oxydations et réductions dans les vins; application B létude du vieilissement et des casses. 2e édition, revue et augmentée.; Bordeaux, Delmas., 1933; pp 86-103.
8. Ribereau Gayon, J. Oxidizable substances of wine. In [Contribution to the study of oxidations and reductions in wines; application to the study of aging and of casses]. Translation at Fresno State University for the Roma Wine Company and M. Turner. Corrections and emendations made by M.A. Amerine.; Bordeaux, Delmas., 1933; pp 63-75.
9. Ribereau Gayon, J.; Peynaud, E. Formations et précipitations de colloides dans les vins rouges. C. R. Seances Acad. Agric. Fr. 1935, 21, 720-725.
10. Bate-Smith, E. C. Paper chromatography of anthocyanins and related substances in petal extracts. Nature 1948, 161, 835-838.
11. Ribéreau-Gayon, P. Recherches sur les anthocyannes des végétaux. Application au genre Vitis; Librairie general de l'enseignement: Paris, 1959.
12. Berg, H. W. Stabilisation des anthocyannes. Comportment de la couleur dans les vins rouges. Annales de Technologie Agricole 1963, 12, 247-257.
13. Mareca-Cortes, I.; Del Amon-Gili, E. Evolucion de la materia colorante de los vinos de La Rioja con el aZejamiento. Anales de la Real Sociedad EspaZola de Física y Química 1956, 52, 651.
14. Somers, T. C. Wine tannins-isolation of condensed flavonoid pigments by gel-filtration. Nature 1966, 209, 368-370.
15. Somers, T. C. The Polymeric Nature of Wine Pigments. Phytochemistry 1971, 10, 2175-2186.
16. Jurd, L. Anthocyanidins and related compounds-XI. Catechin-flavylium salt condensation reactions. Tetrahedron 1967, 23, 1057-1064.
17. Brouillard, R.; Dubois, J. E. Mechanism of structural transformations of anthocyanins in acidic media. J. Am. Chem. Soc. 1977, 99, 1359-1364.
18. Brouillard, R.; Delaporte, B. Chemistry of Anthocyanin Pigments. 2. Kinetic and Thermodynamic Study of Proton-Transfer, Hydration, and Tautomeric Reactions of Malvidin 3-Glucoside. J. Am. Chem. Soc. 1977, 99, 8461-8468.
19. Wulf, L. W.; Nagel, C. W. High Pressure Liquid chromatographic separation of anthocyanins of Vitis vinifera. Am. J. Enol. Vitic. 1978, 29, 42-49.
20. Vivar-Quintana, A. M.; Santos-Buelga, C.; Rivas-Gonzalo, J. C. Anthocyanin-derived pigments and colour of red wines. Analyt. Chim. Acta 2002, 458, 147-155

Chapter 2

Yeast-Mediated Formation of Pigmented Polymers in Red Wine

Jeff Eglinton[1], Markus Griesser[1,3], Paul Henschke[1], Mariola Kwiatkowski[1,2], Mango Parker[1,2], and Markus Herderich[1,2,*]

[1]The Australian Wine Research Institute, P.O. Box 197, Glen Osmond, South Australia 5064, Australia
[2]Cooperative Research Centre for Viticulture, P.O. Box 154, Glen Osmond, South Australia 5064, Australia
[3]Current address: Biomolekulare Lebensmitteltechnologie, Technische Universität Müchen, D–85350 Freising-Weihenstephan, Germany
*Corresponding author: telephone: +618–8303–6601; fax: +618–8303–6601; email: markus.herderich@awri.com.au

To gain full advantage of viticutural techniques that optimise anthocyanin levels in grapes, it is essential to identify factors that contribute to the stabilization of anthocyanins in wine and to characterise the reactions which are involved in the transformation of grape-derived anthocyanins into stable red wine pigments. HPLC analysis was used to determine the relative concentrations of anthocyanins, tannins and pigmented polymers in commercial-scale replicated fermentation trials, in the corresponding red wines during ageing, in small scale model ferments and in chemical model reactions. By closely monitoring the time course of anthocyanin degradation and pigmented polymer formation we were able to identify essential parameters for anthocyanin stability. In addition, studies of analyte profiles and reaction kinetics demonstrated that both condensed tannins and anthocyanins were required for the formation of pigmented polymers. Maximal formation of pigmented polymers was achieved in the presence of fermenting yeast, and soluble yeast metabolites were actively involved in the condensation reaction of anthocyanins with tannins. Finally, the results from the commercial-scale replicated fermentation trials could be confirmed by yeast-mediated biotransformation reactions employing purified substrates, and by chemical model reactions.

The wine industry has recognized the correlation between wine colour density and red wine quality scores. In general, everything else being equal deeper coloured wines are more likely to have greater flavour and body than lighter coloured wines (*1, 2*). In combination with other established compositional measures of grape quality such as total soluble solids, pH, titratable acidity, and glycosylated aroma compounds, methods to determine grape colour are essential for continued improvement of wine quality. As an example, robust and rapid measurements of grape colour based on near infrared spectroscopy are currently being implemented by the wine industry (*2*).

The essential pigments responsible for red grape colour are the anthocyanins. These are, however, not only susceptible to bisulfite bleaching and pH-induced colour changes, but are also quite reactive molecules and degrade quickly in solution (*3*). Consequently, only small amounts of grape anthocyanins that have been extracted from skins during fermentation can be detected in aged red wine (*4*). Anthocyanins are largely responsible for the colour of young red wine and the concentration of individual anthocyanins can reflect grape composition. After less than twelve months of ageing, however, the concentration of anthocyanins has declined substantially and pigments that have been formed after crushing, during fermentation and during ageing are essential to maintain the red colour of wine (*3-5*). Two types of pigments contribute to the colour of red wines together with the anthocyanins: Pigmented polymers, a heterogenous group of pigments formed from anthocyanins and tannins (*5*), and pyranoanthocyanins, which are formed by the addition of vinylphenols or carbonyls to anthocyanins followed by successive cyclisation and oxidation reactions (*6, 7*). Pigmented polymers are of higher molecular weight than pyranoanthocyanins and they are resistant to SO_2-bleaching (*8*). Pigmented polymers can be clearly separated from anthocyanins as well as from other SO_2-resistant pigments like the vitisins by HPLC analysis, but coelute with tannins (*8*). To differentiate by HPLC analysis between pigmented polymers and colourless tannins, tannins are detected at 280 nm, while pigmented polymers are detected at 520 nm. Based on spectrophotometric measurements, pigmented polymers are regarded to contribute as much as 90% to the color of red wine after 2 years of storage (*9, 10, 27*) and according to the commonly accepted paradigm a gradual transition from monomeric anthocyanins through oligomers to polymeric pigments occurs during ageing (*9*).

The focus of many viticultural research activities is to maximize grape colour and grape anthocyanin concentrations. To gain full advantage of enhanced grape anthocyanin levels it is, however, equally important to identify factors that contribute to the stabilization of anthocyanins in wine and to characterise the reactions which are involved in the transformation of grape-derived anthocyanins into stable red wine pigments during winemaking. To address these objectives, we have studied parameters involved in anthocyanin degradation and pigmented polymer formation during replicated commercial-scale winemaking trials and compared the outcomes to complementary model reactions involving purified anthocyanins and condensed tannins.

Materials and Methods

Grapes and Wines. Machine harvested Shiraz grapes from the Coonawarra region of South Australia (20 bins of approx. 400 kg) were used for commercial-scale winemaking in the 2001 vintage. Two randomly selected bins of grapes per fermentation tank (either a 900 L rotary fermenter (*RF*) or a 1100 L Potter fermenter (*PF*)) were destemmed and crushed. The must in each fermenter was individually adjusted to pH 3.3 with tartaric acid prior to inoculation with the active dry wine yeast *Saccharomyces cerevisiae* (Lalvin® EC1118, 0.25 g/L ADWY), that had been previously rehydrated. Triplicated fermentations were conducted at 18°C (*RF-18°C, PF-18°C*) and duplicated fermentations were conducted at 25°C (*RF-25°C, PF-25°C*). Fermenting must was pressed off skins after seven days (≤1° Baumé) and the free run was combined with the pressings. After fermentations were completed (day 14, combined concentration of glucose and fructose ≤ 0.4 g/L), the wines were racked from gross lees and inoculated for malolactic fermentation (MLF) with *Viniflora oenos* (Chr. Hansen). Wines were stored in 200 L stainless steel drums at 20°C and protected from air contact. After MLF was completed, all wines were racked into 80 L stainless steel drums, 30 mg/L SO_2 was added and the wines were stored at 0°C for cold stabilization. After approx. 150 days all wines were adjusted to achieve 30 mg/L free SO_2 and 50 mg/L total SO_2 as measured by the aspiration method (*25*), bottled in 750 mL bottles closed with ROTE closures and stored at 15°C.

Analysis of Anthocyanins and Pigmented Polymers in Samples from Winemaking Trials. Samples were taken from each fermentation tank daily from day 0 (immediately after crushing) up to pressing, after which samples were taken less frequently. Following removal of gross solids, all samples were frozen at -20°C for approximately 6 months, thawed, clarified by centrifugation and analysed batchwise by HPLC as described previously (*8*). Phenolic compounds were detected using a photodiode array detector at 280 nm and 520 nm. Tannin (peak 16 in ref. 8, detected at 280 nm) was quantified at 280 nm, using an external calibration based on catechin hydrate (Aldrich) at 0, 10, 50, 100, 250, 500 and 1000 mg/L, and expressed as catechin equivalents. Malvidin-3-glucoside and pigmented polymers (peaks 5 and 16 in ref. 8) were quantified at 520 nm using an external calibration based on malvidin-3-glucoside hydrochloride (Polyphenols Laboratories) at 0, 17.5, 35, 70, 100, 250 and 500 mg/L.

The spectral properties of wines were measured with a Varian 300 UV/Vis spectrophotometer, and wine colour density (absorbance [420 nm + 520 nm]), hue (absorbance [420 nm / 520 nm]), and non-bleachable pigments (absorbance 520 nm in presence of SO_2) were determined as described by Somers and Evans (*4*). Based on colour and compositional data, one wine that had been made in a Potter fermenter at 18°C was identified as an outlier and was subsequently removed from the trial, leaving us with nine individual wines. Additional data of the wines have been published previously (*11*).

Formation of Pigmented Polymers in Model Experiments. Anthocyanins were purified from a commercially available grape skin extract (Quest International). Grape skin extract (250 g) was extracted twice with 1 L of a methanolic solution of formic acid (3% v/v) at room temperature, the suspension was clarified by centrifugation at 4400g for 5 minutes, and the extract was filtered to remove particulate material (GH Polypro-450 0.45 µm membrane filter, Pall Gelman). Pigmented polymers were then removed from anthocyanins by preparative ultrafiltration using a regenerated cellulose membrane filter (76 mm diameter, Diaflo YM10, Amicon). The solvent was evaporated under reduced pressure (20 mbar) at 40 °C and the volume of the concentrated solution was adjusted to 50 mL with water. This enriched anthocyanin preparation contained 5.4 g/L M3G and no detectable pigmented polymers (i.e. less than 1 mg/L). A commercially available tannin preparation was used without further processing (Tanin Vinification, FERCO).

Model Fermentations. A chemically-defined grape juice medium at pH 3.5, with amino acids as the nitrogen source, was used for the model experiments (*12*). Sugar (D(+)-glucose 200 g/L) was included unless otherwise indicated. For those model fermentations that were conducted in the presence of tannins (1.5 g/L), the medium was sonicated for 15 minutes at room temperature after the tannin addition, and the insoluble material was removed by centrifugation at 48000 g for 15 minutes. In ferments containing added anthocyanins, the enriched anthocyanin preparation was sterile filtered (0.2 µm), and 1.6 mL were added to 47.4 mL of sterile filtered chemically-defined grape juice medium for each model fermentation.

Model fermentations (50 mL final volume) were conducted in triplicate in 250 mL fermentation flasks equipped with air locks and a side arm, which was sealed with a rubber septum to enable aseptic anaerobic sampling. Yeast suspension (1 mL *Saccharomyces cerevisiae* AWRI 838 in chemically-defined grape juice medium) was added to each ferment to yield a cell density of 5×10^6 cells/mL at the start of fermentation (approx. 0.4 OD_{650}). Chemically-defined grape juice medium (1 mL) was added to those control ferments that did not contain yeast. The fermentation flasks were flushed with sterile high purity nitrogen gas for 5 minutes immediately after inoculation, in order to remove as much oxygen as possible from the headspace. Fermentations were performed at 25 °C with shaking at 160 rpm. For model experiments with a high number of non-fermenting cells the yeast inoculum was prepared by fermentation in chemically-defined grape juice medium at 25 °C as above, then isolating the late-exponential phase cells by centrifugation at 2000g for 10 minutes. The cells were washed once with saline (0.85% aqueous NaCl, 10 mL) to remove traces of extracellular sugar and other fermentation metabolites, re-suspended in saline, and re-inoculated into medium lacking sugar with anthocyanins and tannins to yield an OD_{650} of approximately 23, which was an estimated maximum value based on a preliminary experiment.

Cell-free Experiments. Cell-free extracts were prepared from fermentation samples (12 mL) by centrifugation (2700g for 10 minutes) of the clarified supernatant. Tannin (15 mg) and enriched anthocyanin (400 μL, adjusted to pH 3.5) preparations were added to the cell-free supernatant (10 mL, final volume 10.4 mL), any insoluble material was removed by centrifugation at 48000g for 15 minutes, and the medium was sterile filtered (0.2 μm) into sterile 10 mL culture tubes to prevent the growth of microorganisms. The headspace of the tube was then flushed with sterile N_2 for 30 seconds to minimise the amount of O_2 present, and the tubes were incubated at 20°C in an anaerobic hood filled with 5% hydrogen in nitrogen. All subsequent sampling was performed under anaerobic conditions.

Model Reactions between Anthocyanins and Tannins in Presence of Acetaldehyde. Tannin (37 mg, Tanin Vinification, FERCO) was dissolved in chemically-defined grape juice medium and sonicated for 5 minutes, 0.8 mL anthocyanin stock solution was then added, the pH was adjusted to pH 3.53 using 5M NaOH, and the final volume was adjusted to 25 mL with chemically-defined grape juice medium. The reaction mixture was then sterile filtered using a 0.2 μm filter (Schleicher and Schuell), and divided into two vials. Into one vial containing 10 mL reaction mixture 1 mg freshly distilled acetaldehyde was added, and the medium was filtered again using a 0.2 μm filter (Schleicher and Schuell). The reaction mixtures with 100 ppm acetaldehyde, and the control reaction mixtures without acetaldehyde were then divided into nine 1 mL aliquots each, and placed in sterile 1.5 mL plastic vials (18 vials in total). A fresh solution of potassium metabisulphite was used to adjust the level of total sulphur dioxide to 50 ppm or 200 ppm in some vials. The same volume of sterile water was added to the flasks without sulphur dioxide. The flasks were then capped and placed into the anaerobic hood for 7 days at room temperature. Samples (100 μL) were taken from each vial on days 2, 4, and 7 and analysed by HPLC as described below using an Agilent 1100 HPLC system (Agilent). The column used was a Phenomonex Synergi Hydro-RP (4 μm particle size, 80Å pore size, 150 x 2 mm), at 25°C. Solvents were A) 1% acetonitrile, 1.5% phosphoric acid in water; B) 20% solvent A, 80% acetonitrile for gradient elution at a flow rate of 0.4 mL/min: 0 min (14.5% solvent B), 18 min (27.5% solvent B), 20 min (27.5% solvent B), 21 min (50.5% solvent B), 22 min (50.5% solvent B), 26 min (100% solvent B), 28 min (100% solvent B). The injection volume used was 20 µL. Phenolic compounds were detected by the photodiode array detector at 280 nm (tannins) and 520 nm (anthocyanins and pigmented polymers) and quantified as described above. The same HPLC method was used to analyse samples from the model fermentations.

Results and Discussion

Extraction and Stability of Anthocyanins

After crushing of grapes, anthocyanins were extracted from grape skins during alcoholic fermentation. A peak concentration of 295 mg/L of the dominating anthocyanin malvidin-3-glucoside (M3G) was achieved after seven days, after which the skins were removed by pressing (Table 1).

Table 1. Average Concentration of Malvidin-3-glucoside (M3G) and Pigmented Polymers in Red Wine during Alcoholic Fermentation, MLF and Bottle Storage.

		day 7	***day 97***	***day 480***
M3G	average (mg/L)	295±42	189±13	90±10
	range (mg/L)	234–372	170–200	76–102
Pigmented Polymers	average (mg/L)	42±10	70±21	72±15
as M3G	range (mg/L)	29–63	44–103	51–103
	n	9	5*	8

n: number of ferments analyzed; *: ferments which had completed MLF by day 97.

The difference between the M3G concentration in individual ferments (Table 1) could be partially attributed to the bin-to-bin variation of grapes having different maturity and might have been influenced by the choice of tank type and fermentation temperature (*11*). It remains to be established whether the peak M3G concentration reflected the maximum possible concentration of extractable anthocyanins under the conditions studied, or whether it was limited by the physical skin removal and could have been further increased by extended maceration on skins.

After racking off gross lees, two different phases of M3G disappearance were evident: Initially, the M3G concentration declined rapidly by approximately 30% (100 mg/L) within 60 days, followed by a second period of slower anthocyanin disappearance with the loss of a further 100 mg/L M3G over the subsequent 14 months. In total, approximately 70% of M3G was lost after 16 months and the wines retained an average concentration of 90 mg/L M3G. In addition, the difference between M3G concentration in individual ferments was reduced from a maximum range of 138 mg/L at pressing to 26 mg/L after 16 months. The concentration-dependent rate of the degradation reaction with the fastest loss occurring in young wine at high M3G concentration, and the smaller differences between the M3G concentration of individual ferments at later time points, are consistent with first order kinetics for M3G loss as described previously (*10, 13*). In addition, the rapid decline in M3G concentration during the first months after fermentation and the transition into the second phase of slower degradation coincided with the progress and end of MLF. It is reasonable to assume that substantial anthocyanin losses were caused by the elevated storage temperature at 20°C and lack of SO_2 protection, that are required for successful conduct of MLF. While it has been previously observed that *Oenococcus oeni* can degrade phenolic compounds including anthocyanins (*26*), it remains to be clarified whether the bacteria involved in MLF contributed actively to the degradation of anthocyanins.

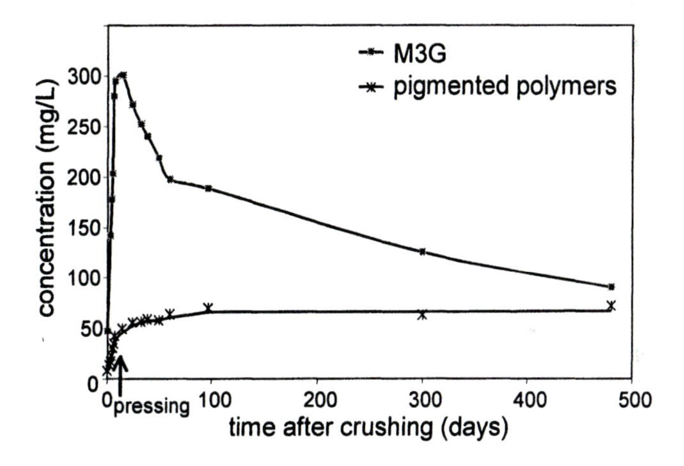

Figure 1. Average concentration of malvidin-3-glucoside (M3G) and pigmented polymers in red wine during alcoholic fermentation, MLF and bottle storage.

Formation of Pigmented Polymers during Commercial Scale Winemaking Trials

HPLC analysis demonstrated that pigmented polymers were formed in all ferments during fermentation at an average rate of 6 mg/(L x day) until pressing (Figure 1), when an average concentration of 42 mg/L was observed. In contrast to anthocyanins such as M3G, which declined rapidly thereafter, the concentration of pigmented polymers continued to increase for the following 90 days at a slower rate of approximately 0.3 mg/(L x day), resulting in an average concentration of 70 mg/L after three months. Over the remaining 13 months, the pigmented polymer concentration remained almost unchanged with a maximum increase of only 11 mg/L observed.

Precursors involved in Formation of Pigmented Polymers. The concentration of pigmented polymers in the four treatments at pressing is shown in Figure 2. It was closely correlated to the concentrations of M3G. This observation served as indirect evidence that anthocyanins, not unsurprisingly, were required as chromophores for the formation of pigmented polymers (*11*).

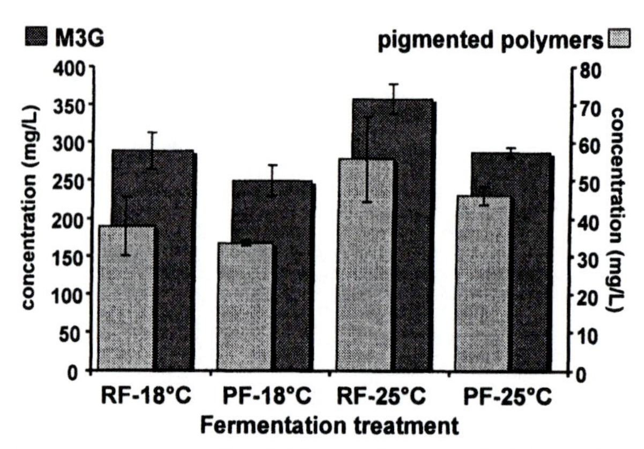

Figure 2. Concentration of malvidin-3-glucoside (M3G) and pigmented polymers in red wine at pressing.

The formation of pigmented polymers was correlated to the concentration of tannin. As an example, the rotary fermenters at 25 °C, which yielded the highest concentration of pigmented polymers, also extracted with 875 mg/L the most tannins (*11*). In addition, the observation that pigmented polymers showed tannin-like properties (8) and co-eluted with tannins during HPLC analysis, and that both tannin and pigmented polymer concentrations remained largely unchanged after pressing suggested that grape tannins were, together with anthocyanins, involved in the formation of pigmented polymers. The substantial increase of pigmented polymers during fermentation until pressing indicated that fermenting yeast could play a crucial role in their formation. The relevance of yeast-catalyzed reactions or yeast metabolites for linking anthocyanins to tannins was further supported by the observation that the rate of pigmented polymer formation was significantly slower after the wine had been fermented to dryness, even though anthocyanins and tannins were still available at high concentrations. In addition, the chemical degradation of anthocyanins that coincided with MLF resulted in significant reduction in anthocyanin levels, while at the same time only a small increase in pigmented polymer concentration could be observed.

Formation of Pigmented Polymers in Model Fermentation Experiments

In order to investigate the requirement for fermenting yeast for the formation of pigmented polymers, we developed a model system to study yeast mediated biotransformation reactions in the presence of purified anthocyanins (174 mg/L M3G) and commercially available grape tannins (1.5 g/L). Inoculation of sterile chemically-defined grape juice medium with *S. cerevisiae* AWRI 838 resulted in a much greater rate of formation of pigmented polymers from anthocyanins and tannins after the fermentation was started by addition of sugar to the medium (Figure 3).

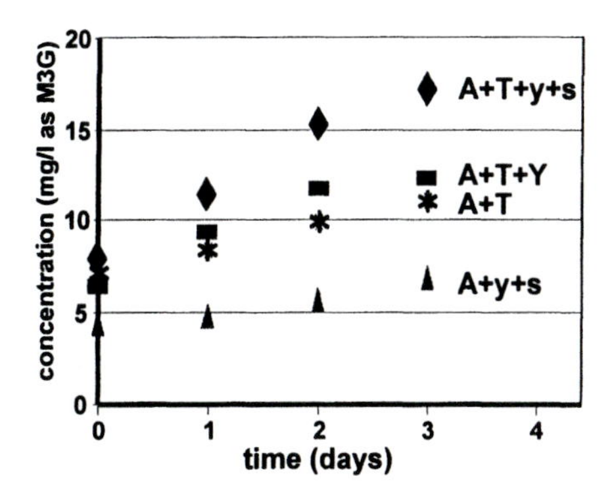

Figure 3. Formation of pigmented polymers in chemically-defined grape juice medium. Average concentration by HPLC, values are the mean of triplicate fermentations, A: anthocyanins, T: tannins, s: sugar (200 g/L glucose), y: standard inoculation with yeast (0.4 OD_{650}), Y: inoculation with high number of yeast cells (23 OD_{650}).

For all control experiments, the lack of fermentation activity in the absence of sugar was confirmed by constant refractometric readings during the incubation period. As all controls were essentially sterile cultures, we could exclude involvement of other microorganisms and thus clearly associate the formation of pigmented polymers with the presence of fermenting *S. cerevisiae*. From the controls it also was clear that the presence of fermenting yeast cells and anthocyanins, in the absence of tannins, resulted in a small increase in the concentration of pigments which coeluted with pigmented polymers (Figure 3, 'A+y+s'). Spontaneous condensation of tannins with anthocyanins accounted for approximately 50% of the pigmented polymers in the model reactions (Figure 3, 'A+T'). The presence of non-fermenting yeast cells at a cell density similar to a fermenting sample had a small effect on the formation of pigmented polymers in comparison to the reaction of tannins with anthocyanins in absence of yeast (Figure 3, 'A+T+Y'). In summary, these model experiments demonstrated that pigmented polymer formation was optimal when both anthocyanins and tannins were present. The maximum concentration of pigmented polymers was clearly achieved in the presence of actively fermenting yeast.

The colour properties of the model reactions were determined by a repeat experiment conducted at pH 3.2, and the results are shown in Figure 4. Non-bleachable pigments that retained their red colour in presence of an excess of SO_2 (Figure 4b, non-bleachable pigments, 'A+T+y+s') were significantly enhanced when compared to the controls. The significantly higher hue of the fermented samples (Figure 4a, 'A+T+y+s') correlates with the loss of approximately 40% of all anthocyanins after 4 days (data not shown), causing a reduction in absorbance at 520 nm. It may also reflect the colour properties of the pigmented polymers to some extent.

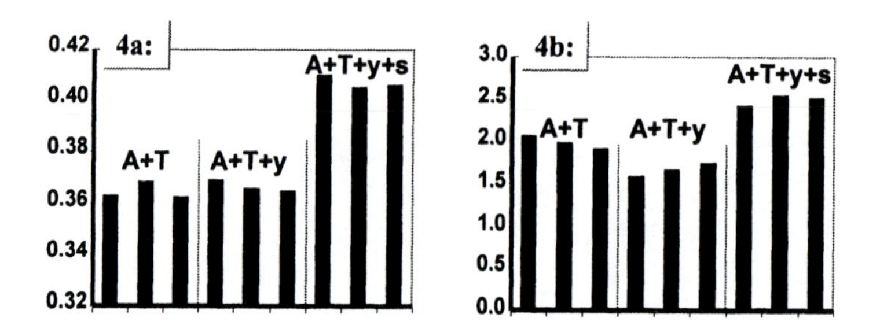

Figure 4. Colour properties of anthocyanin-tannin model reactions. a: hue, b: non-bleachable pigments, samples measured spectrophotometrically immediately after fermentation was complete, A: anthocyanins, T: tannins, s: sugar (200 g/L glucose), y: standard inoculation with yeast (0.4 OD_{650}).

Formation of Pigmented Polymers in chemical Model Systems

To determine whether enhanced formation of pigmented polymers from anthocyanins and tannins required physical contact with yeast cells, or could be mediated by yeast metabolites and other soluble factors, chemically-defined grape juice medium was fermented with *S. cerevisiae* AWRI 838 in presence of sugars. Samples were taken at day 0 (prior to inoculation), at day 2 during the exponential phase of cell growth (when approximately 50% of the carbohydrates had been utilized), and at day 4 after fermentation was completed. Upon removal of yeast cells by centrifugation and filtration, anthocyanins and tannins were added to each of the samples and HPLC analysis was performed at intervals during storage under anaerobic conditions for 6 days. Formation of pigmented polymers from anthocyanins and tannins in all three yeast-free media was noted. Interestingly, the concentration of pigmented polymers increased by 8 mg/L after 6 days in the samples prepared with medium taken before or after fermentation, while the medium from day 2 caused a significant increase in the pigmented polymer concentration to 17 mg/L after 6 days. The results from these model experiments containing yeast metabolites reinforced the relevance of fermenting yeast in the formation of pigmented polymers, and identified yeast metabolites as linking partners during the condensation reaction of anthocyanins with tannins.

While the identity of the relevant yeast metabolites in the fermented medium sampled at day 2 needs to be clarified, previously published data have provided some evidence about the role of acetaldehyde-mediated condensation of catechin with M3G (*13-20*). We therefore aimed to extend these model studies and to confirm chemical formation of pigmented polymers from condensed tannins, which are commercially used in red winemaking, and anthocyanins. The model reactions were conducted with varying concentrations of acetaldehyde and SO_2 as shown in Table 2 and analysed by HPLC after 2, 4 and 7 days. After 7 days visible precipitation of unidentified material started to occur in presence of acetaldehyde and the reactions were discontinued.

The data clearly demonstrated the protective effect of SO_2 on anthocyanins, as the M3G concentration remained essentially unchanged in the absence of acetaldehyde and in the presence of surplus SO_2. At a constant SO_2 concentration, formation of pigmented polymers was significantly enhanced in the presence of acetaldehyde. In contrast, increased SO_2-protection resulted in reduced pigmented polymer concentration in the presence of a constant acetaldehyde concentration. The effects of SO_2 and acetaldehyde on anthocyanin and pigmented polymer concentrations described here are consistent with previously published data from winemaking trials (*21*). The observation that tannin levels were not affected to a large degree during our short-term model experiments, while the coeluting pigmented polymers increased by almost 600%, provided further support for the concept that tannins were directly converted into pigmented polymers. The model reactions with acetaldehyde also confirmed that anthocyanins were the likely chromophores involved in pigmented polymer formation as anthocyanin degradation could be directly correlated to pigmented polymer concentration (Figure 5).

The model experiments involving condensed tannins used for winemaking thus provided confirmatory evidence to preceding research in which catechin or oligomeric procyanidins had been employed to characterize the newly formed pigments and colour stabilisation in the presence of acetaldehyde (*13-20*).

Summary

With this study we applied HPLC analysis to quantify anthocyanins, tannins and pigmented polymers in multiple samples from commercial-scale replicated

Table 2. Formation of Pigmented Polymers (PP) from Malvidin-3-glucoside (M3G) and Tannin in the Presence of Acetaldehyde and SO_2.

		0 ppm Acetaldehyde			***100 ppm Acetaldehyde***		
		M3G	***PP***	***Tannin***	***M3G***	***PP***	***Tannin***
Initial Concentration		130 mg/L	4 mg/L	739 mg/L	130 mg/L	4 mg/L	739 mg/L
day 7	***0 ppm SO_2***	75%±2%	344%±4%	88%±1%	49%±5%	588%±5%	85%±3%
	50 ppm SO_2	92%±2%	250%±3%	87%±1%	56%±2%	561%±13%	89%±7%
	200 ppm SO_2	104%±7%	136%±3%	71%±6%	96%±1%	257%±3%	95%±0%

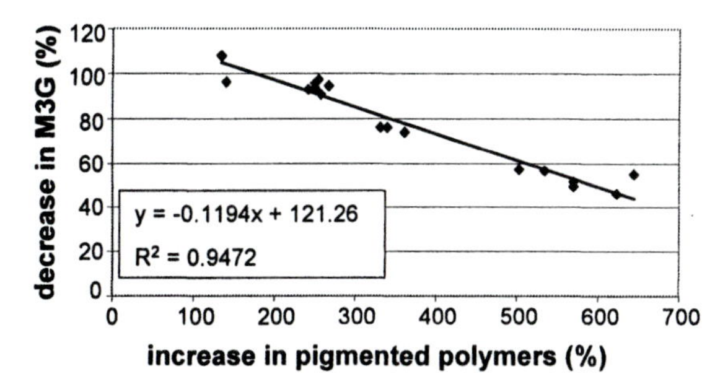

Figure 5. Correlation of loss of malvidin-3-glucoside (M3G) with pigmented polymer formation in model reactions with anthocyanins, tannins, acetaldehyde and SO_2.

fermentation trials, model ferments and chemical model reactions. Analyte profiles and analysis of the reaction kinetics confirmed that both condensed tannins and anthocyanins were required for substantial formation of pigmented polymers. This is consistent with previous research which has demonstrated chemical formation of pigmented polymers after anthocyanins and tannins had been added to a white wine (*22*). Most importantly, our data revealed that yeast metabolites were actively involved in the condensation reaction of anthocyanins with tannins and that fermenting yeast cells and metabolites derived therefrom were required for enhanced formation of pigmented polymers. A similar reaction has been described recently for the formation of the 'small' anthocyanin-derived pigment vitisin-A during fermentation and, while vitisin-A concentrations in wine appear to be much smaller when compared to levels of pigmented polymers, this further corroborates the relevance of yeast-mediated reactions for wine pigment composition (*23*). Our study also confirms and extends the results of early fermentation trials which employed indirect spectrophotometric measurements and a limited number of samples taken after pressing and ageing only, in which it was shown that a large proportion of 'non-anthocyanin pigments' had been formed during fermentation on skins (*10, 24*).

Our research provides robust evidence that wine composition is affected by yeast-mediated reactions in addition to grape composition and extraction parameters. It remains to be established, however, whether yeast metabolites act as linking partners, or accelerate the condensation of anthocyanins with tannins. From a wine industry perspective, wineries could target the few days of alcoholic fermentation to achieve enhanced pigmented polymer concentrations. This represents an important extension of the commonly accepted concept that formation of pigmented polymers is the result of a gradual transition from monomeric anthocyanins through oligomers to polymeric pigments during ageing (*9*). In addition, further studies are needed to characterize structural and sensory differences, if any, between pigmented polymers that have been formed by fast yeast-mediated reactions or relatively slow chemical condensation of anthocyanins and tannins. Still, it can be predicted that a better understanding and greater control of the reactions responsible for pigmented polymer formation during fermentation could contribute to increased colour stability in red wine. Our research also provides an explanation for the anecdotal evidence that addition of grape-derived oenotannins before or during fermentation is likely to support the formation of stable colour in red winemaking.

Acknowledgments

We acknowledge essential advice and assistance with winemaking by C. Day and S. Clarke, both at the University of Adelaide, and P. Godden, and thank Z. Peng and A. Pollnitz for assistance with sampling and HPLC analysis. The grapes for the commercial-scale winemaking trial were provided by Southcorp and L. Lurton from FERCO provided the sample of oenotannins. We thank Professors P. Høj and S. Pretorius for their comments and encouragement. This research was supported by Australia's grapegrowers and winemakers through their investment body the Grape and Wine Research and Development Corporation, with matching funds from the Federal government, and by the Commonwealth Cooperative Research Centres Program. The work was conducted by the Australian Wine Research Institute and forms part of the research program of the Cooperative Research Centre for Viticulture in Australia.

References

1. Somers, T.C.; Evans, M.E. *J. Sci. Food Agric.* **1974,** *25*, 1369-1379.
2. Gishen, M.; Iland, P.G.; Dambergs, R.G.; Esler, M.B.; Francis, I.L.; Kambouris, A.; Johnstone, R.S.; Høj, P.B. In: Proceedings of the eleventh Australian wine industry technical conference; Blair, R.J.; Williams, P.; H¢j, P.B.; Eds.; Adelaide, 2002, 188-194.
3. Mazza, G. *Crit. Rev. Food Sci. Nutr.* **1995**, *35*, 341–371.
4. Somers, T.C.; Evans, M.E. *J. Sci. Food Agric.* **1977**, 28, 279–287.
5. Somers, T.C. *Nature* **1966**, *209*, 368–370.
6. Fulcrand, H.; Benabdeljalil, C.; Rigaud, J.; Cheynier, V.; Moutounet, M. *Phytochemistry* **1998**, *47*, 1401–1407.
7. Håkansson, A.E.; Pardon, K.; Hayasaka, Y.; de Sa, M.; Herderich, M. *Tetrahedron Lett.* **2003**, *44*, 4887-4891.
8. Peng, Z.; Iland, P.G.; Oberholster, A.; Sefton, M.A.; Waters, E.J. *Aust. J. Grape Wine Res.* **2002**, *8*, 70–75.

9. Somers, T.C. *Phytochemistry* **1971**, *10*, 2175-2186.
10. Bakker, J.; Bridle, P.; Bellworthy, S. J.; Garcia-Viguera, C.; Reader, H. P.; Watkins, S. J. *J. Sci. Food Agric.* **1998**, *78*, 297-307.
11. Kwiatkowski, M.J.; Peng, Z.K.; Waters, E.J.; Godden, P.W.; Day, C.J.; Clarke, S.J.; Herderich, M.J. In: Proceedings of the eleventh Australian wine industry technical conference 2001; Blair, R.J.; Williams, P.; Høj, P.B.; Eds.; Adelaide, 2002, 150-151.
12. Jiranek, V.; Langridge, P.; Henschke, P.A. in: Proceedings of the international symposium on nitrogen in grapes and wine 1991; Rantz, J.M.; Ed.; American Society for Enology and Viticulture, Seattle, 2002, 266-269.
13. Baranowski, E.S.; Nagel, C.W. *J. Food Sci.* **1983**, *48*, 419-429.
14. Nagel, C.W; Baranowski, E.S.; Baranowski, J.D. In: Condensation reactions of flavonoids; Grape and Wine Centennial; Davis, 1980, 235-239.
15. Bakker, J.; Picinelli, A.; Bridle, P. *Vitis* **1993**, *32*, 111-118.
16. Garcia-Viguera, C.; Bridle, P.; Bakker, J. *Vitis* **1994**, *33*, 37-40.
17. Rivas-Gonzalo, J.C.; Bravo-Haro, S.; Santos-Buelga, C. *J. Agric. Food Chem.* **1995**, *43*, 1444-1449.
18. Dallas, C.; Ricardo da Silva, J.M.; Laureano, O. *J. Agric. Food Chem.* **1996**, *44*, 2402-2407.
19. Escribano-Bailon, T.; Alvarez-Garcia, M.; Rivas-Gonzalo, J.C.; Heredia, F.J.; Santos-Buelga, C. *J. Agric. Food Chem.* **2001**, *49*, 1213-1217.
20. Es-Safi, N.-E.; Fulcrand, H.; Cheynier, V.; Moutounet, M. *J. Agric. Food Chem.* **1999**, *47*, 2096-2102.
21. Dallas, C.; Laureano, O. *Vitis* **1994**, *33*, 41-47.
22. Singleton, V.L.; Trousdale, E.K. *Am. J. Enol. Vitic.* **1992**, *43*, 63-70.
23. Asenstorfer, R.E.; Markides, A.J.; Iland, P.G.; Jones, G.P. Australian Journal of Grape and Wine Research *Aust. J. Grape Wine Res.* **2003**, *9*, 40-46.
24. Bakker, J.; Preston, N.W.; Timberlake, C.F. *Am. J. Enol. Vitic.* **1986**, *37*, 121-126.
25. Rankine, B.C.; Pocock, K.F. *Aust. Wine Brew. Spirit Rev.* **1970**, *88 (8)*, 40-44.
26. Vivas, N.; Lonvaud-Funel, A.; Glories, Y. *Food Microbiology* **1997**, *14*, 291-300.
27. Schwarz, M.; Quast, P.; von Baer, Dietrich; Winterhalter, P. *J. Agric. Food Chem.* **2003**, *51*, 6261-6267.

Chapter 3

Color and Phenolic Compounds of Oak-Matured Wines as Affected by the Characteristics of the Barrel

E. Gómez-Plaza[1], L. J. Pérez-Prieto[1], A. Martínez-Cutillas[2], and J. M. López-Roca[1]

[1]Dept. Tecnología de Alimentos, Nutrición y Bromatología, Facultad de Veterinaria, Universidad de Murcia, 30071 Murcia, Spain
[2]Instituto Murciano de Investigación y Desarrollo Agroalimentario. Ctra. La Alberca s/n, 30150 Murcia, Spain

The content and nature of wine polyphenols change during the maturation in oak barrels. The most important changes that occur are due to the polymerization and degradation of anthocyanins and tannins, reactions that deeply affect wine color and its stability. We have studied the influence of the oak origin, barrel size and barrel age on the extent of these reactions and on the final color and polyphenol content of wine after six months in the barrel, followed by one year in bottle. Wines from grapes c.v. Monastrell were matured in medium-toasted American and French oak barrels. Three different barrel sizes were assayed: 220, 500 and 1000 liters. One lot of barrels was being used for the first time while the other lot of barrels were three years old. The results showed that the largest differences in wine color were observed when analyzing wines aged in new and used barrels. Anthocyanin/tannin condensation reactions occurred in a larger extent in wines aged in new barrels resulting in greater wine color stability.

Introduction

The polyphenolic reactions involved in wine aging are basically described as anthocyanin/tannin combinations that can either be direct, generating xanthylium salts, or involve acetaldehyde, leading to purple pigments (*1,2*). Other reactions where tannins (or procyanidins) participate are, on one hand, acid-catalyzed bond-making and bond-breaking processes (characteristics of procyanidin chemistry) and, on the other hand, oxidation reactions leading to browning *(3,4)*. All these reactions are likely to be involved in the color and taste changes observed during wine aging. In particular, the new pigment species arising from anthocyanin reactions generally appear more stable with respect to hydration and sulfite bleaching *(3)*, ensuring the color of aged wines (*5*).

Oak barrels are commonly used in the aging of wine and spirits because of their positive effects, which include increased color stability, spontaneous clarification and a more complex aroma.

The evolution of color may be affected by the characteristics of the oxidative process, that is, the barrel which has been used. Oak barrels are porous recipients that allow oxygen to enter continuously. The dissolved oxygen in wines matured in new barrels is higher than in used barrels because prolonged use causes a progressive colmatation of wood pores and a consequent decrease in the wine oxygen content. After the barrel has been used three to five times, the quantity of dissolved oxygen in wines will be very close to that of wines stored in tanks, so different wine color characteristics will be obtained *(6)*.

Other differences in the wine oxygen content may be due to the oak origin. The wood porosity and permeability change between species and will so influence the rate of oxygen consumption *(7)*.

Barrel volume may also influence the oxidative process. In this way, the favorable surface/volume ratio of 220 liters barrels may facilitate the phenolic polymerization. The four kinds of maturation effects which are attributed to barrels (evaporation, extraction, oxidation and component reaction) would all be intensified by a greater wood surface in contact with a unit of beverage *(8)*. Although much of the literature focuses on the value of maturing wine in small oak barrels, many fine wines are aged in mid-size to large (more than 1000 liters) oak barrels *(9)*.

However, another important factor, which must be taken into account when a winery has to decide on the kind of barrel it should buy, is the economic factor. French oak barrels cost twice or more the price of American oak barrels. The difference is partly due to the greater losses involved in coopering French oak because of its more irregular grain structure and higher wood porosity. For this reason, the wood must be split instead of sawn. With splitting it is impossible to obtain more than 25% of staves from a log, whereas sawing can give more than 50% . Large oak barrels can also be used for a longer time whereas small barrels are usually replaced after a few years.

The variety Monastrell represents 90% of the vines destined for D.O. Jumilla wines. It is the second most cultivated variety for red wines in Spain and its importance is increasing in other countries such as Australia and France. The resulting wines usually have problems withstanding long aging periods, they lose the vivid color towards orange tones very soon *(10)*. The objective of this study was to ascertain the effect of wood origin, barrel age and barrel size on the color of Monastrell wines, in order to obtain wines with the best color characteristics.

Material and Methods

The twelve new and twelve used barrels used in this experiment were made of American white oak (Quercus alba, fine grain) or French oak (Quercus petraea from the Allier forest in France), and were obtained from the same cooperage firm in Spain with the same specifications (medium toast level). The used barrels had been used three times for the aging of Monastrell wine and they were sanitatied before use by burning sulphur inside. The following barrels were used in the experiment: 220 liter French oak barrel (3 new barrels and 3 used barrels); 220 liter American oak barrels (3 new barrels and 3 used barrels); 500 liter American oak barrels (3 new barrels and 3 used barrels); and 1000 liter American oak barrels (3 new barrels and 3 used barrels).

The wine used in this experiment was a 1998 Monastrell red wine from Bodegas San Isidro in Jumilla, Murcia. Table I shows the analytical data of the wine prior to wood aging. All the barrels were filled with the same wine. A control wine was kept in a stainless steel tank. The wines were matured in the oak barrels for six months and stored during one year in the bottle, to meet the specifications for "crianza" wines. Wine samples were taken after three and six months of oak aging and after one year of bottle storage. Duplicate samples were taken from each barrel.

Color determination

Absorbance measurements were made in a Helios Alpha (Unicam, UK) spectrophotometer with 0.2 cm path length glass cells and the results were multiplied by 5 to express them as 1 cm path lenght glass cells. The samples were clean and contained no CO_2 which was eliminated by ultrasound and stirring. Color density (CD) was calculated as the sum of absorbance at 620 nm, 520 nm, and 420 nm *(11)*. Other variables calculated were the red, yellow and blue percentages, according to Glories *(11)*. PVPP index, which measures the extent of combination between anthocyanin and tannins, was determined following the method described by Ribereau-Gayon *et al* (*12*).

Table I. Parameters of the initial wine

Monomeric anthocyanins (mg/L)	152
Ellagitannins (mg/L)	0
Monomeric procyanidins (mg/L)	51.98
Oligomeric procyanidins (mg/L)	426.23
Polymeric procyanidins (mg/L)	551.43
Color density	13.19
% Yellow	34.15
% Red	53.87
% Blue	11.98
PVPP index	23.16

Monomeric anthocyanins represent the sum of the five anthocyanin-3-monoglucosides. Wine samples were analyzed by direct injection in the HPLC system (Waters 2690, USA, equipped with a DAD detector). The identification and quantification procedures have been previously described *(13)*.

Ellagitannins were analyzed by measuring the free ellagic acid after a hydrolysis with methanol/HCl 6N (8.4:1.6) following the method described by Vivas et al. *(14)*.

Procyanidin fractionation and quantification were made following the methods described by Sun et al. *(15,16)*.

Significant differences among wines and for each variable were assessed with an analysis of variance (ANOVA). To determine statistically significant differences among the means the LSD method has been used. A principal component analysis (PPC) was also performed, both statistical analyses using Statgraphics 2.0 Plus.

A sensory descriptive analysis was also carried out, using a panel of well-trained judges from the Jumilla Wine Council. The intensity of each attribute was rated on a scale of zero to nine. A score of zero indicated that the descriptor was not perceived, while a score of nine indicated a high intensity.

Results and Discussion

Figure 1 shows the evolution of monomeric anthocyanins during oak aging and bottle storage. A substantial decrease in these compounds was observed, especially during the wood aging period, because these compounds took part in condensation and polymerization reactions *(7,17)*. The age of the barrel, rather than the volume of the barrels or the origin of the wood was the main cause of differences among wines. Martinez-Garcia (*18,19*) found similar evolution patterns, not finding differences in the final concentration of monomeric anthocyanins regarding the origin of the wood.

The decrease in the monomeric anthocyanins content in wines stored in used barrels was lower and the evolution of these phenolic compounds in all the wines was slower in the bottle than in the barrel, as also found by Revilla and Gonzalez-San Jose *(20)*.

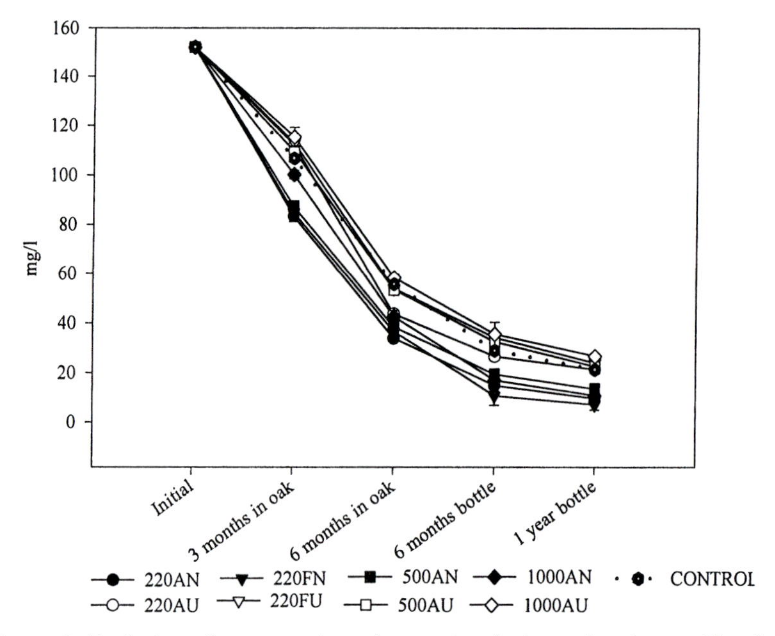

Figure 1. Evolution of monomeric anthocyanins during oak aging and bottle storage

This decrease in the monomeric anthocyanins content was not reflected in color density (Table II). Whereas the concentration of monomeric anthocyanins in wines after six months of wood aging was only 25% of the initial value in wines stored in new barrels, and 36% in wines stored in used barrels, in the same period of time, the color density only decreased by 5.5% in wines from new barrels and by 16% in wines stored in used barrels. After one year in the bottle, the decrease in CD value was 30% and 36% in wines stored in new and used barrels, compared with the initial wine, while the decrease in monomeric anthocyanins accounted for 93% and 85%, respectively.

It is clear that monomeric anthocyanins are not the main responsible for color density in aged wines but, rather, the formation of new compounds, mainly condensation products between anthocyanins and tannins. Such conclusions were also reached by Revilla et al. *(21)* After one year in the bottle, wines that have been matured in new French barrels had the highest values.

Table II. Evolution of color density (CD) and PVPP index during oak and bottle aging

	3 months in oak		*6 months in oak*		*One year in bottle*	
	CD	*PVPP index*	*CD*	*PVPP index*	*CD*	*PVPP index*
Control	14.7g	42.5b	13.1c	37.7a	8.9cd	46.8a
220AN	12.3e	44.1c	12.0c	52.2c	8.9cd	67.5d
220AU	10.8b	35.6a	11.6b	44.7b	8.7bc	59.7b
220FN	12.1d	48.8d	12.7c	58.2d	9.6f	75.0e
220FU	10.5a	38.6ab	10.9ab	35.8a	8.4ab	57.15b
500AN	12.7f	45.7b	12.8c	51.5c	9.3e	61.1bc
500AU	11.2c	39.7ab	11.1ab	46.1b	8.5ab	52.6a
1000AN	12.7f	43.6b	12.4c	54.7cd	9.1de	65.4cd
1000AU	10.7ab	36.5a	10.6a	46.0b	8.2a	49.7a

Different letters within the same column means significant differences ($p<0.05$)

The decrease in monomeric anthocyanins was greater in new barrels while the decrease in color density was more pronounced in wines aged in old barrels. This can be explained assuming that the color density of aged wines is essentially due to polymerized anthocyanins and is not related to free anthocyanins. Also, Revilla and Gonzalez-San José *(20)* found that wines stored in used barrels had a lower color density and higher tint values.

The PVPP index increased during storage for all the wines but more so in wines from new barrels. After six months in oak, wines from new barrels had significantly higher values of this index than wines in used barrels (Table II). The more favorable dissolution of oxygen in wines aged in new barrels is probably the origin of the easier polymerization of phenolics. The presence of oxygen leads to the formation of oxidized compounds, among them, acetaldehyde, the pivoting point of condensation between anthocyanins and tannins. At the end of the studied period, the lowest value was found in the control wine and the highest value in wine stored in new French barrels. The explanation for the difference in the PVPP index between wines stored in French and American oak barrels may be related with the structure of Quercus alba wood, which makes it more difficult for the wine to impregnate the wood and may reduce the penetration of oxygen into the wine, with consequent lower levels of oxidative condensation processes *(7,22)*.

The proportion of red, yellow and blue changed during aging and storage (Table III). Reactions resulting in yellow and bluish pigments take place during the entire aging period *(23)*, decreasing the percentage intensity of the red color in wines. The increase in the blue component was caused by the formation of new compounds that shift the maximum wavelength towards higher values *(21)*. In new barrels, color density was higher as was the absorbance at 620 nm, a finding also mentioned by Revilla and Gonzalez-San Jose *(20)*.

The greatest increase in the yellow percentage of the color took place in the bottle, as also observed by Del Alamo et al. *(23)*, with wines from used barrels exhibiting significant higher values.

However, the role of ellagitannins on the evolution of oak-matured wine color must not be overlooked. The evolution of ellagitannins in the wines is shown in Figure 2.

Ellagitannins, phenolic compounds extracted from wood, enhance color stability and reduce the astringency of red wine by favoring anthocyanin-tannin

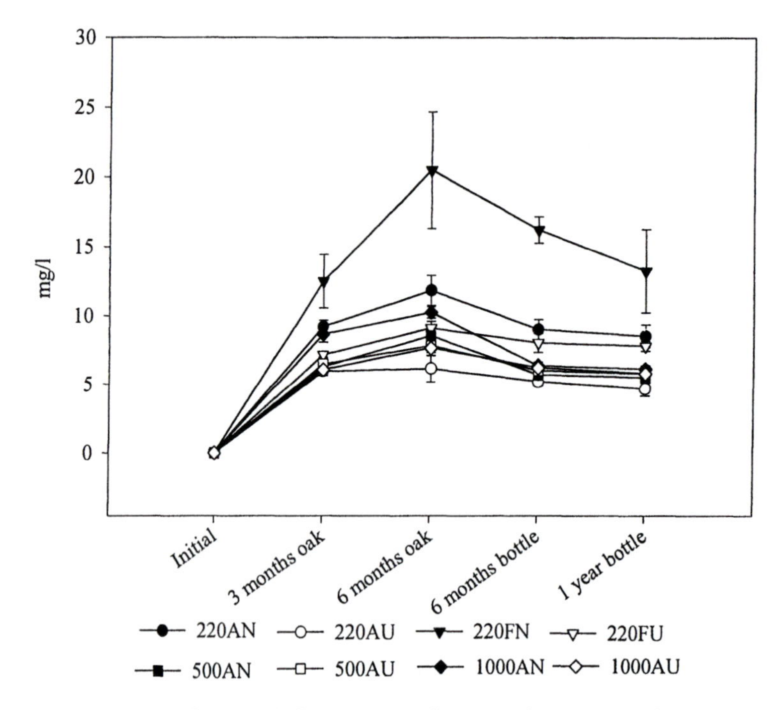

Figure 2. Evolution of ellagitannins during oak aging and bottle storage

Table III. Evolution of the percentages of yellow (%Y), red (%R), and blue (%B) in wine color

	3 months in oak barrel			*6 months in oak barrel*			*One year in bottle*		
	%Y	*%R*	*%B*	*%Y*	*%R*	*%B*	*%Y*	*%R*	*%B*
Control	34.08a	50.22a	15.70f	35.93a	51.58b	12.49c	39.98a	48.58b	11.45de
220AN	35.05bc	52.64bc	12.41d	36.28ab	51.37b	12.65cd	40.12a	48.66bc	11.23de
220AU	36.24e	52.06a	11.70c	36.68c	52.13cd	11.19b	40.75b	48.92de	10.32c
220FN	35.10c	52.46b	12.46d	36.37b	50.77a	12.85d	40.39a	48.11a	11.50e
220FU	36.26e	52.19a	11.56b	36.89de	52.06c	11.04b	40.80bc	48.91cd	10.28bc
500AN	34.90bc	52.61bc	12.49d	36.43b	50.89b	12.67cd	40.18a	48.51b	11.31de
500AU	35.65d	52.81c	11.55b	36.80cd	52.22d	10.98b	40.82bc	49.09e	10.09b
1000AN	34.75b	52.58b	12.67e	36.12ab	51.24b	12.64c	40.41a	48.50b	11.10d
1000AU	36.03e	52.59b	11.38a	37.24e	51.87c	10.88a	41.07c	49.06de	9.87a

Different letters within the same column means significant differences according to LSD test ($p<0.05$)

condensation reactions and by speeding up the condensation of tannins while limiting degradation processes, such as the precipitation of condensed tannins and anthocyanin destruction *(24)*. Their concentration increased during oak aging and slightly decreased during bottle storage.

Puech et al *(25)* stated that ellagitannins are not very stable in hydroalcoholic solutions and the fall in their concentration has been shown to increase with greater oxygen availability and in the presence of metal cations. The high oxidative ability of ellagitannins generates peroxides and therefore, large quantities of acetaldehyde could be formed.

Other studies have described similar results to those found in our research, that is, that ellagitannins are more abundant in French oak *(26)* and therefore, in wines matured in French oak barrels. So, the presence of ellagitannins enhances the color of wine and increases absorbance at 620 nm by favoring anthocyanin-procyanidin type tannin condensations with acetaldehyde (purple pigments) and that fact also helps to explain the higher PVPP index values of wines from new French oak barrels. They also prevent the development of brick-yellow color by preventing the oxidation of phenolic compounds *(24)*.

Table IV shows the evolution of monomeric, oligomeric and polymeric procyanidins, measured at the end of the wood aging period and after one year in the bottle. Some increases in monomeric procyanidins were detected during oak aging, perhaps as a result of the breakdown of oligomeric procyanidins or the extraction of monomeric compounds from the wood.

Table IV. Evolution of monomeric (MP), oligomeric (OP) and polymeric procyanidins (PP) during oak and bottle aging (mg/L)

	6 months in oak barrel			*One year in bottle*		
	MP	*OP*	*PP*	*MP*	*OP*	*PP*
Control	71.0de	332.8a	949.5a	49.1e	250.7bc	754.5a
220AN	48.8ab	345.0ab	1324.5b	31.0a	200.6a	917.9bc
220AU	66.7d	389.2bc	1282.1b	40.6bc	260.8c	1010.3bc
220FN	46.4a	355.3ab	1511.0c	36.1b	194.1a	880.3ab
220FU	75.1e	452.1d	1255.6b	44.6cd	250.9bc	1094.2cd
500AN	58.3c	393.1c	1397.4bc	30.1a	239.7b	918.4bc
500AU	67.3d	396.6c	1302.3b	38.7b	282.6d	1104.6d
1000AN	55.8bc	391.8c	1408.0bc	39.2bc	237.1b	858.5ab
1000AU	72.7de	452.3d	1275.6b	46.3de	284.1d	1123.0d

Different letters within the same column means significant differences ($p<0.05$)

The greatest increase during barrel aging was detected in polymeric procyanidins in wines aged in new barrels, an increase that was also detected in wines aged in used barrels and in the control wine, although to a lesser extent. The oxidative process of oak aging led to a high polymerization of procyanidins and similar results have been found by Del Alamo et al. *(23)*.

During bottle storage, all the different procyanidin fractions decreased, particularly the polymeric procyanidins in wines aged in new barrels. The evolution of these large quantities of phenolic compounds towards homogeneous polymerization until the precipitation of the polymeric procyanidins, phenomena described by Riberau-Gayon et al. *(12)*, could explain part of these reduction. But also, Vidal et al *(27)* demonstrated that at wine pH conditions, polymeric tannins undergo spontaneous cleavage of their interflavanic bonds. The process results in a reduction of average tannin chain length.

After one year in the bottle, a principal component analysis (PPC) was carried out (Figure 3).

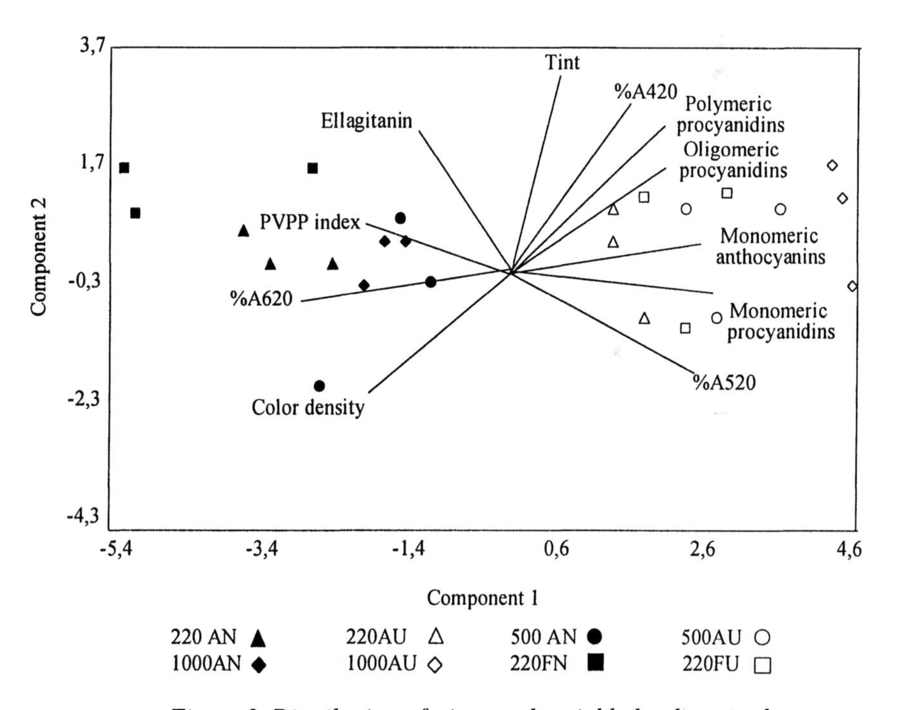

Figure 3. Distribution of wines and variable loadings in the two-dimensional coordinate system defined by the two first principal components

Two groups were observed, oak matured wine samples were separated along Component 1, according to whether new or used barrels had been used. From the variable loadings on each axes, it can be seen that wines stored in new barrels had a higher PVPP index and absorbance at 620 nm (indicating greater condensation between anthocyanins and tannins), and color density. Wines aged in used barrels had higher absorbance at 420 nm, and higher monomeric and oligomeric tannin and monomeric anthocyanin content.

The results of the descriptive analysis (Figure 4) showed that the origin of the wood caused little differences in most of the descriptors, only some were found in those descriptors related to aroma. A previous study showed that the concentration of cis-oak lactone was the most significant difference between wines matured in American oak and French oak barrels *(28)*. The age of the barrel produced the greatest differences in all the descriptors, especially as regards the intensity of the vanilla and woody character. When different volumes were compared, wines stored in 220 liter barrels had the highest scores in the aroma descriptors.

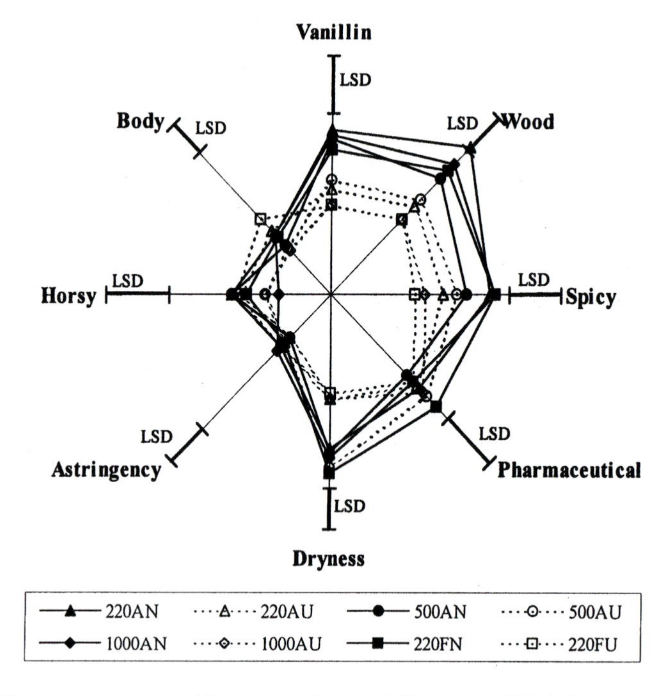

Figure 4. Sensory scores and least significant differences (LSD) for wines stored six months in oak barrels and one year in bottle.

Enologists must have available to them as much information as possible when buying barrels to produce high quality matured wines, although economic factors must not be forgotten. French oak barrels cost almost twice the price of American oak barrels and large volume barrels save space in the winery.

The chemical analysis showed that polymerization and color stability seemed to be favored in small and new barrels. The results of the sensory analysis showed that the origin of the wood did not lead to important sensorial differences and that wines held in 500 and 1000 L barrels had a significantly lower score, wines from 220 liter barrels being, in general, preferred.

So, we can conclude that 220 liter barrels will give better color characteristics and this color will be more stable and the taste and aroma of these wines are the preferred by the panelist. Regarding the origin of the wood, the large presence of ellagitannins in French oak promotes in a larger extent the polymerization of anthocyanins and tannins giving wines with deeper color, although many consumers prefers the aroma of the wines matured in American oak barrels.

Acknowledgments

The authors wish to acknowledge the financial assistance of the Fundación Séneca for this research (Project AGR/1/FS/99) and the cooperation and assistance of Bodegas San Isidro.

Abbreviations

AN: New American oak barrels
AU: Used American oak barrels
FN: New French oak barrels
FU: Used French oak barrels

References

1. Somers, T.C. *Phytochemistry* **1971**, *10*, 2175-2186.
2. Timberlake, C.F.; Bridle,P. *Am. J. Enol. Vitic.* **1976**, *27*, 97-105.
3. Cheynier, V.; Remy, S.; Fulcrand, H. *Proceedings of the ASEV. 50th Anniversary Annual Meeting,* Seattle, WA, **2000**, pp. 337-344.
4. Vidal, S.; Cartalade, D.; Souquet, J.M.; Fulcrand, H.; Cheynier, V. *J. Agric. Food Chem.* **2002**, *50*, 2261-2266.
5. Remy, S.; Fulcrand, H.; Labarbe, B.; Cheynier, V.; Moutounet, M. *J. Sci. Food Agric.* **2000**, *80*, 745-751.

6. Vivas, N. *J. Sci. Tech. Tonnellerie* **1995**, *1*, 9-16.
7. Fernández de Simón, B.; Hernández,T.; Cadahía, E.; Dueñas,M.; Estrella, I. *Eur. Food Res. Technol.* **2003,** *216*, 150-156.
8. Pérez-Coello, M.S.; Sanz, J.; Cabezudo, M.D. *Am. J.Enol. Vitic.* **1999**, *50*, 162-165.
9. Jackson, R.S. *Wine Science. Principles, practice and perception*, Academic Press: Washington D.C., **2000**.
10. Díaz Plaza, E.V.; Reyero, J.R.; Pardo, F.; Alonso, G.L.; Salinas, M.R. *J. Agric. Food Chem.* **2002**, *50*, 2622-2626.
11. Ruiz Hernández, M. *Vitivinicultura* **1991**, *9*, 24-28.
12. Ribéreau Gayon, P.; Glories, Y.; Maujean, A.; Dubourdieu, D. *Traité d'Oenologie.2. Chimie du vin. Stabilisation et traitements*, Editorial Dunod: Paris, **1998**.
13. Gómez Plaza, E.; Gil Muñoz, R.; López Roca, J.M.; Martínez, A. *J. Agric. Food Chem.* **2000**, *48*, 736-741.
14. Vivas, N.; Glories,Y.; Bourgeois, G.; Vitry, C. *J. Sci. Tech. Tonnellerie* **1996**, *2*, 25-49.
15. Sun, B.S.; Ricardo da Silva, J.M.; Spranger, M.I. *J. Agric.Food Chem.* **1998**, *46*, 4267-4274.
16. Sun, B.S.; Leandro, M.C.; Ricardo da Silva, J.M.; Spranger, M.I. *J. Agric. Food Chem.* **1998**, *46*, 1390-1396.
17. Gómez Cordobés, C.; Gonzales San José, M.L. *J. Agric. Food Chem.* **1995**, *43*, 557-561.
18. Martínez García, J. *Viticultura/Enología Profesional* **1998**, *57*, 48-56.
19. Pomar, M.; González-Mendoza, L.A. *J. Int. Sci. Vigne Vin* **2001**, *35*, 41-48.
20. Revilla, I.; Gonzales San José, M.L. *Eur. Food Res. Technol.* **2001**, *213*, 281-285.
21. Revilla, E.; González San José, M.L.; Gómez-Cordobes, C. *Food Sci. Technol. Int.* **1999**, *5*, 177-181.
22. Chatonnet, P.; Dubourdieu, D. *Am. J. Enol. Vitic.* **1998**, *49*, 79-85.
23. del Álamo Sanza, M.; Bernal Yagüe, J.L.; Gómez-Cordobes, C. *J. Agric. Food Chem.* **2000**, *48*, 4613-4618.
24. Vivas, N.; Glories, Y. *Am. J. Enol. Vitic.* **1996**, *47*, 103-107.
25. Puech, J.L.; Feuillat, F.; Mosedale, J.R. *Am. J. Enol. Vitic.* **1999**, *50*, 469-478.
26. Haba, M.; Chirivella, C.; Méndez, J. *Viticultura/Enología Profesional* **1995**, *37*, 32-37.
27. Vidal,S.; Cartalade,D.; Souquet, J.M.; Fulcrand, H.; Cheynier, V. *J. Agric. Food Chem.* **2002,** 50, 2261-2266.
28. Pérez Prieto, L.J.; López Roca, J.M.; Martínez Cutillas, A.; Pardo Minguez, F.; Gómez Plaza, E. *J. Agric. Food Chem.* **2002**, *50*, 3272-3276.

Chapter 4

The Variation in the Color Due to Copigmentation in Young Cabernet Sauvignon Wines

Joanne Levengood[1,2] and Roger Boulton[1]

[1]Department of Viticulture and Enology, University of California, Davis, CA 95616–8749
[2]Current address: Manatawny Creek Winery, 227 Levengood Road, Douglassville, PA 19518

This study measured the variation in the color due to copigmentation in 69 Cabernet Sauvignon wines from 12 wineries in the Napa Valley of California. The range in the level of color due to copigmentation is almost three-fol and the wine color is more strongly correlated with this form than either the total anthocyanin or polymeric pigment contributions. An analysis of the variation in color due to copigmentation found that among several reported viticultural measures (crop level, vine age, Brix at harvest, irrigation frequency, vine vigor and vine age) together could only account for 37% of the total variation. A similar set of winemaking variables (prefermentation, fermentation and extended contact times, sulfur dioxide addition, fermentation temperature) could only explain 34% of the variation. Together these viticultural and enological measures explained 66% of the total variation. The wide variation between wines from the same winery suggests that variation in initial composition influences the levels of this color contribution in young wines more than does the winery contacting practices.

Copigmentation is defined as the enhancement of color due to formation of complexes between pigments (anthocyanins) and cofactors (other non-colored compounds). Anthocyanins, when they are associated in these complexes, are more highly colored than when in the free, monomeric form. The phenomenon of copigmentation is responsible for the purple hue and intensity of color in young red wines and contributing to short-term color stability. It permits more extensive partitioning of anthocyanins between the grape skins and the wine during contacting and permits a larger pool of anthocyanins to be available for the development of polymeric pigment during aging. The variation in red wine color can be due to biochemical differences in both anthocyanin and cofactor synthesis and retention in developing grapes, as well differences in extraction conditions during winemaking. A review of studies of copigmentation over the past 30 years (*1*), includes an extensive bibliography. This study assessed the variation in the color due to copigmentation in wines from several wineries, using grapes of the same cultivar, season and region, but with different vineyard and winemaking practices

Pigment Assays and Red Wine Color

For more than 30 years the color of red wine has been characterized into two forms, monomeric anthocyanins and polymeric pigments (*2-6*). The monomeric fraction has generally been distinguished by its dependence on pH and its bleaching by sulfur dioxide. These features have been the basis on most color and pigment assays. While several authors have made reference to the non-linear effects of dilution (and the possibility of self-association and copigmentation), none have incorporated modifications into their color and pigment assays (*8-11*) to account for this. Until recently, the ability to relate wine color to these pigment measurements (or even with component concentrations of the 20 major phenols and anthocyanins) was far from acceptable with between 30 to 40% of the observed color of young wines, unaccounted for.

Color and Anthocyanin Assays

The measurements proposed by Somers and Evans (*8*) and others, which are of relevance to the current research are 1) absorbance of the sample at 520 nm with SO_2 added (A^{SO2}), 2) absorbance of the sample at 520 nm with acetaldehyde added (A^{Acet}) and 3) The dilution in buffered ethanol (*9*).

There are two basic problems with all of the previous spectral methods. The first is that pH is uncontrolled and since the copigmented form and free anthocyanin display quite different pH response (*12*), the estimate of the anthocyanin content and degree of ionization are unreliable. The second problem is that color of the anthocyanins in the copigment complex is included in the measurement at wine pH, but not in the lower pH (or diluted) measurement. This leads to an overestimate of the anthocyanin content in young wines.

The Copigment Complex

Copigmentation is known to exist because the anthocyanin concentration in young red wines is higher than that obtained in model wine studies. It is defined as color enhancement resulting from complexing of an anthocyanin with a cofactor, thought to be a wide variety of compounds including flavonoids, polyphenols, amino and phenolic acids and alkaloids. Asen et al. (*13*) developed a numerical sorting of the cofactors based on their enhancement of color and the shift in the maximum wavelength when present a three times the concentration of cyanidin 3,5-diglucoside. Other studies (*14-17*) have considered other anthocyanin and cofactor pairs. In wines, gallic acid, the cinnamates, caffeic and caftaric acids, the flavonoids, catechin and epicatechin, the flavonols, quercetin, kaempferol and myrecetin are thought to be the most significant copigmentation cofactors (*1*). The complexes can be thought of as either vertically stacked molecular aggregates involving π-π interaction or more perpendicular alignments based on CH-π interactions. Weak molecular forces are involved including hydrophobic interactions; phenolics associate with each other to minimize interactions with relatively polar, highly-ordered water (*15*). There are also likely contributions of counter ions such as bitartrate whe., un-ionizable cofactors are involved. The copigmentation process increases the proportion of colored forms at equilibrium since the stacking of one phenolic against another is easier for colored forms of anthocyanins than for less planar colorless forms (*15*).

Analysis of Copigment Complex

To date, there has been no known work done to actually measure the quantity of anthocyanin in the copigment complex of wines. Boutaric et al. (*18*) in 1937 and others (*9*) more recently, suggested approaches involving dilution and correction to quantify the amount of color that was due to non-Beer's law effects. The amount of copigmentation is dependent on the concentration of the anthocyanins and the cofactor compounds, as previously discussed. Therefore, as a young red wine is diluted and the concentration of pigment and cofactor declines, the amount of copigment complex that can form also declines. When this occurs, the absorbance values will decrease since anthocyanins are more highly colored when they are in the complexed form compared to when they are in the free form. This is the basis for the method.

This work is a survey of the color due copigmentation in young Cabernet Sauvignon wines and some simple viticultural and enological measures concerning the grapes used and the winemaking practices. Sixty-nine Cabernet Sauvignon wines from the 1995 harvest were assayed in an attempt to quantify the variance in the color due to this complex formation. It was initiated because no published work could be found quantifying the amount of color due to the complex and no recent survey of the color components in young red wines.

Experimental Methods

The Wines

The wines were produced from only Cabernet Sauvignon grapes grown during the 1995 season in locations within the Napa Valley appellation. The 69 wine samples were collected from 12 wineries who have established reputations for their Cabernet Sauvignon wines and who have control over the vineyards practices. The samples were of single wine lots from single vineyards, and samples were collected before any blending.

The measurements were completed when the wines were between 5.5 and 6.5 months of age. The variation in their age was approximately two weeks, being the variation in the harvest date.

Spectrophotometric Measurements

The method used to measure the amount of the copigment complex in a red wine, and the other measures, is referred to as the copigmentation assay. This method measures the absorbance of a sample altered by dilution together with those resulting from additions of acetaldehyde or sulfur dioxide (*7, 8*). The procedure differs from previous methods in that all readings are made at a standardized pH (3.6) and at 12% ethanol and makes no assumptions about the extinction coefficients, ionization or the color shifts. It also subtracts the color due to copigmentation from wine color in order to make the anthocyanin estimate.

The first step is to adjust the wine samples to a constant pH. Either hydrochloric acid or sodium hydroxide, as appropriate, was added to the wine samples to obtain a pH of 3.6. The second step is to filter the wine sample using a 0.45 μm pore size filter. Following completion of these two steps, four absorbance readings are obtained on the spectrophotometer at a wavelength of 520 nm as defined below:

A^{Acet}: 20μL 10% acetaldehyde solution is added to 2 mL of wine sample in a 10 mm cuvette. After 45 minutes, the wine sample is placed in a quartz cuvette with 2 mm pathlength and the absorbance is measured at 520 nm. The reading is corrected for the cuvette pathlength by multiplying by 5. 1 mm pathlength may need to be used with highly colored wine samples.

A^{20} : The wine sample is diluted 1/20 by placing 100 μL wine and 1900 μL buffer in a cuvette with 10 mm pathlength. The absorbance is measured at 520 nm after 10 minutes. The reading is corrected for dilution by multiplying by 20. (The Diluting Buffer solution was prepared as follows: 24 mL pure ethanol is added to 176 mL distilled water. 0.5 grams potassium bitartrate is dissolved in the solution. The solution is then adjusted to pH 3.6 with HCl or NaOH as appropriate). Higher dilutions of 1/30 or 1/40 will usually be required to completely dissociate the complex in darker wines.

A^{SO2} : 160 μL 5% SO_2 solution is added to 2 mL of wine sample in a cuvette with 10 mm pathlength. The absorbance is measured at 520 nm after 10 minutes.

The replicate variation (n=5) for the color due to copigmentation was 0.7%. That for the color due to anthocyanin was 0.6% and that for the color due to polymeric pigment was 0.3%

Calculations

The following equations were used to calculate the color contributions:

Color due to Copigmentation (AU) $[C] = (A^{Acet} - A^{20})$

Color due to Anthocyanins (AU) $[A] = [(A^{20} - A^{SO2})$

Color due to Polymeric Pigment (AU) $[P] = A^{SO2}$

Correlations

The principal component analyses (PCA) and partial least squares (PLS) regressions were made using the Uncrambler software (Version 7.6 SR-1, Camo Asa, Oslo, Norway).

Results and Discussion

Variation in the Color Components.

The range, mean, standard deviation and coefficient of variation of each component of the wine color, across all wines are shown in Table I. It can be seen that there is at least a two-fold variation within these wines in all color components. In the cases of the color due to copigmentation and the polymeric pigment, this variation is three-fold.

Table I. The range, mean, standard deviation and coefficient of variation for each of the color contributions in the Cabernet Sauvignon wines (n=69).

	Range (AU)	Mean (AU)	Standard Deviation (AU)	Coefficient of Variation (%)
Wine Color (AU@520nm)	5.47 to 12.3	8.24	1.79	21.7
Copigmentation (AU@520nm)	1.81 to 5.67	3.34	0.92	27.7
Anthocyanins (AU@520nm)	1.20 to 3.23	2.01	0.45	22.2
Polymeric Pigment (AU@520nm)	1.64 to 4.88	2.90	0.73	25.0

The distributions of wine color within the wines of each of the wineries and that for all samples are shown in Figure 1. The dotted curve shows the group distribution of all 69 wines. Wineries 1, 5 and 10 have higher than average colored samples and Winery 9 has a wide variation but this is based on only 3 samples. The mean color of the wines from a given winery is not significantly different between the wineries. The variation in color within most wineries is similar to that for the group (n=69) indicating that winery practices do not have a major effect on it.

The variation in the color due to copigmentation is shown in Figure 2. The wines of Winery 5 are significantly higher in the color due to copigmentation than those of Winery 3, 7 and 11, but the wines from the rest of the wineries are not significantly different in this measure. Like the wine color, the variation in the color due to copigmentation is similar for all wineries, despite the use of significantly different contacting conditions.

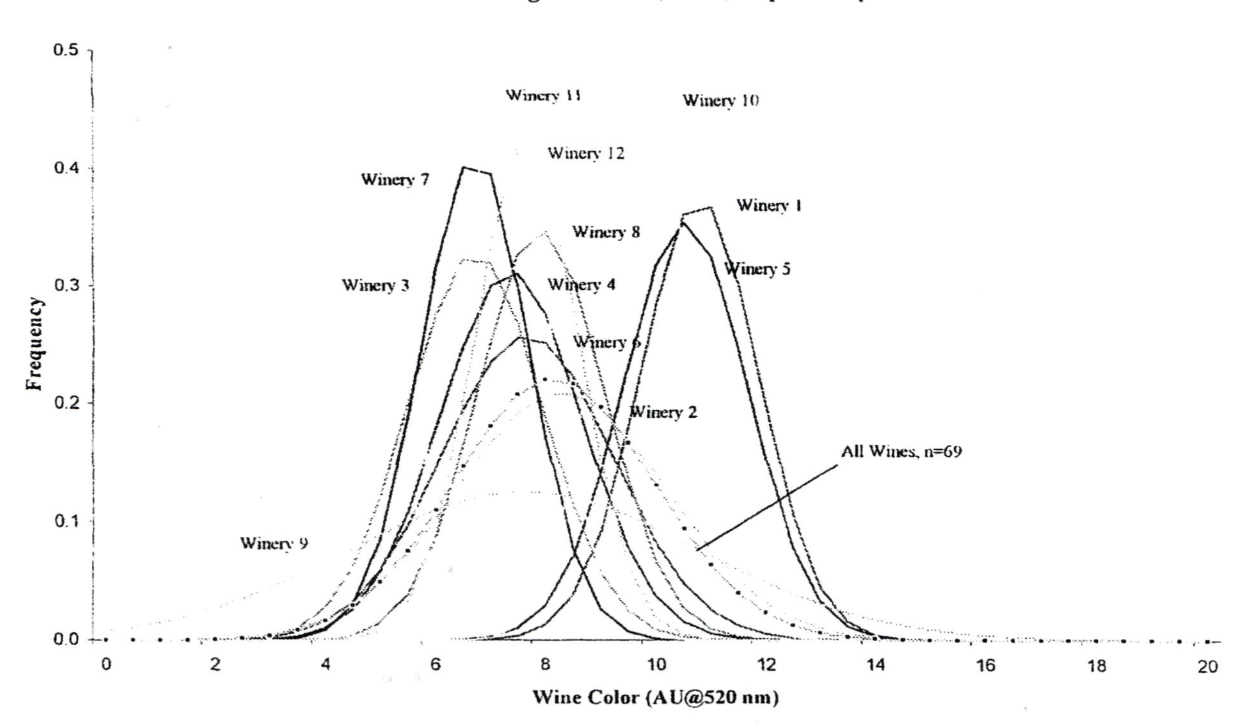

Figure 1. The distribution of wine color by winery and across all samples.

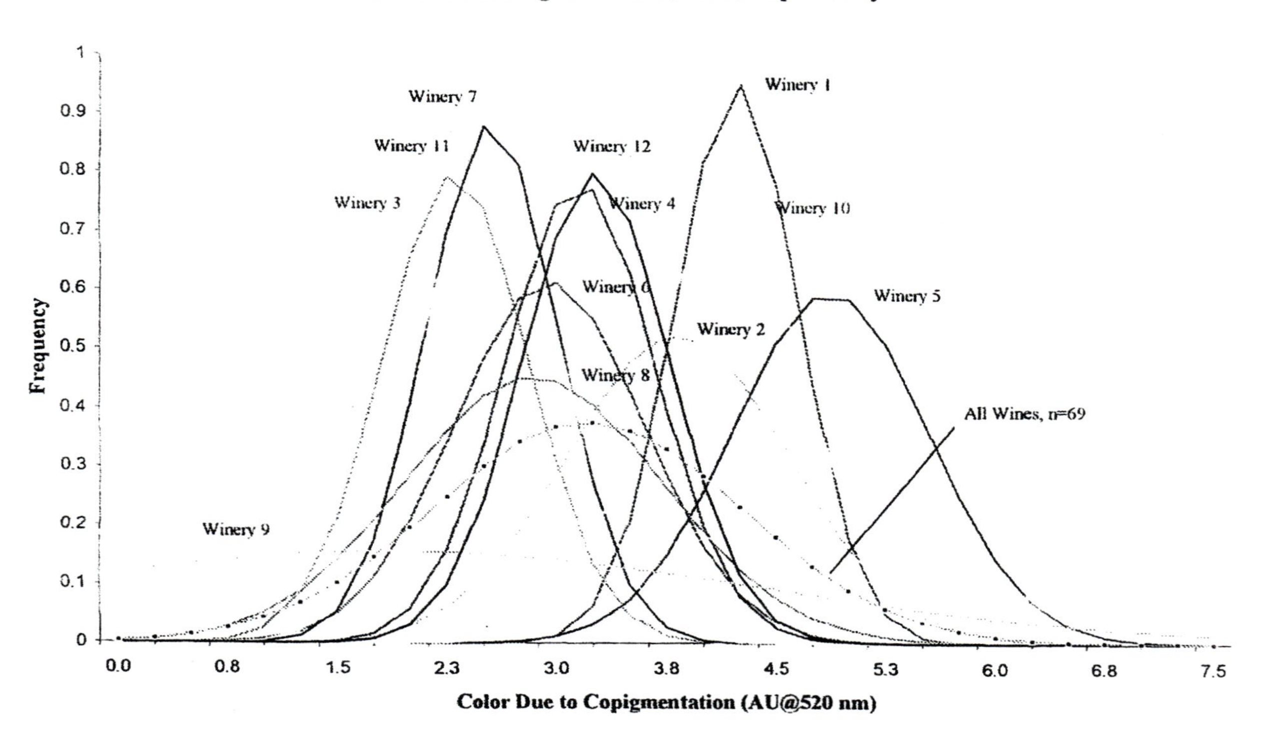

Figure 2. The distribution of color due to copigmentation by wineryand across all samples.

Wine Color at pH 3.6

The wine color, measured at pH 3.6 in the absence of SO_2, ranged from 5.5 to 12.3 AU with a mean of 8.24 and a coefficient of variation of 22%. This variation is quite large given that it is for wines of the same cultivar, season and appellation and the effects of SO_2 and pH on wine color have been removed.

Color due to the Anthocyanin Content

The color due to the monomeric anthocyanins, [A], ranged from 1.20 to 3.23 AU which a mean of 2.0 and a coefficient of variation of 22%. The anthocyanins are contributing approximately one quarter of the wine color. These values indicate that there is also a wide range of free anthocyanins in these 6-month old Cabernet Sauvignon wines.

Color due to Copigmentation

The range of the color due to copigmentation was from 1.8 to 5.7 with a mean of 3.3 and a coefficient of variation of 28%. This variation is the highest of all the color measures, considerably more variation than the anthocyanin content and this must be due to the variation between the wines in their cofactor content. This wide variation within wines of a single cultivar and region would have been underestimated if the traditional measures of color had been used.

Color due to the Polymeric Pigment

The average value for color of anthocyanin in the polymeric form, [P], measured was 2.90 AU with a coefficient of variation of 25%. The range for this parameter was from 1.64 to 4.88 AU, which corresponds to a range ratio of 3.0. Again, the wide range of color due to polymeric pigment had greater variation than the anthocyanin content, presumable due to the additional variation in tannin content of the wines.

Wine Color Correlations

The correlation of wine color with the anthocyanin color, the color due to copigmentation and the color of the polymeric pigments, individually, are straight-line relationships. More important is the fact that a higher correlation exists for the color contribution of the copigment complex, (R^2 = 0.77), Figure 3, than for the color due to anthocyanins (R^2 = 0.57) or polymeric pigments (R^2 = 0.66), data not shown.

The percentage of the total color at pH 3.6 due to the anthocyanins in the copigment complex ranges from 25 to 50% with an average of 40%. This

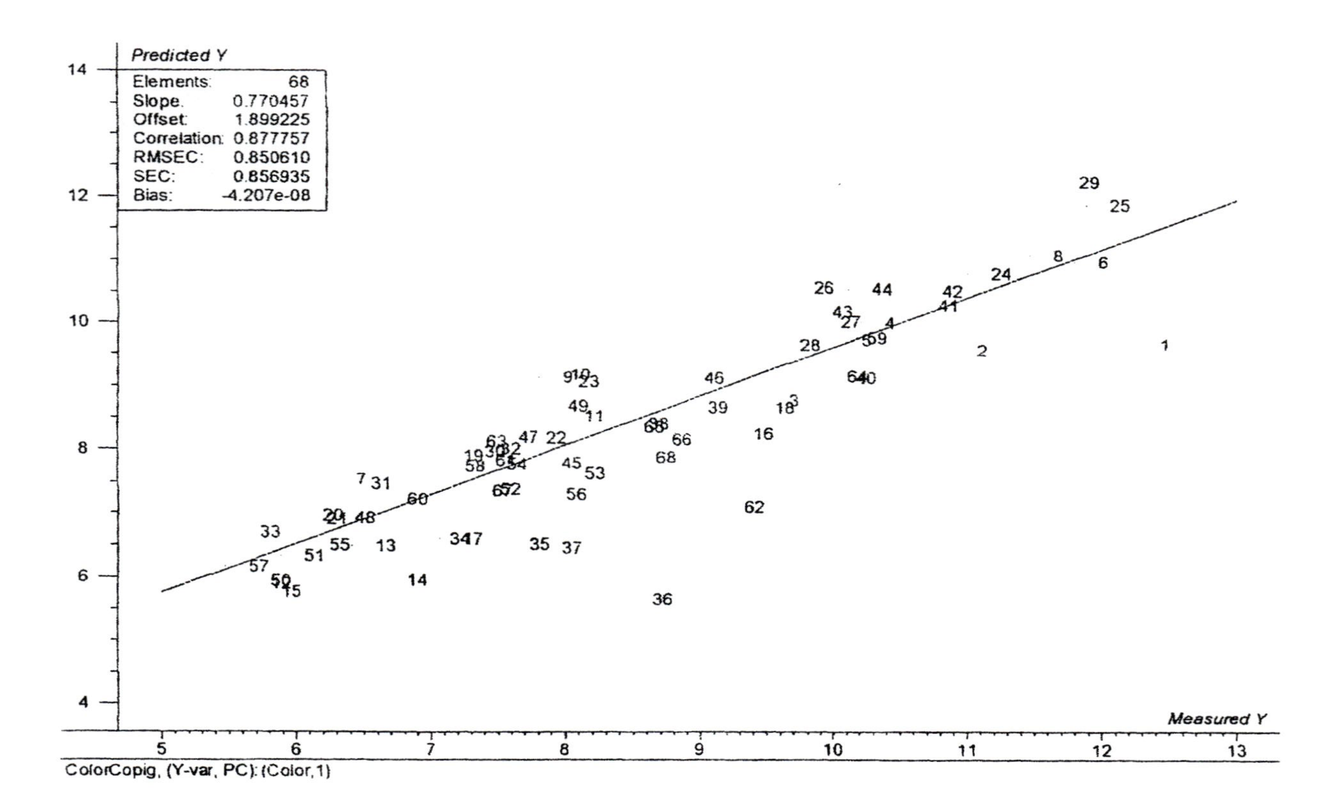

Figure 3. Correlation between wine color and color due to copigmentation.

means that the color of a 6-month old Cabernet Sauvignon wine is most influenced by the amount of the complex. Wide ranges also exist for fraction of wine color due to monomeric and polymeric anthocyanins. This shows that there is a wide variation in extent of polymerization in wines of the same age, more than would be expected in a very similar group of single cultivar wines from the same season and appellation.

Relationships Between Copigmented Color and Viticultural/Enological Factors

Principal Component Analysis (PCA) and Partial Least squares (PLS) regression were used to investigate relationships that might exist between various vineyard and winemaking practices and the color due to copigmentation.

The viticultural parameters evaluated were Vine Age, Trellis Type (a numerical vigor score of 1 to 3 (highest), Rootstock (a numerical vigor score), Hillside Location (a numerical vigor score), Irrigation Frequency, Crop Level in both ton/acre and pound/vine, Vigor Level (a numerical score of the vines within the site). The enological parameters evaluated were Initial Brix, Initial pH, Length of Prefermentation Contact, Length of Fermentation, Length of Extended Maceration, Fermentation Temperature, Percentage of Fruit Crushed, Time until SO_2 added, SO_2 Concentration after addition. More details of the actual scores can be found elsewhere (12).

Sixty-eight out of the sixty-nine Cabernet Sauvignon samples discussed previously had an assigned value for each of these viticultural and enological parameters. The first component accounted for only 23% of the variation in the measured variables and 41% of the variation in copigmentation. It was weighted positively with Length of Fermentation and negatively with Crop Level. The second component accounted for only 10% of the variation in the measurement variables and 17% of the variation in copigmentation color (PCA not shown). It was not significantly weighted with any of the variables.

Many of the sets of samples from the same winery fall in groups on this PCA graph. This is not unexpected since a winery may perform the same viticulture and enological practices for much of their wine. In some cases, however, the sets of samples are scattered across the plot, indication variation from sources other than those included. It could be useful for a winery to examine the location of their sample points on this figure to determine differences among samples and potential causes.

Prediction functions, based on the principal components, were used to see how much of the variation in the copigmentation measures could be accounted for by the vineyard and winery practices, using PLS methodologies. When all of the vineyard measures and winery conditions were included, 66% of the variation could be accounted for. The predicted color due to copigmentation is plotted against the actual values in Figure 4. Four components were used and the correlation coefficient is 0.82.Note that wine number 50 seems to be an outlier in this correlation.

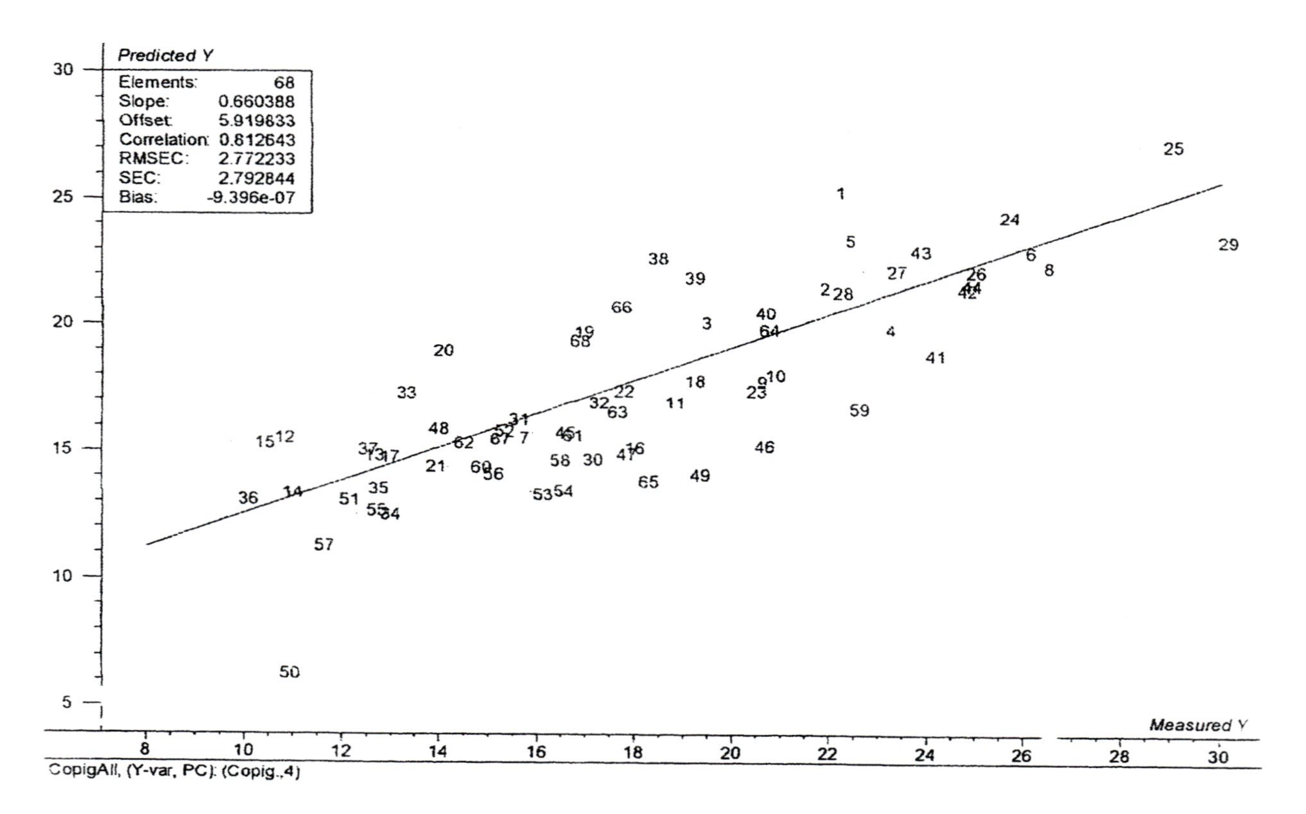

Figure 4. The correlation of color due to copigmentation with all variables.

Similar PLS models were developed for the vineyard variable and the winery variable separately. Figure 5 shows the poor relationship when all of the vineyard variables are considered. Only 37% could be accounted for and there is a significant non-zero offset.

Figure 6, shows the simple regression between the crop (as pounds per vine) and the color due to copigmentation. The crop level alone accounts for 57% of the variation with a significant offset. Note that the correlation does not mean that higher crop loads lead to higher color.

Two viticultural measures which are thought by some to be related to wine color (and other quality factors) are high sugar levels at harvest, and vine age. Figure 7, shows the correlation between degrees Brix and the color due to copigmentation. The sugar content at harvest, even for samples as high as 29° Brix, can only account for 3% of the variance and clearly it is of no significance in these color measurements. The corresponding regression between the color due to copigmentation and vine age shows that it accounts for only 2% of the total variation in color and is insignificant.

The extent to which the winemaking measures can account for the variation in color due to copigmentation is shown in Figure 8. Together they are responsible for only 34% of the variation in the color due to copigmentation

The primary conclusion reached in this research is that none of the viticultural or enological factors examined affected the color due to copigmentation of the anthocyanins. .

Conclusions

The color due to copigmentation, along with the other color parameters evaluated, varied greatly within this set of wines, even though the wines were all the same cultivar from the same appellation. It was found that the copigmentation contributes a large percentage of the color in 6-month old Cabernet Sauvignon wines and in fact, the component is the most important factor in the prediction of wine color.

The conclusion reached in the evaluation of relationships between the color due to copigmentation and various viticultural and enological factors was, under this set of experimental conditions, that no significant relationships exist

Acknowledgments

We would like to thank the following wineries for providing the wine samples and the associated viticultural and enological information used in this research: Beringer Vineyards, Cain Cellars, Cakebread Cellars, Franciscan Vineyards, La Jota Vineyard Co., Markham Vineyards, Robert Mondavi Winery, Newton Vineyards, Raymond Vineyards, Spring Mountain Vineyards, Stag's Leap Wine Cellars and Trefethen Vineyards.

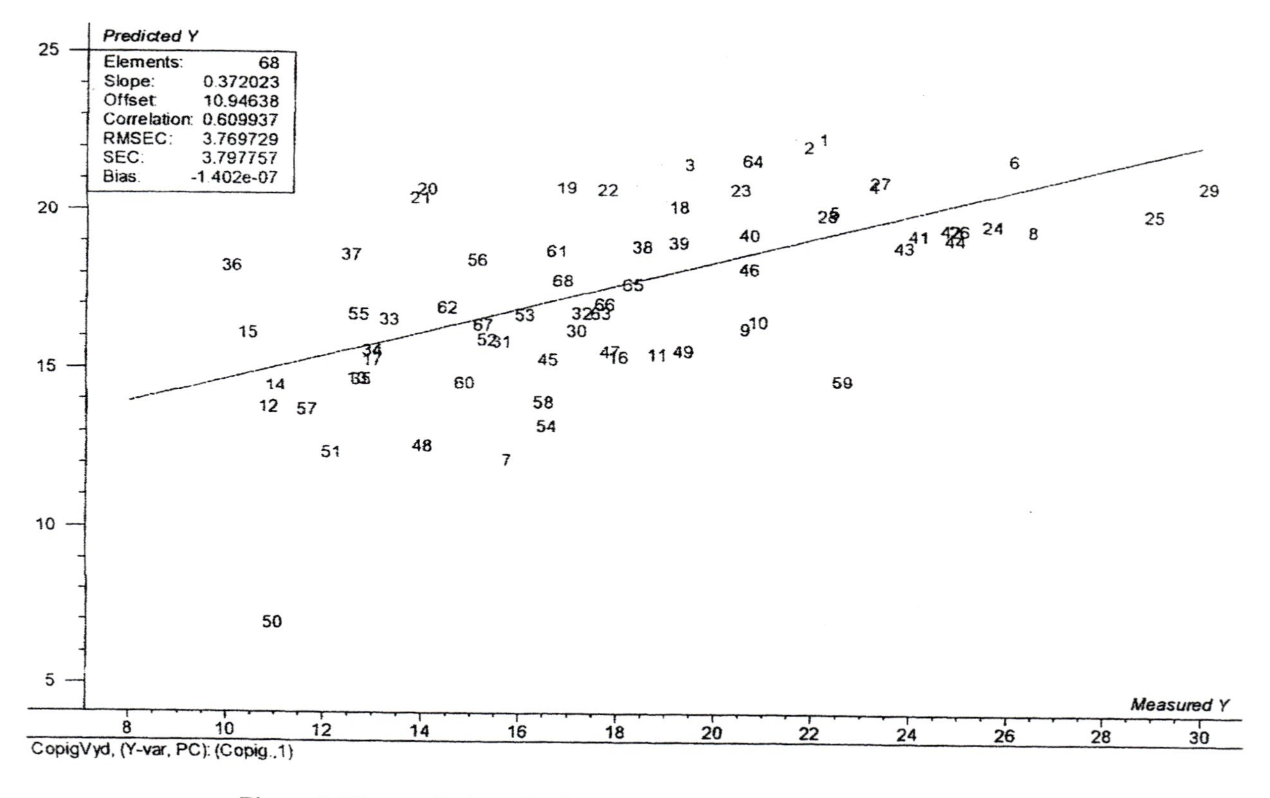

Figure 5. The prediction of color due to copigmentation from all vineyard variables.

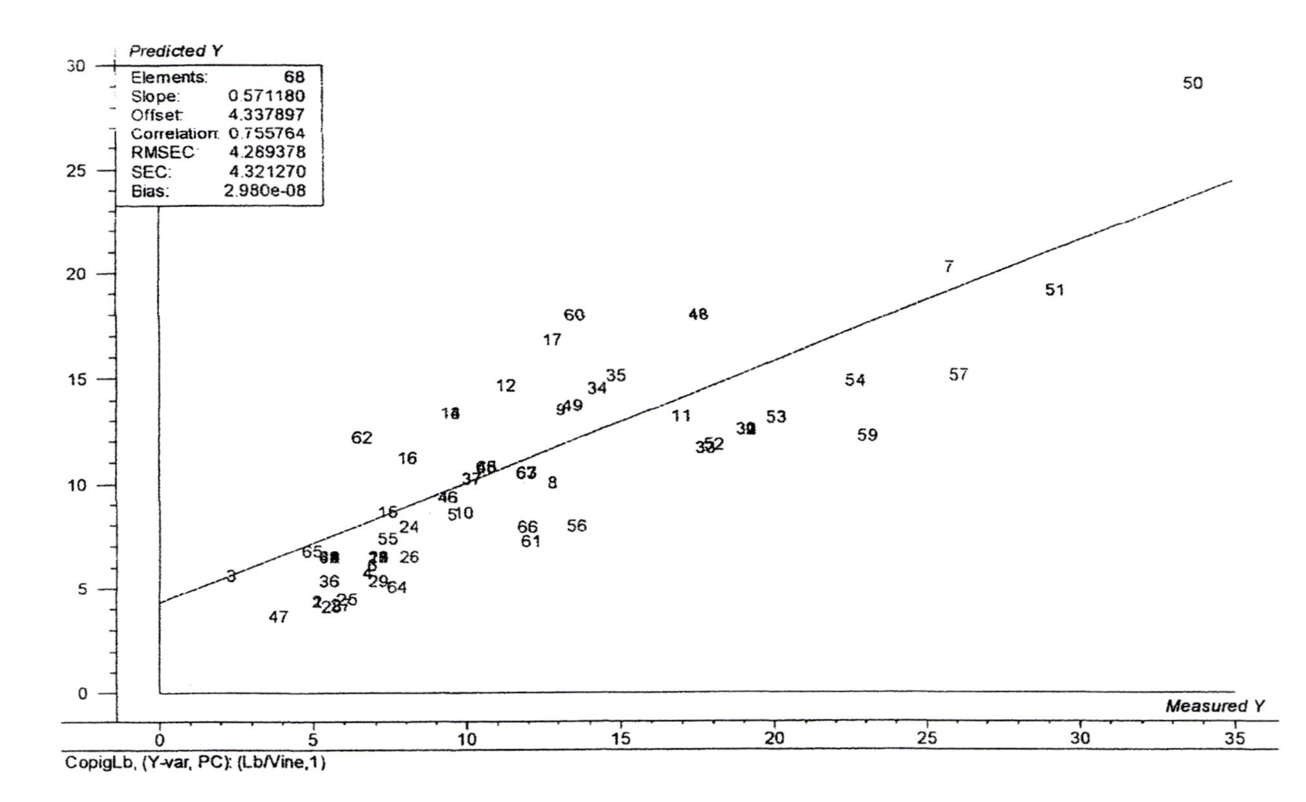

Figure 6. The correlation of color due to copigmentation with crop level.

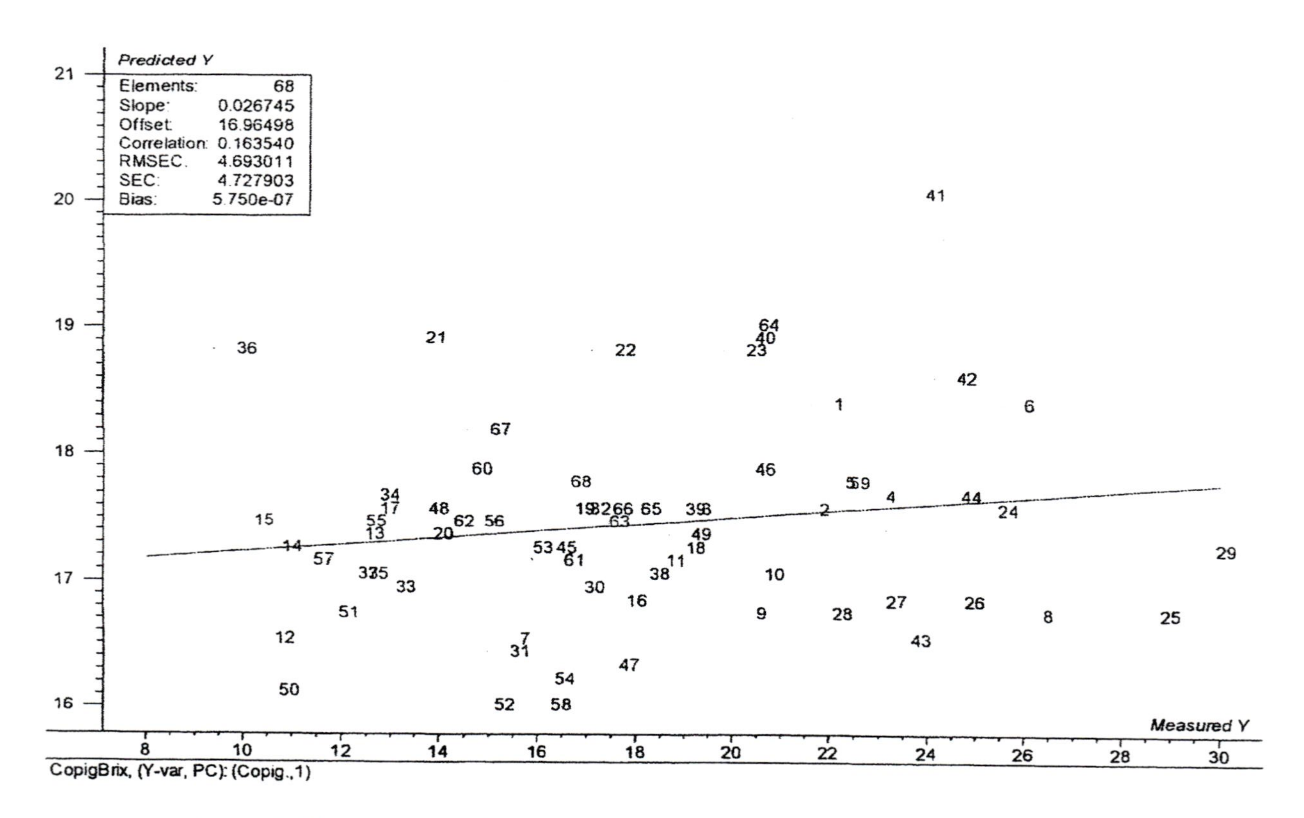

Figure 7. The prediction of color due to copigmentation from degrees Brix at harvest.

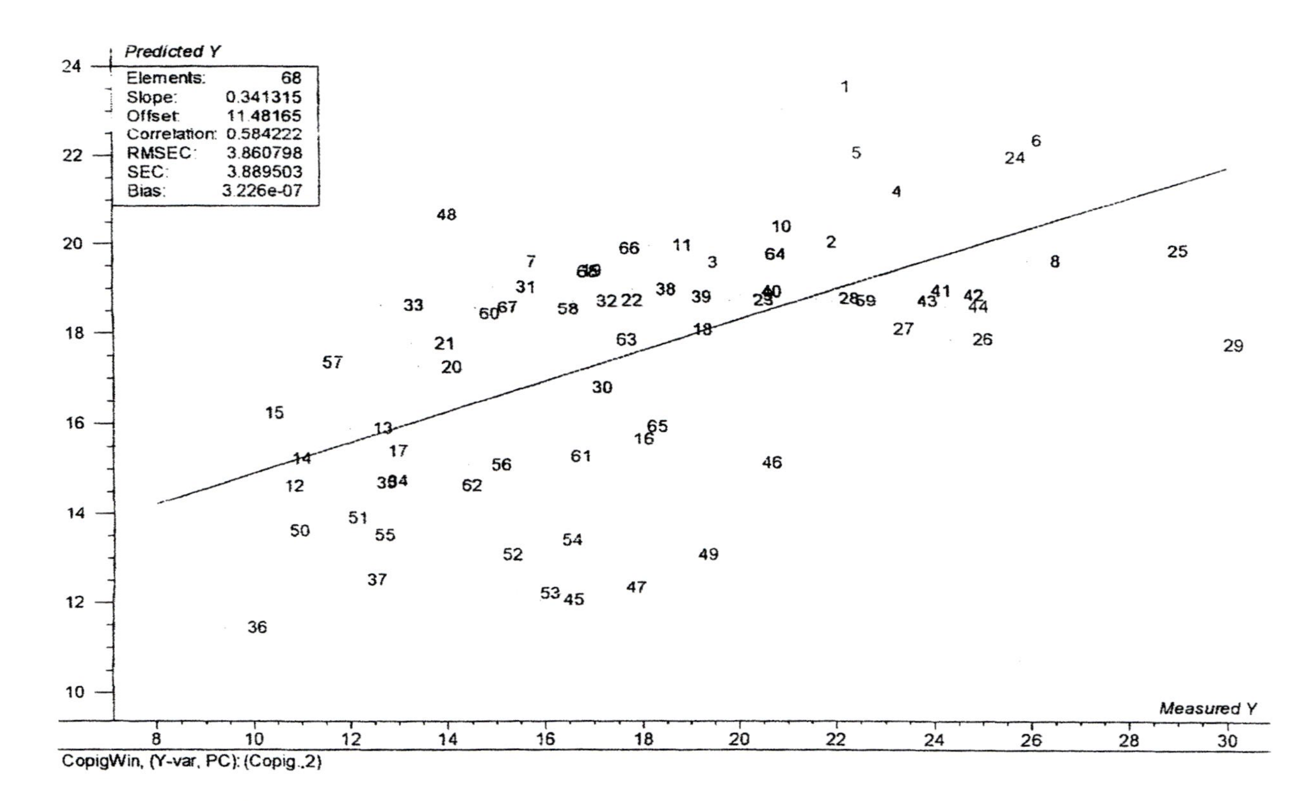

Figure 8. The prediction of color due to copigmentation from all winery variables.

References

1. Boulton, R. B. *Am. J. Enol. Vitic.* **2001,** 52, 67-87
2. Aubert, S. *Ann. Fals. Exp. Chim.* **1970,** 63, 107-117.
3. Berg, H. W. *Ann. Technol. Agric.* **1965,** 12, Suppl, 247-259.
4. Berg, H. W; Akiyoshi, M. A. *Am. J. Enol. Vitic.* **1975,** 26, 134-143.
5. Ribereau-Gayon, P.; Stonestreet, E. *Bull. Soc. Chim. France* **1968,** 47, 2649-2652.
6. Somers, T. C. *Phytochem.* **1971,** 10, 2175-2186
7. Somers, T. C.; Evans, M. E. *J. Sci. Food Agric.*, **1974,** 25,1369-1379.
8. Somers, T. C.; Evans, M. E. *J. Sci. Food Agric.*, **1977,** 28,279-287.
9. Somers, T. C.; Verette, E. In: *Modern Methods of Plant Analysis, New Series, Vol. 6, Wine Analysis;* Linskens, H. F.; Jackson, J. F., Eds. Springer-Verlag, Berlin. 1988, 219-257.
10. Somers, T. C. *The Wine Spectrum.* Winetitles, Adelaide, SA, 1998
11. Timberlake, C. F. *Proc. University of California, Davis, Centennial Symp.* 1981, 240-254.
12. Levengood, J. S. M.S. thesis, University of California, Davis, CA, 1996.
13. Asen, S.; Stewart, R. N.; Norris, K. H. *Phytochem.* **1972,** 11, 1139-1145.
14. Davies, A. J.; Mazza, G. *J. Agric. Food Chem.*, **1993,** 41,716-720.
15. Mazza, G.; Brouillard, R. *Phytochem.* **1990,** 29, 1097-1102.
16. Miniati, E.; Damiani, P.; Mazza, G. *Ital. J. Food Sci.* **1992,** 2, 109-116.
17. Scheffeldt, P.; Hrazdina, G. *J. Food Sci.* **1978,** 43,517-520.
18. Boutaric, A.; Ferre, L.; Roy, M. *J. Annal. Falsific.* **1937,** 30, 196-209.

Chapter 5

The Prediction of the Color Components of Red Wines Using FTIR, Wine Analyses, and the Method of Partial Least Squares

Andrea Versari[1], Roger Boulton[2], and John Thorngate III[2]

[1]Department of Food Science, University of Bologna, Cesena, 47023, Italy
[2]Department of Viticulture and Enology, University of California, Davis, CA 95616–8749

The ability to predict the levels of the free anthocyanins, copigmented anthocyanins, the polymeric pigment fraction and actual red color at 520 nm and Infrared (IR) spectra was investigated. Twenty wines ranging in red color from aımost zero to 10 AU at 520nm and mostly one year old, were analyzed by HPLC, protein precipitation and a copigmentation assay for their levels of the three color fractions. The mid-IR spectrum from 5012 to 926 wave numbers (corresponding to the wavelengths of 2 to 10.8 μm) was recorded for each sample. The ability to predict the levels of each color fraction from the spectra was tested using the method of Partial Least Squares (PLS) regression. The groups of wavelengths which were best able to predict the contribution of each color fraction were determined. The selections were made based on the use of valid PLS factors and the variance of the color measure that was explained by the factors and their weightings.

Infra-red (IR) spectroscopy, first discovered by Sir William Herschel in 1800, measures the absorption at different frequencies of infra-red radiation (*1*). Infra-red radiation is electromagnetic energy located in the wavelength range between the visible light (800 nm) and the shorter microwaves (100 μm). The area of infra-red light is generally divided into near-infra-red, NIR, (0.8–2.5 μm), mid-infra-red, MIR, (2.5–15 μm) and far infra-red, FIR, (15–100 μm). Molecular absorption of electromagnetic radiation in the infra-red range relates to the uptake of energy. Infrared radiation promotes the transition to rotational or vibrational levels from the ground electronic energy state. Different types of vibrations and rotations result in absorption of energy at different wavelengths. Fourier transform infra-red spectroscopy (FTIR) it is based on interferometry, which allows the measurement all wavelengths simultaneously. The interferogram is converted into a conventional spectrum using a Fourier transform algorithm. In the majority of cases spectral normalizing is used to eliminate the disturbances caused by any air present in the optical path. In particular, carbon dioxide and water strongly interfere (*2*).

FTIR spectra contain so much information that the same spectra may easily be used to calibrate for many types of constituents. Spectroscopy signal usually comprise rather large data sets but because they are co-linear there is a need to apply such methods as partial least square (PLS) regression to handle them (*3*). The X-variables are wavelengths and the X-values can be the absorbance, reflectance or transmission reading. The Y-variables (often called constituents or properties) may be chemical concentrations or physical parameters. The most widely-used application of PLS in spectroscopy is indirect measurement and calibration, where the aim often is to replace costly reference measurements with prediction from fast and inexpensive spectroscopic measurements, preferably with smaller uncertainty. The use of PLS is well established today within NIR, because this application often requires methods based on many wavelengths due to the abundance of non-specific wavelengths. PLS-applications continue to be developed in other wavelength ranges, like IR, UV or VIS, where information can remain hidden and univariate methods or multiple linear regression (MLR) were the only analysis options. PLS is preferred since it focuses directly on the Y-values and uses the information in Y to find the Y-relevant structure in X (*4*).

The development of infra-red (IR) spectrometers in combination with chemometric methods, make the IR spectroscopy an interesting tool for research, routine analysis and process control. The FTIR technology provides rapid, reproducible, nondestructive, multiconstituent analysis of food sample, with minimal or no sample preparation. The mid-IR spectrometry is being used for compositional analysis or discrimination purposes of several foods, including wine (*5-12*), fruit juices and soft drinks (*13*), soy sauce (*14*), purees (*15*), coffee (*16*), milk (*17*). Typical analytes that have been measured in juices and wines are ethanol, pH, organic acids, sugars, glycerol, SO_2, free amino nitrogen and ethyl carbamate. Few studies are available on its application for polyphenols (*10, 18*) and anthocyanins (*19*).

The purpose of this application is to use fast, inexpensive FTIR measurement of red wine color for quality control. The most important colored component of red wine are free anthocyanins, copigmented anthocyanins, the polymeric pigment fraction and actual red color at 520 nm. These parameters are currently analyzed by using selective UV-Vis spectrophotometric methods, often single wavelength methods. As an alternative, FTIR can collect a global fingerprint of the sample and used in conjunction with PLS regression it should be possible to extract reliable and relevant information to predict the color components of red wine.

Materials and Methods

Wine Samples

Twenty young red wines were selected for a wide range of color density. These were two Sangiovese, 2000 (Atlas Peak Winery), four Merlot, 2001 (Robert Mondavi Winery, Trefethen Vineyards), two Cabernet Sauvignon, 2001 (Robert Mondavi Winery, Trefethen Vineyards), one Cabernet Franc, 2001 (Trefethen Vineyards), nine Pinot Noir, 2001 (Robert Mondavi Winery, Saintsbury, UCDavis Winery), one Cagnulari, 2001 and one Cannonau, 2001 red wines (Santa Maria la Palma). At the time of these analyses eighteen wines were 11 months old and two wines were 23 months old. Before analysis the wine pH of each was adjusted to 3.6 and filtered through a 0.45 μm poly-tetrafluoroethylene (PTFE) membrane filter (Acrodisc CR, 25 mm, Pall Corporation, Corvina, CA).

Reference Measurements

For each wine the following parameters (Y-variables) were measured: total color (Aacet), total anthocyanins (TA), copigmentation (Copig), total polymeric pigment (EpP) at pH=3.6 (*20*), and small and large polymeric pigment (*21*). Values are reported as Absorbance units at 520 nm.

FTIR Measurements and Apparatus

FTIR analyses were carried out in the absorbance mode using a WineScan FT120 spectrometer (Foss, Denmark) equipped with a pyro-electric detector. Measurements were carried out at 40°C using a liquid flow-trough cell equipped with CaF_2 windows and a 37 μm optical path-length was used. The spectra were recorded during the passage of sample through the cell, with an optical resolution of 1 pin number. A HeNe frequency laser with a 632 nm signal beam was used with a reference interferometer. The spectra were obtained in duplicate for each sample and the mean of the two measures was used in all calculations.

The WineScan FT 120 interferometer gives a spectrum from pin number 240 to pin number 1299, corresponding to 926–5012 cm^{-1}, resulting in a total of 1060 data points per spectrum, Figure 1.

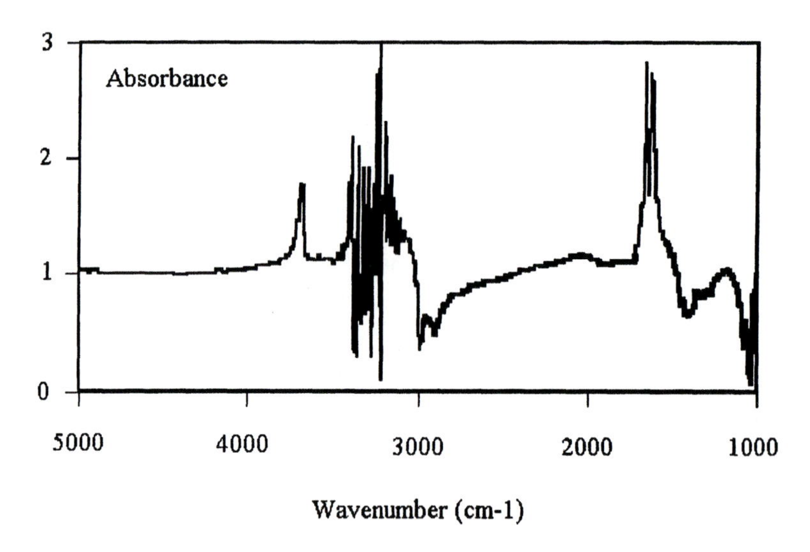

Figure 1. An example of the full MIR spectrum of wine.

In order not to use areas of the spectra that introduce noise (i.e. random variation) to the calibration, only the following areas were used for filter selection: 1543 – 965cm^{-1} (PIN 400 – 250); 2280 – 1717 cm^{-1} (PIN 589 – 445) and 2971 – 2435 cm^{-1} (PIN 770 – 631). The two regions, 1717 – 1543 cm^{-1} (PIN 445 – 400) and 3627 – 2971 cm^{-1} (PIN 940 – 770) are water absorption areas and the area from 5012 – 3627 cm^{-1} (PIN 1299 – 940) contains very little interesting information. Thus, the data set was reduced to 436 wave numbers (X-variables), Figure 2. Wave numbers, also called reciprocal centimeters (cm^{-1}), are more often used than wavelengths, in the mid and far infrared literature. The conversion is $cm^{-1} = 10,000/\mu$. (*22*).

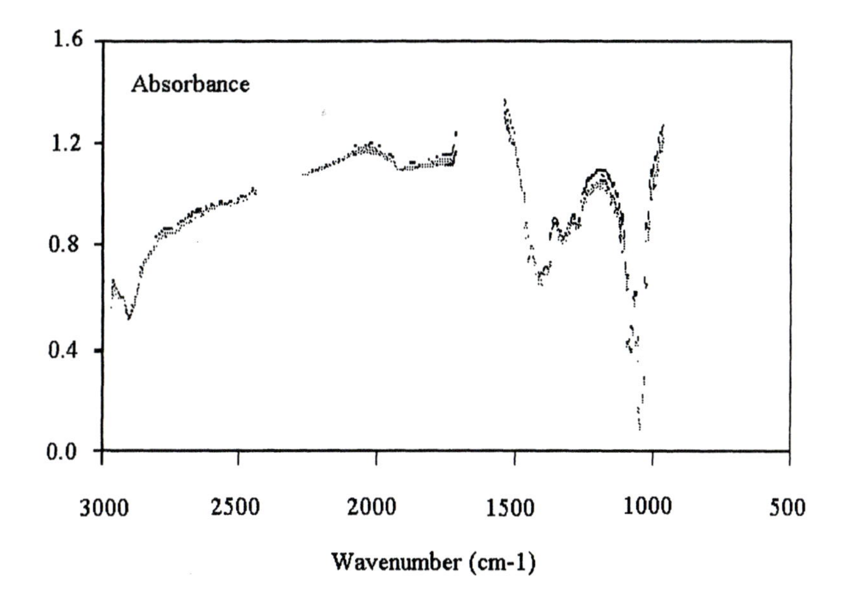

Figure 2. Selected MIR wavenumber ranges for wines.

Statistical Analysis

The data set consisted of a matrix of 20 samples × 436 absorances (X-variables). Partial-least square (PLS) regression (Uncrambler 7.6 SR-1, CAMO ASA, Oslo, Norway) was chosen since it is among the most commonly used and usually the best performing multivariate calibration method for the evaluation of IR spectra (*3*). In this study, PLS was used to predict the three color parameters of wines (Y-variables) by using the FTIR spectra (X-variables). Each raw spectrum was mean-centered prior to the statistical analyses. The raw absorbance, as well as first and second derivatives of the IR spectra were computed using a well-established approach (*23*). As the number of samples is rather limited, the estimation of the prediction ability involved the use of the entire data-set in a procedure called full-cross-validation (*24*). The method of cross-validation was used to determine the maximum number of significant factors (PCs), to ensure the predictive ability and to avoid over-fitting of the data. The wave numbers of interest were selected using a specific test option available within the Unscrambler software. Significant regions in the spectrum are chosen based on (i) a strong correlation between analyte concentration and absorbances in the sample is found, and (ii) the variance of the spectral intensity is high. The optimum number of wavenumbers for inclusion in the calibration equations was determined by comparing the regression results in terms of correlation coefficient (r) and the root mean squared error of prediction (RMSEP).

Results and Discussion

The data set must contain enough relevant information to allow the for an acceptable regression. When building a calibration set, the following parameters should be taken into consideration: the number of samples; the concentration range to be covered by the samples and the distribution of samples within this range (*25*). A total of 20 red wines mostly one year old, so that significant levels of polymeric pigment. They included a series of Pinot noir wines made with skin contact time of between 0 and 7 days, to provide a reference "white wine" with essentially no color, and several low colored samples. These were analyzed for the color measures and their compositional ranges were as follow: Total color (Aacet @ 520 nm): almost zero to 10 AU; Total Anthocyanins (TA @ 520 nm): almost zero to 3.4 AU; Copigmentation (C @ 520 nm): almost zero to 1.9 AU; and Polymeric pigment ([P] @ 520 nm): almost zero to 5.4 AU. Athough the total number of samples was limited, the wines covered a wide range of values that are representative of most of the commercial young red wines.

To get an initial overview of the data distribution it is important to plot the raw data. The selected parameters of wine color showed some deviation from the Gaussian distribution.. Increasing the number of samples may improve the population to a more normal distribution. The Y-variables were kept in their original units for a simpler interpretation of the results.

The shape of the spectrum was easily identified for all of the wines, Figures 1,2. No obvious outliers, grouping of samples or scattering deviation were evident. In data analysis it is the variation between the samples that is of most interest. In our case, the main differences visually, were located in the early region of the spectra between the wavenumber 1500 – 1000 cm^{-1} , Figure 2. The MIR spectra of phenolic compounds that were extracted with methanol_from red wines show various absorption bands around 1450, 1340, 1280, 1230, 1200, 1145, 1110, and 1060 cm^{-1}. Most remarkable are the differences at 1145 and 1340 cm^{-1} that can be considered as characteristic for each cultivar (*10*). The spectra of flavanoid-based tannins in aqueous solution shows a peak around 1285 cm^{-1} (*18*). Generally in the region 1600 – 900 cm^{-1} numerous bands originating from wine phenols can be found, including anthocyanins in solid state (*19*). The bands at 1600 and 1520 cm^{-1} can be assigned to C=C bond vibrations, which are typical of aromatic systems (*26*). A strong contribution of

OH deformation vibration can be found in the region 1410 – 1260 cm^{-1}. Strong C–O valence vibrations between 1150 and 1040 cm^{-1} overlap with aromatic fingerprint bands at 1225 – 950 cm^{-1}. CH_3 symmetric deformation vibrations occur in the region 1370 – 1190 cm^{-1}. Other authors ascribe the 1049 cm–1 region to C-OH and C–C stretching mode (*27, 28*). However, it is not excluded that the FTIR spectra may contain useful information located in too weak peaks not visible by direct visual inspection. More specific chemical and spectroscopic application knowledge is required to fully interpret this plot.

The practice of weighting variables by the reciprocal of their standard deviation is common in many applications, especially where the variables have different units. In contrast this is seldom used in spectroscopy calibration, because the variable are of the same type and units, and information is usually considered to be related to broader peaks (*4*). A preliminary trial showed that X-variable standardization amplified the noise of the PLS model (data not shown). Since the 436 X-variables are all in the same units (ABS @ 520 nm) with presumably similar noise levels, they were not weighted. The Y-variables were also be kept unweighted, since the relative noise levels between X and Y variables are more or less irrelevant (*3*).

When working with FTIR data, an important decision is whether a pre-processing method is necessary. The specific spectroscopic transformations of X includes: derivative, smoothing and Multiplicative Scatter Correction. In particular, first or second derivative spectra enhance the spectral differences between similar compounds and eliminate baseline drift effects (*13*). A constant background is removed by transforming the original spectra into first derivative spectra, a linear background by transforming them into the second derivative spectra. The drawback of numerical differentiation is that it amplifies the noise in the results. Therefore, it is necessary to smooth the data beforehand (*29*). However, it is a good practice to start by modeling raw data and then try transformations if the model needs them (*4*).

In spectroscopy applications it is often appropriate to refine the model by deleting wavelengths. In practice, the B-coefficients (regression coefficients) give the accumulated picture of the most important wavelengths (*4*). The early region of the spectra, 1539 – 926 cm^{-1} (PIN 399–240), Figure 3, provided useful information, whereas after 1891 cm^{-1} (PIN 490)] there is little important information. A new PLS model has been calculated in the signal range 1891 – 965 cm^{-1} (PIN 490 – 250) by using an interval every 4 X-variables (i.e. the wavelengths between 1891 – 965 cm^{-1} [PIN 490 – 250] were sampled every 4 intervals).

After several trails, the best prediction result, in terms of the RMSEP, was obtained under the following conditions. The Y-variable was the original copigmentation values (ABS @ 520 nm); the X-variables were smoothed by the

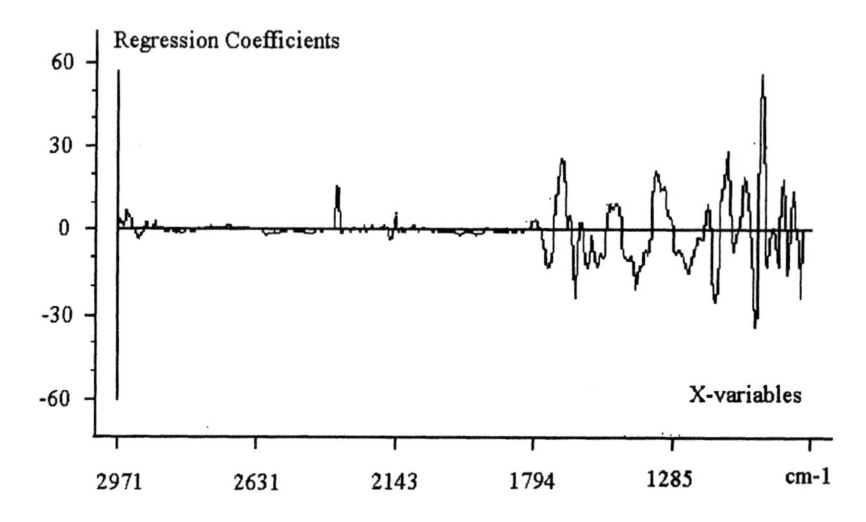

Figure 3. An example of increasing B-coefficients at certain wavenmubers.

Savitzky Golay method, using the first derivative (left and right point = 1; polynomial order = 1) of FTIR selected spectra. The spectra used were from three regions: pins 250 – 400; pins 445 – 589; pins 631 – 770; a total of 436 X-variables. The PLS analysis used every fourth wavelength from pins 250 to 490 (total of 48 channels); with mean centering, full-cross validation with 7 PCs, an outlier warning limit of 4.0 and the bilinear PLSR model.

An outlier is a sample which looks so different from the other that either is not well described by the model or influences the model too much. In regression, there are many ways for a sample to be classified as an outlier. It may be outlying according to the X-variables only, or to the Y-variables only, or to both. It may also not be an outlier for either separate set of variables, but become an outlier when you consider the (X,Y) relationship. Outliers can be detected in the Unscrambler software using X-Y relation outliers, Y-residual vs. predicted Y, the influence plot, the score plots, Y residuals, leverages and normal probability plot.

Outliers are atypical objects but there must always be a good reason to remove them from the data set, otherwise a valuable information may unwittingly be removed (*4*). If they are the result of erroneous measurements they must be removed. In our case, the sample number 20 shows the highest deviation, whereas other samples (#8, 9, 12, 14) influence more than other to the model. However, the Y-residual vs. predicted Y distribution did not show any obvious outlier. Moreover, the Y-variance decrease regularly. Thus, all samples were kept for further analysis.

Table I. Wavenumber filters selected for each color measure in PLS.

Total Color	Anthocyanin Color	Copigmentation Color	Polymeric Pigment Color
1524 – 1451	1470-1451	1528-1516	1489-1451
1408-1358	1404-1358	1374-1300	1385-1354
1335-1304	1331-1308	----	1335-1304
1269-1246	----	1265-1235	1269-1250
1188-1150	1188-1150	1188-1150	----
1115-1103	1115-1103	1115-1103	1115-1107
1084-1065	1084-1069	1088-1065	1084-1069
7 Filters	6 Filters	6 Filters	6 Filters
71 Wavenumbers	46 Wavenumbers	54 Wavenumbers	43 Wavenumbers

The PCA model, Figure 4, shows that there is a spectral structure, such that the wines tend to be grouped according to their cultivar. Red wines are well spread over the first two PCs and the score plot shows the presence of three clusters: (1) the Pinot noir wines with varying skin contact from Davis; (2) the two wines from Italy; and wines from Napa Valley. The Cabernet sauvignon, Merlot and Cabernet Franc are grouped together, as are the Pinot noir wines and the Sangiovese are together separately. This finding suggests that copigmentation varies depending on several factors, including grape cultivar,

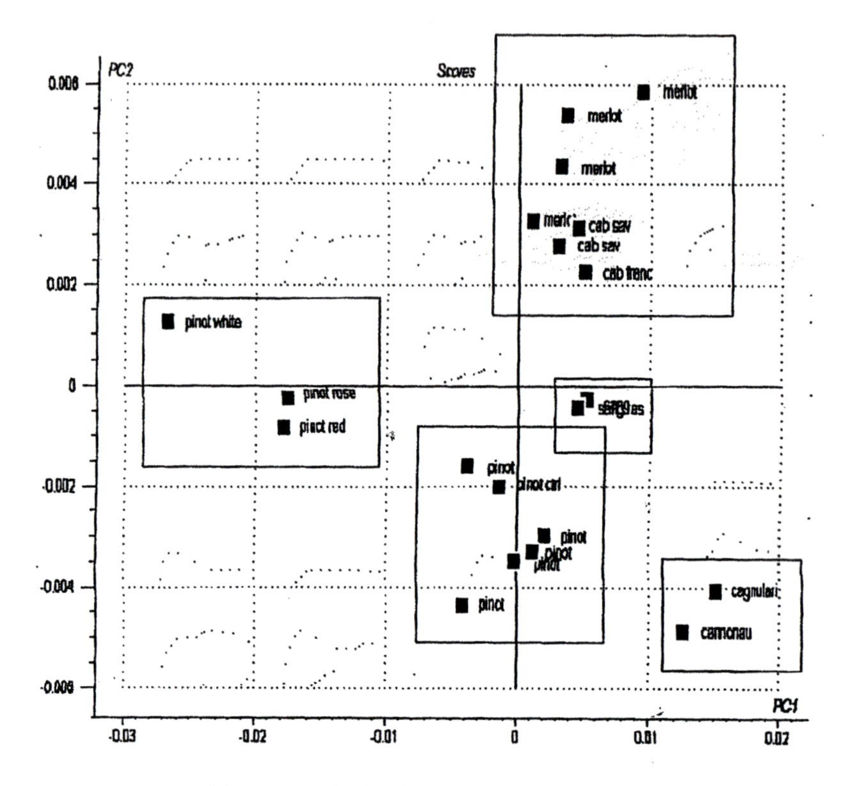

Figure 4. Principal Component Diagram.

origin and winemaking conditions. The first PC component use about 69% of the X information to explain 41% information on Y.

Thus, there is little reduntant/irrelevant information in X, with respect to Y. In other words, the FTIR measure is low-selective and the eventual attempt to reduce the number of X-variables could be problematic. The regression coefficient spectrum (top right) is often useful in understanding the chemistry of the application. Large values imply wavelengths in which there is significant absorption related to the constituent of interest. The overall shape of the regression coefficient spectrum is rather complicated.

No single wavelength shows any selective signal for copigmentation in red wine. This plot demonstrated the ability of multivariate calibration to find a set of positive and negative weights to ascribe to the different wavelengths in order to remove several interferences in the FTIR data. Only through the empirical

Table II. Variance explained by 1 PC and 2 PC models.

Parameter	PCs	Variance Explained in X (%)	Variance Explained in Y (%)
Total Color	1 2	84 89 (5)	32 89 (37)
Anthocyanin Color	1 2	83 86 (3)	27 92 (65)
Copigmentation Color	1 2	81 89 (8)	35 93 (58)
Polymeric Pigment Color	1 2	87 92 (5)	33 89 (56)

combination of several wavelength channels it become possible to attain selectivity (*3*)

The residual Y-variance (bottom left) for both calibration and validation decrease regularly by increasing the number of PCs. A similar trend is showed by the prediction error in y, RMSEP = 0.162 AU. For practical purposes an optimal model includes 7 PCs; the two later PCs were very small and the prediction ability improve very little. After the 9-PC the Y-variance increase, thus indicating that some problem occurred. PLS is very sensitive to overfitting, so finding the optimal number of PLS-components is crucial (*4*). In PLS any increase in the validation variance is a very bad sign, usually because of the presence of outliers, noise, non-linearities (4). Since PLS focuses on Y, the Y-relevant information is usually expected already in early components. There are, however, situations where the variation related to Y is very subtle, so many components will be necessary to explain enough of Y. For example, modeling protein in wheat from NIR-spectra for example may require 8 to 18 components, because the dominating variations in the data may be related also to grain size, packing, chemistry, etc. (*4*). Up to 7 PCs are required if there is little knowledge

Table III. Calibration statistics for all wine color measures.

Parameter	Range (AU)	Filters	PCs	Slope	Offset	Coefficient r	RMSEC	SEC	Bias
Total Color	0-10	7	2	0.898	0.510	0.947	1.004	1.030	--
			6	0.982	0.088	0.991	0.418	0.428	--
Anthocyanin Color	0-3.4	6	2	0.919	0.122	0.959	0.276	0.283	--
			6	0.975	0.038	0.987	0.154	0.158	--
Copigmentation Color	0-1.9	6	2	0.926	0.067	0.962	0.159	0.163	--
			6	0.985	0.013	0.993	0.071	0.073	--
Polymeric Pigment Color	0-5.4	6	2	0.891	0.280	0.944	0.536	0.550	--
			4	0.939	0.155	0.969	0.399	0.410	--

Table IV. Prediction statistics for all wine color measures.

Parameter	Range (AU)	Filters	PCs	Slope	Offset	Coefficient r	RMSEP	SEP	Bias
Total Color	0-10	7	2	0.779	1.068	0.906	1.335	1.370	-0.033
			6	0.966	0.071	0.968	0.793	0.808	-0.099
Anthocyanin Color	0-3.4	6	2	0.790	0.305	0.926	0.373	0.383	-0.011
			6	0.959	0.046	0.962	0.267	0.274	-0.015
Copigmentation Color	0-1.9	6	2	0.811	0.176	0.928	0.221	0.227	-0.005
			6	0.960	0.018	0.960	0.166	0.169	-0.018
Polymeric Pigment Color	0-5.4	6	2	0.797	0.507	0.909	0.678	0.695	-0.015
			4	0.894	0.231	0.935	0.579	0.593	-0.042

about how many physical phenomena varied in the chosen calibration data set (*4*).

The error variance (bottom right) shows the apparent predictive ability of the model: the copigmentation predicted with 7 PCs during cross-validation (ordinate) is plotted against the measured copigmentation (abscissa) Given the limited number of sample calibrated for this study, with little attempt to improve X-data normality distribution and non-linearity, this is considered a satisfactory predictive performance. Addition of more components may improve the modeling fit but reduces the prediction ability (*4*). The optimum number of terms (PCs) for inclusion in a PLS calibration equation was selected based on the standard errors of cross-validation, which should be minimized. Bias is the average difference between predicted and measured Y-values for all samples in the validation set. Bias measures the accuracy of a prediction model. Bias also indicates the presence of a systematic difference between the average values of the training set and the validation set. If there is no such difference, the Bias will be zero (*4*). In our model there is a very small bias (–0.019) so the root mean square error of prediction (RMSEP = 0.162) and the standard error of prediction (SEP = 0.162) are about equal and showed a satisfactory correlation, $r = 0.962$ ($r^2 = 0.925$, i.e. 92% Y-variance explained). The RMSEP expresses the error expected in future predictions (*4*). Therefore, RMSEP is an estimated precision; e.g. 0.91 AU ± 0.162 AU calculated on the average value of wine copigmentation. The relative error is thus max. 0.162/0.17 = 95% on the low value levels and max. 0.162/1.89 = 8% on the high level. The copigmentation of wine showed an approximate reliability range of the mean, estimated as ±2 standard errors of the mean, that is (0.91 – 0.16; 0.91 + 0.16) or from 0.75 to 1.07 AU.

References

1. Günzler, H.; Gremlich, H.-U. *IR spectroscopy.* 2002, Wiley-VCH, Weinheim, Germany.
2. Reh, C. *In-line and off-line FTIR measurements.* 2001, Woodhead Publishing Limited and CRC Press LLC.
3. Martens, H.; Martens, M. *Multivariate analysis of quality. An introduction.* 2001, John Wiley and Sons, Ltd, New York, NY.
4. Esbensen, K.; Schönkopf, S.; Midtgaard, T.; Guyot, D. *Multivariate analysis in practice.* 1998, CAMO ASA, Oslo, Norway.
5. Schindler, R.; Vonach, R.; Lend, B.; Kellner, R. *Fresenius J. Anal. Chem.*, **1998**, 362, 130–136.
6. Patz, C.-D.; David, A.; Thente, K.; Kürbel, P.; Dietrich, H. *Vitic. Enol. Sci.*, **1999,** 54, 80–87.
7. Dubernet, M.; Dubernet, M. *Rev. Fr. d'Œnol.*, **2000,** 181, 10–13.

8. Gishen, M.; Holdstock, M. *Aust. Grapegrower Winemaker*, Annual Technical Issue, **2000,** 75–81.
9. Pérez-Ponce, A.; Garrigues, S.; de la Guardia, M. *Química Analítica,* **2000,** 19, 151–158.
10. Edelmann, A.; Diewok, J.; Shuster, K. C.; Lendl, B. *J. Agric. Food Chem.,* **2001,** 49, 1139–1145.
11. Manley, M.; van Zyl, A.; Wolf, E. E. H. *Sth. African J. Enol. Vitic.*, **2001,** 22, 93–100.
12. Palma. M.; Barroso, C. G. *Talanta,* **2002,** 58, 265–271
13. Rambla, F. J.; Garrigues, S.; Ferrer, N.; de la Guardia, M. *Analyst,* **1998,** 123, 277–281.
14. Iizuka, K.; Aishima, T. *J. Food Comp. Anal.,* **1999,** 12, 197–209.
15. Defernez, M.; Kemsley, E. K.; Wilson, R. H. *J. Agric. Food Chem.,* **1995,** 43, 109–113.
16. Kemsley, E. K.; Ruault, S.; Wilson, R. H. *Food Chem.,* **1995,** 54, 321–326.
17. Šašic, S.; Ozaki, Y. *Anal. Chem.,* **2001,** 73, 64–71.
18. Edelmann, A.; Lendl, B. *J. Am. Chem. Soc.*, **2002,** 124, 14741–14747.
19. Merlin, J. C.; Cornard, J. P.; Stastoua, A.; Saidi-Idrissi, M.; Lautie, M. F.; Brouillard, R. *Spectrochimica Acta* **1994,** 50A, 703–712.
20. Boulton, R. B. *Am. J. Enol. Vitic.,* **2001,** 52, 67–87.
21. Adams, D. O.; Harbertson, J. F. *Am. J. Enol. Vitic.,* **1999,** 50, 247–252.
22. Williams, P. C.; Norris, K. H. *Near-infrared technology in the agricultural and food industries.* 1987, Am. Assoc. Cereal Chem. St. Paul, MN.
23. Savitsky, M.; Golay, J. E. *Anal. Chem.* **1964,** 36, 1627–1639.
24. Martens, H.; Naes, T. *Multivariate calibration.* 1989, Wiley-Chichester, New York, NY, pp. 254–257.
25. Wiedemann, S. C. C.; Hansen, W. G.; Snieder, M.; Wortel V. A. L. *Analysis Magazine,* **1998,** 26, 38M–43M.
26. Hesse, M.; Meier, H.; Zeeh, B. *Spektroskopische Methoden in fer Organischen Chemie.* 1991, Thieme, Stuttgart, Germany.
27. Wilson R. H.; Belton P.S. . *Carbohydrates Res.,* **1988,** 180, 339–344.
28. Bellon-Maurel, V.; Vallat, C.; Goffinet, D. *Appl. Spectroscopy,* **1995,** 49, 556–562.
29. Candolfi, A.; De Maesschalck, R.; Jouan-Rimbaud, D.; Hailey, P. A.; Massart, D. L. *J. Pharm. and Biomed. Anal.,* **1999,** 21, 115–132.

Chapter 6

The Fate of Anthocyanins in Wine: Are There Determining Factors?

Hélène Fulcrand*, Vessela Atanasova, Erika Salas, and Véronique Cheynier

UMR Sciences Pour l'Œnologie, 2, place Viala, 34060 Montpellier, France
***Corresponding author: phone: 33 4 99 61 25 84, email: fulcrand@ensam.inra.fr**

After extraction, grape anthocyanins undergo many reactions during winemaking and ageing. Relative amounts of anthocyanins and tannins, dissolved oxygen, pH, and presence of yeast metabolites (e.g. acetaldehyde, pyruvic acid) influence these reactions. pH controls reactivity because at low pH the electrophilic flavylium ion is the most abundant form of anthocyanin whereas at higher pH the nucleophilic hydrated is the most abundant form. Tannin to anthocyanin ratio determines the proportions of tannin derivatives, tannin anthocyanin adducts and anthocyanin polymers. Wine exposure to oxygen favors ethanol oxidation to acetaldehyde and susbsequent reactions involving acetaldehyde. Acetaldehyde-induced reactions first lead to ethyl-linked species, including flavanol oligomers, anthocyanin oligomers and tannin-anthocyanin adducts. Acid catalysed cleavage of the labile ethyl-bonds between flavanol units followed by addition of the resulting species onto flavylium ions then generates pyroanoanthocyanins. Current knowledge on the impact of wine-making processes on formation of anthocyanin-derived pigments as well as their impact on wine organoleptic properties is discussed.

Introduction

Anthocyanins are water-soluble pigments, providing a wide range of colors to living plant tissues. However, structures associated with the colored forms of these pigments are not stable at the pH encountered in plants. Thus, the red primary structures referred to as flavylium cations (AH^+), are predominant only in acidic media (pH<2). Above this value, these cations are converted into other anthocyanin forms through proton transfer, hydration and tautomerisation reactions. As the pH is raised, fast deprotonation of flavylium cations leads to the blue quinonoidal bases (A), while water addition on flavylium cations progressively yields the hydrated hemiketal forms (AOH), which have partly converted into chalcones (*1*). The relative proportion of each form is determined by the pH. In the plant pH range (3 to 7), anthocyanins are mainly under colorless hydrated forms. Therefore, nature has developed very efficient mechanisms such as formation of molecular complexes to preserve anthocyanin color.

Quinonoidal bases and cations stack vertically into hydrophobic planar structures that exclude. The association can involve identical anthocyanin molecules (self-association), a part of the structure itself (usually one of the aromatic acyl group substituents) in a process then referred to as intra-molecular copigmentation or, finally, another planar molecule (inter-molecular copigmentation). Recent reviews describe in details the mechanism of these associations (*2*), (*3*), (*4*). The resulting complexes are more stable than the individual anthocyanin, so that the balance is displaced from hydrated forms towards the red flavylium cations (figure 1). In this way, the color is maintained in the plant pH range. Metallic complexation can also take place, especially with aluminium, producing a similar effect. However, when anthocyanins are extracted from plants, their high reactivity overcomes their stabilization processes. In wine-making, anthocyanins are extracted together with other phenolic compounds with which they react in many different ways. Copigmentation may be the first step to covalent bonding (*5*).

This paper reviews recent findings on the products and mechanisms of these reactions, the major factors determining their relative importance and their consequences on wine colour and colour stability.

Anthocyanin Reactivities

Structural identification of anthocyanin derived pigments and the studies carried out on their mechanisms of formation allowed us to bring out a general pattern of the flavonoid reactivities (figure 2). Basically, flavonoids are built on a three-ring core consisting of “A” and “B” aromatic systems bearing hydroxyl

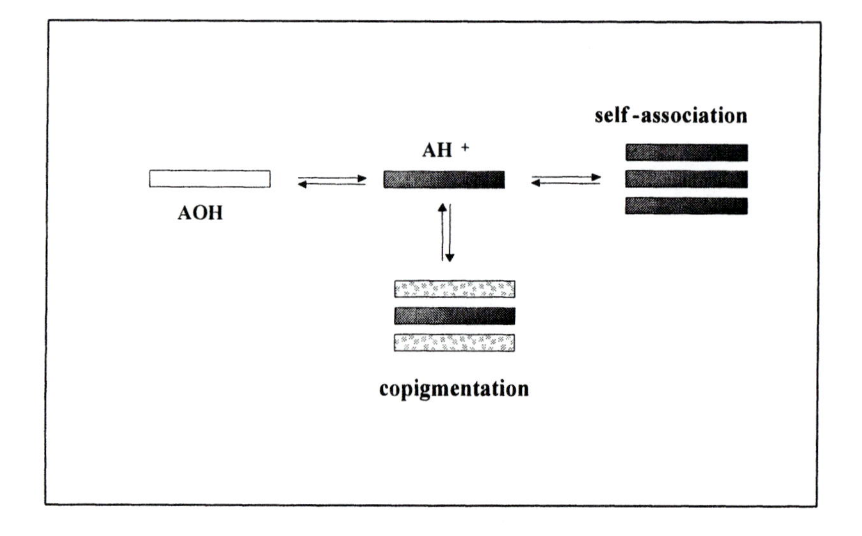

Figure 1. Copigmentation mechanism

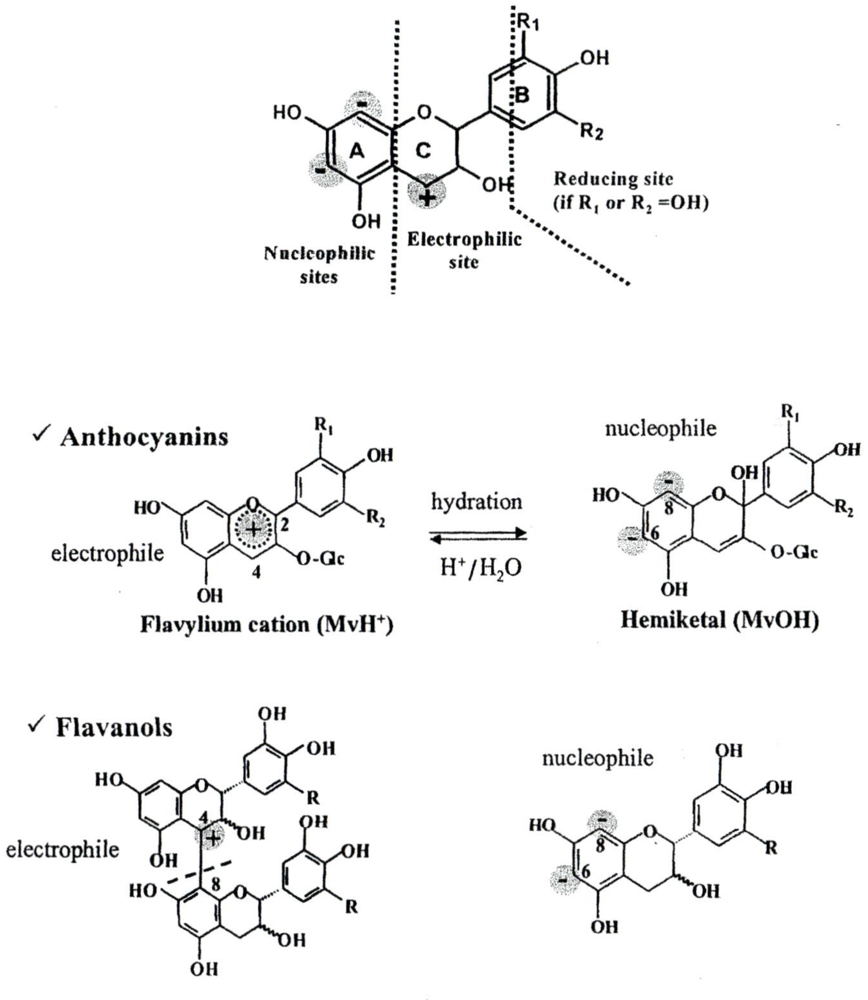

Figure 2. Flavonoid reactivities

substituents connected by a central pyran ring, "C".. Different flavonoid classes are differentiated according to the oxidation state (degree of unsaturation) of the pyran ring. Except flavanol monomers and polymers (classically named tannins) that have a fully saturated C-ring, all major flavonoid molecules are pigments, absorbing in higher wavelengths as the unsaturation increases (resulting in an extent of electron delocalization): from yellow for flavones and flavonols to red and blue for anthocyanins. A type of reactivity can feature each ring.

When the B-ring possesses a catechol type structure, this nucleus acts mostly as a reducing agent. Once oxidized, it can undergo either coupled oxidation or Michael addition as a powerful electrophile, both mechanisms restore the catechol structure. The main grape anthocyanins, as their B-ring lacks the catechol hydroxylation pattern.

The A-ring shared by all grape flavonoids is a phloroglucinol type structure possessing two nucleophilic sites activated by three hydroxyl groups, two located in *ortho* position and one in *para* position. Therefore, the A-ring acts as a nucleophilic reagent and undergoes electrophilic aromatic substitution. For anthocyanins, the nucleophilic character of A-ring is provided by their hydrated forms.

The central C-pyran ring tends to have a cationic charge, natural for the flavylium forms of anthocyanins, or formed by acidic cleavage of the interflavan bonds connecting the flavanol units of tannins. Therefore, the C-ring reacts as an electrophilic moiety and undergoes nucleophilic addition.

Another kind of reaction has been demonstrated for the cationic forms of anthocyanins, resulting in the formation of a second pyran ring. The mechanism involves both the electron deficient C-4 and the 5-hydroxyl group of the anthocyanin with compounds possessing a polarizable double bond (*6*). The new pigments thus formed are referred to as pyranoanthocyanins.

The general reactivity of the A, B, and C rings of anthocyanins and other flavonoids results in a large number of potential products that may undergo further reactions. Given the large structural diversity of grape tannins (*7*) (*8*) (*9*)), a quasi-infinite number of reaction products, colored or not, may be formed in wines. Therefore, the conversion pathways of native components and their impact on wine sensory properties is difficult to define.

Different Types Of Pigments Identified In Wines And Wine-Like Solutions

Numerous pigments have been characterised in wines and wine–like model solutions in the last decade. The structures newly identified can be classified into three groups with respect to their formation pathways (figure 3). Reactions involved in the conversion of grape anthocyanins to more stable pigments are

✓A^+-F type pigments

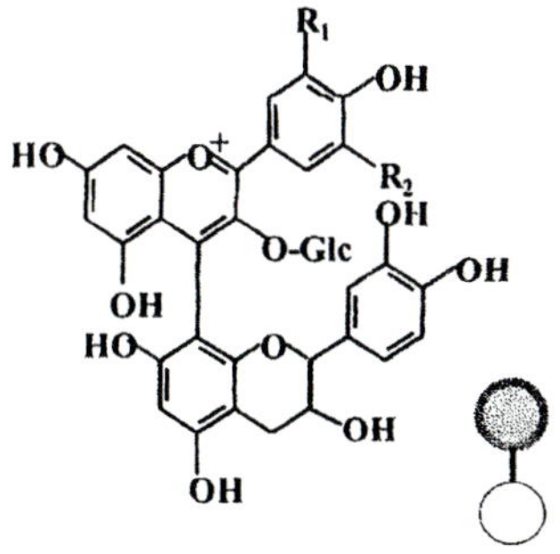

✓A(-O-)F type adducts

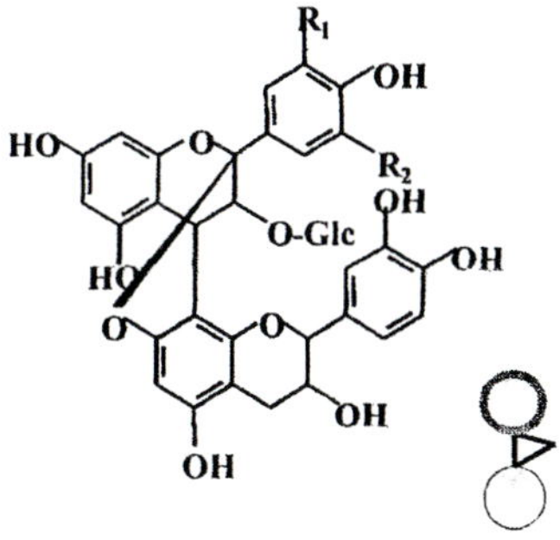

✓F-A^+ type pigments

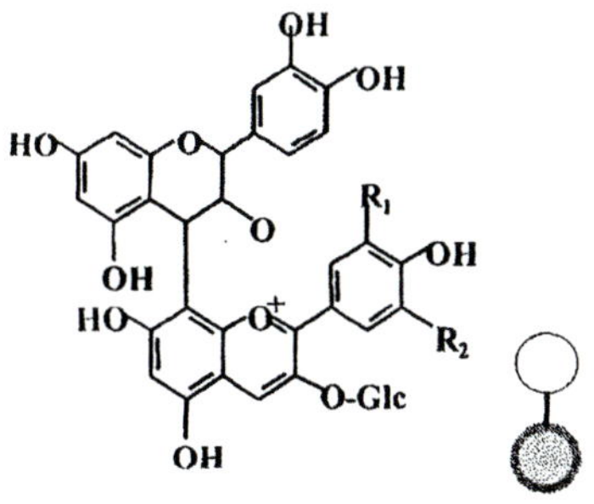

✓A^+-AOH type pigments

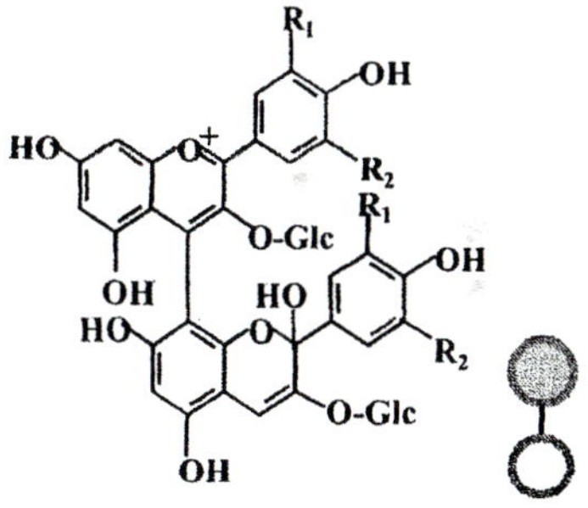

Figure 3a. Products issued from direct reactions

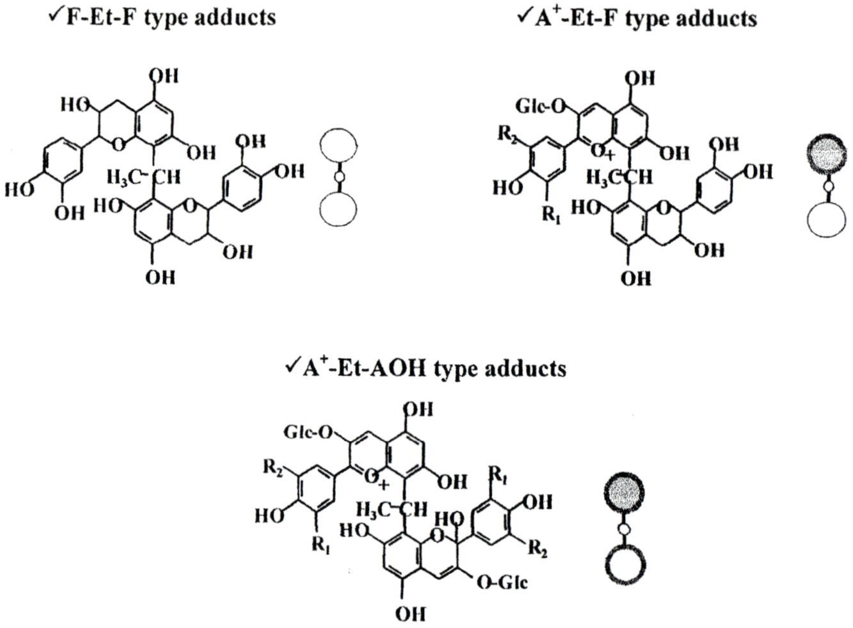

✓F-Et-F type adducts

OH HO OH OH H_3C-CH OH HO OH HO OH OH OH

✓A^+-Et-F type adducts

OH Glc-O R_2 O+ OH OH H_3CCH OH HO R_1 HO O OH OH

✓A^+-Et-AOH type adducts

OH Glc-O R_2 O+ OH R_1 OH H_3C-CH HO HO HO R_1 O R_2 O-Glc OH

Figure 3b. Products issued from acetaldehyde-induced polymerization

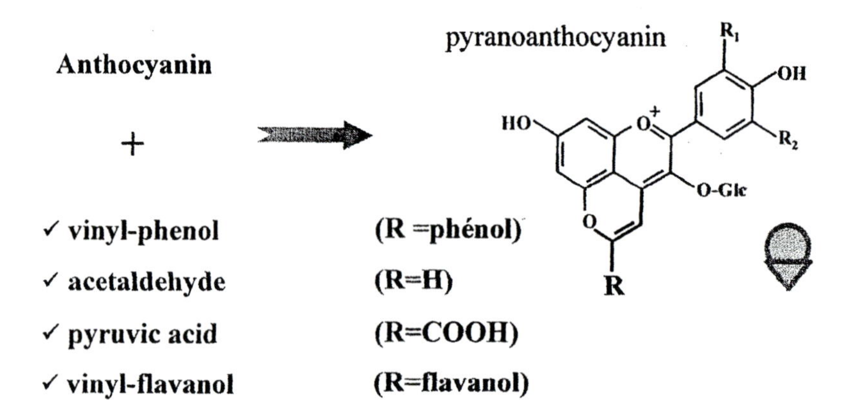

Figure 3c. Pigments issued fron cycloadditions

classically described as anthocyanin–flavanol reactions. They can be direct, leading to a first group of products (fig. 3a), including F-A$^+$ or A$^+$-F pigments as well as A(-O-)F colorless compounds that contain an additional ether linkage and a structure analogous to that of A-type proanthocyanidins. F-A$^+$ structures arise from nucleophilic addition of the anthocyanin hemiketal form onto the flavanol carbocation (F$^+$), followed by dehydration of resulting F-AOH adduct. The other two structures, in which the anthocyanin is in the upper position, result from the reverse reaction, *i.e.* addition of the nucleophilic flavanol onto the flavylium carbocation. The intermediate flavene adduct then either oxidizes to the A$^+$-F pigment, or proceeds to A(-O-)F.

Interestingly, recent findings in characterization of grape skin tannins suggest that pigmented tannins can already form during grape berry ripening (*10*). However, no clear evidence can support whether these structures are covalently bound or just non-covalently associated nor if they result from an artifact during extraction.

Both flavanol-anthocyanin pigments and colorless A(-O-)F have been detected in wines (*11*). Mass signals do not allow to distinguish between F-A$^+$ and A$^+$-F derivatives. However, dimeric anthocyanins consisting of one unit under flavylium cation and the other one under hydrated hemiketal form (A$^+$-AOH) were characterized by mass spectrometry in wine like solutions (*12*). The existence of such structures arising from direct reactions between anthocyanin molecules confirms that they react both as nucleophiles and as electrophiles in wine.

The second type of reaction involved in conversion of grape anthocyanins to wine pigments is a polymerization process involving an aldehyde component (fig.3b). Acetaldehyde-induced reactions have been thoroughly studied with respect to oxidation processes, as acetaldehyde is believed to result from ethanol oxidation (*13*). They involve protonation of the aldehyde followed by nucleophilic addition of the flavonoid onto the resulting carbocation (*14*). Ethyl-linked products, including ethyl-linked flavanols (F-Et-F) (*15*), (*16*) and ethyl-linked anthocyanin-flavanol pigments (A$^+$-Et-F) (*16*) have been detected in wines. The formation of ethyl-linked anthocyanin oligomers (A$^+$-Et-AOH) was also recently shown to occur both in model solution and in wine (*17*). The ethyl-linked 8,8-malvidin 3-glucoside dimer was characterized by NMR under bi-flavylium cation forms (*17*). However, physico-chemical studies carried out on this pigment showed that the dimer under monoflavylium cation is the most abundant form at wine pH since the first hydration constant ($pK_h = 1.8$) is lower than that of its precursor and the second one much higher ($pK = 4.6$) (*18*). Analogous oligomers have been detected by mass spectrometry. Other aldehydes including glyoxylic acid (*19*) and furfural (*20*) were shown to react in the same way. However, products resulting from flavanol reactions with these aldehydes undergo further rearrangement to xanthylium pigments (*21*).

The third group of wine pigments corresponds to pyranoanthocyanins (fig. 3c) formed, as described in the anthocyanin reactivity section, by reaction of genuine anthocyanins with yeast metabolites released during fermentation (*6*) (*22*). Pyruvic acid leads to the major pyranoanthocyanins currently found in wines, *i.e.* carboxy-pyranoanthocyanins (R=COOH) ((*6*), sometimes referred to as vitisins A (*23*). As well, acetaldehyde yields pyranoanthocyanins (R=H). When flavanols are present together with anthocyanins and acetaldehyde, flavanyl-pyranoanthocyanins (R=flavanol monomer through tetramer) were shown to occur both in model solutions and in wines ((*24*) (*25*) (*26*)). They can be generated following two reaction pathways: either reaction between anthocyanins and ethyl-linked flavanol dimers or reaction between pyranoanthocyanins and flavanols (*12*). Besides, further reactions of carboxy-pyranoanthocyanins with flavanols and acetaldehyde yield blue pigments based on flavanol and pyranoanthocyanin units linked through a vinyl bridge (*27*).

Pyranoanthocyanins are exceptionally stable pigments towards sulfite bleaching and pH variations, due to substitution of the flavylium C-4.

More complex structures consisting of both malvidin 3-glucoside and flavanol units connected by different types of linkages, including direct, ethyl and A-type bonds have been partially characterized by mass spectrometry in model solutions (*12*). Additionally, some of these compounds also contain pyranoanthocyanin moiety.

Thus, the huge diversity of structures formed in wines and wine–like solutions display a great variety of properties, contributing differently to the wine quality. The influence of vine growing conditions and wine-making processes on the relative importance of these products is still largely unknown. However, some factors may favor specific reaction pathways and induce selectivity in the formation of wine pigments, thus leading to a particular trend in wine quality, as detailed below.

Factors Influencing The Type Of Pigments Yielded In Wines

From a chemistry standpoint, the nature and amounts of end products (in this case, wine pigments) depend obviously on the relative proportion of the precursors, the relative kinetics of the reactions and finally the relative stability of the reaction products. In addition to these primary factors, physico-chemical parameters like pH and temperature can affect the type and/or kinetics of the reactions. In addition, the presence of oxygen and co-factors, like metal ions, may induce catalytic activities and favor one kind of process. The following section will focus on factors that vary according to grape growing and wine making conditions (pH, anthocyanin to tannin ratio and oxygen availablility) and their impact on final wine composition and quality.

pH Effect On Wine Pigment Composition

Based on the flavonoid reactivity pattern (figure 2), the dichotomy between electrophilic and nucleophilic characters is strongly related to pH. The more acidic the pH is, the more cationic charges prevail. Thus, the balance between flavylium cations (AH^+) and hydrated forms (AOH) is pH-dependent as well as the interflavanic bond cleavage yielding the flavanol cations (F^+). In addition, acetaldehyde protonation is also controlled by pH. Therefore, it can be expected that reactions involving flavylium, flavanol or protonated acetaldehyde cations are favored at lower pH values. Based on carbocation reactivity, these cationic species can be classified as follows:

protonated acetaldehyde (CH_3CHOH^+)> flavanol cation (F^+) > flavylium (A^+).

Acetaldehyde reactions should thus be prevalent provided acetaldehyde is present. As regards direct reaction, it can similarly be expected that formation of F-A^+ exceeds that of A^+-F and A^+-A species at lower pH values.

In fact, studies performed in model solutions in the pH range 2.2 to 4.0 showed acetaldehyde-induced polymerization is faster at lower pH values (*28*). This indicates that acetaldehyde induced-polymerizations are limited by protonation of the aldehyde reagent, even when the process involves anthocyanins. Thus, the availability of hydrated anthocyanins (*29*) required in this case is not the limiting factor. On the other hand, products resulting from acetaldehyde induced-polymerization are not very stable. Ethyl-linked flavanol oligomers are highly susceptible to acid-catalysed cleavage and thus labile in wine (C-8-Et-C8>C8-Et-C6>C6-Et-C6) (*28*) whereas ethyl-linked anthocyanin derivatives (*i.e.* A^+-Et-AOH and F-Et-A^+) are more resistant. Therefore, the latter are expected to be predominant at lower pH values and after longer storage. In fact, cleavage of the ethyl-bonds results in rearangement to other ethyl-linked species (*28*) or to flavanyl-pyranoanthocyanins (*30*) (*12*).

Similarly, direct reactions between anthocyanins and a flavanol dimer (namely epicatechin-epicatechin 3-gallate) were investigated in model solutions at pH 2 and pH 3.8 (*31*). Compounds with mass values corresponding to epicatechin-anthocyanin adducts and to anthocyanin-epicatechin-epicatechin 3-gallate, were detected at pH 2 and pH 3.8, respectively. Given the structure of the flavanol precursor, the former can be interpreted as F-A^+ arising from cleavage of the flavanol dimer followed by addition of the anthocyanin AOH form onto the epicatechin carbocation whereas the latter is presumably an A^+-F species formed by nucleophilic addition of the flavanol dimer onto the flavylium cation. Anthocyanin dimers (A^+-AOH) were detected in model solutions at pH 3.2 and 3.8, suggesting that A^+-F can similarly form in the wine pH range. These results indicate that F-A^+ reaction is a prevalent reaction at pH 2, meaning that availability of anthocyanin under hemiketal form is not a limiting factor at this pH value. Cleavage of the interflavanic bonds also required for this reaction was shown, besides, to prevail at pH 2 and to still occur at pH 3.2 (*32*) but appeared limited at pH 3.8 where flavanols mostly reacted as nucleophiles.

Formation of pyranoanthocyanins is faster at lower pH values since the mechanism requires anthocyanins to be under the flavylium cation form and ketones to be in the enol form, both of which are favoured in acidic media. Formation of flavanyl-pyranoanthocyanins should be further facilitated in acidic media due to enhanced cleavage of the ethyl bridges. However, even at higher pH values, pyranoanthocyanins are expected to accumulate progressively in wines because of their stability.

Impact Of Anthocyanin To Tannin Ratio On Wine Pigment Composition

The nature and relative proportions of anthocyanins and tannins are also of prime importance for phenolic composition changes occuring during ageing and resulting wine quality. Thus, anthocyanins and tannins usually compete as nucleophilic species in direct reactions and acetaldehyde-induced polymerizations. Their relative proportions lead to selectivity between the three types of reaction products containing F and A units (FA$_+$ type, including F-A$_+$, A-F and F-Et-A$_+$ products), F units only (FF, including F-F and F-Et-F) and A units only (A$_+$A, including A$_+$-AOH and A$_+$-Et-AOH), respectively. Consequently, a high concentration in anthocyanins (A>>F) is expected to increase significantly the amount of anthocyanin polymers A$_+$A with respect to other wine products. On the other hand, excess of tannins should favor FF reactions, including aldehyde induced-polymerization, direct reaction and oxidation, some of which may result in browning (*33*) (*34*) (*20*). When a well-balanced ratio between anthocyanins and tannins occurs in the early stages of wine making, all kinds of products, namely FA$_+$, FF and A$_+$A, can be potentially formed in wines. The respective nucleophilic strength of A-ring, provided by anthocyanin hydrated forms on one hand, and flavanol units on the other hand, can further modulate the ratio between FA$_+$, FF and A$_+$A type products whether they arise from direct or acetaldehyde-induced reaction. In addition, the relative stability of FA$_+$, FF and A$_+$A type products determines the final wine composition.

The relative stability of these products was deduced from results obtained in model solutions. As a general rule, it was found that bonds connecting flavanol units directly or through ethyl groups are less stable than those linking homologous FA$_+$ or A$_+$A pigments. In particular, F-A$_+$, F-Et-A$_+$, and A$_+$-Et-AOH were shown to be partly resistant to acidic cleavage under thiolysis conditions. This may result from occurrence within these structures of flavylium nuclei that are unable to add a proton required for initiating acidic cleavage. Therefore, FA$_+$

and AA^+ are expected to be the most stable among wine pigments. Nevertheless, the ethyl-linked pigments are not as stable as those arising from direct reactions.

Hydration of A-Et-A^+ type pigments was investigated with dimeric species. The greater resistance towards hydration established for the dimer can be extended to larger anthocyanin polymers that should display even greater resistance. Thus, anthocyanin polymers can be considered as chemically stable pigments.

Characteristics of FA^+ and AA^+ pigments were studied in wine-like model systems containing anthocyanins in the presence and absence of equimolar amounts of procyanidin dimer (B2 3'gallate) (*31*). Disappearance of anthocyanins was increased in the presence of the flavanol both at pH 2 and at pH 3.8. Decay in the values of absorbance at 520 nm was also slightly faster in solutions containing the flavanol dimer but appeared rather limited in both solutions compared to the loss of anthocyanins, confirming that most of this loss is due to conversion to FA^+ and AA^+ pigments (as shown in figure 4 for the pH 3.8 solutions). Pigments formed in the solution containing tannins were found to be more resistant to sulfite bleaching.

Finally, the relative amounts of anthocyanin-flavanol pigments and anthocyanin-anthocyanin pigments obviously depend on the anthocyanin to flavanol ratio. However, the kinetics of anthocyanin polymerization and flavanol-anthocyanin reactions as well as the quantities of the resulting pigments in wine remain to be established. Consequently, the importance of these processes with respect to other anthocyanin reactions (hydration and subsequent degradation, cycloadditon with yeast metabolites) and the impact of resulting polymeric anthocyanins on wine quality are speculative.

Oxygen Impact On Wine Composition And Color Stability

Oxidation reactions involving phenolic compounds are extremely complex and not fully elucidated. In fact, molecular oxygen (O_2) cannot directly react with most organic compounds owing to its electronic configuration (triplet state). However, O_2 needs to become reactive by enzymatic, metal ions, or free radical catalysis. In wines, enzymatic oxidations prevail in the early stages of winemaking. Once enzymatic activities are no longer sustainable (after fermentation), chemical oxidations take over. Unexpectedly, chemical oxidations are not principally addressed to phenolic compounds but rather to other major wine components, like ethanol and tartaric acid. In fact, acetaldehyde has been proposed to result from ethanol oxidation by hydrogen peroxide, itself preliminary formed by radical reactions between phenolic compounds and O_2 (*13*). Although acetaldehyde is also a major yeast metabolite, it can be considered as an oxidation marker. Consequently, pigments arising from acetaldehyde reactions can also serve as markers of wine oxidation.

Moreover, all cycloaddition pathways require an oxidation step to recover the flavylium moiety within the final structures, as well as direct reactions yielding A^{+}-F species. In fact, it has been shown that direct reactions may lead to either A(-O-)F product or A^{+}-F pigment from the same flavene intermediate. The former was detected only in wines carefully protected from oxygen exposure (*11*) and obtained in model solutions under anaerobic conditions (*35*), suggesting that molecular oxygen may participate in the formation of the latter.

Modulation Of Wine Pigment Composition Through Vine Growing And Wine Making Practices

Relative proportions of anthocyanins and flavanols in grapes depend primarily on the variety but environmental and cultural conditions that affect grape maturity can further modify grape composition as tannin composition remains stable after veraison whereas anthocyanins continue to accumulate. Winemaking practices also modulate extraction of phenolic compounds and their subsequent reactions in wine. In particular, maceration conditions greatly affect the relative proportion of anthocyanins and flavanols extracted into the wine. Basically, diffusion kinetics depend both on the respective solubilities of polyphenolic constituents and on their localization in the grape berry. Anthocyanins are usually quickly extracted while the concentration of tannins progressively increases with time. Tannins from skin are extracted faster than seed tannins and polymers with highest molecular weight diffuse slower than those of lower molecular weight. Fermentation temperature can modify slightly extraction of anthocyanins and tannins. When a high temperature is maintained at the end of fermentation, phenolic extraction and especially that of tannins is more important. Introduction of maceration phases is more effective. Pre-fermentation maceration at low temperature enhanced extraction of both anthocyanins and tannins, without deeply modifying the tannin to anthocyanin ratio. On the other hand, a three-week maceration performed after fermentation increased tannin extraction and reduced anthocyanin content, presumably owing to their involvement in tanin-anthocyanin reactions, so that the tannin to anthocyanin ratio increased 2 to 3 fold. (figure 5)

Along with changes in anthocyanin and tanin proportions, winemaking practices can bring additional components in wines through selection of fermentation yeasts. Enzymatic activities of particular yeast strain influence the release of volatile phenols, pyruvic acid and acetaldehyde and, therefore, control the relative proportions of the corresponding pyranoanthocyanins.

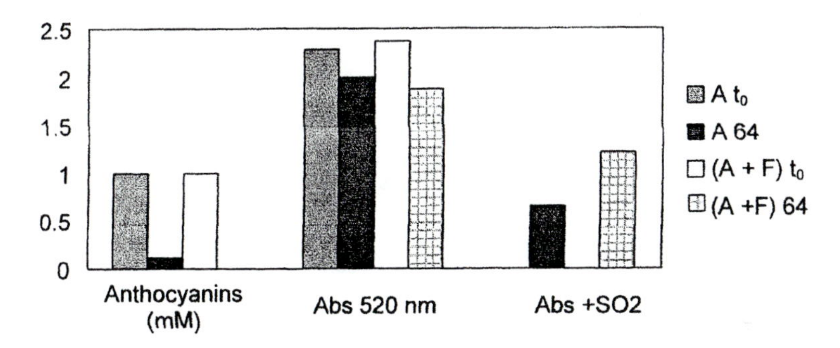

Figure 4. Anthocyanin and color changes in wine-like model

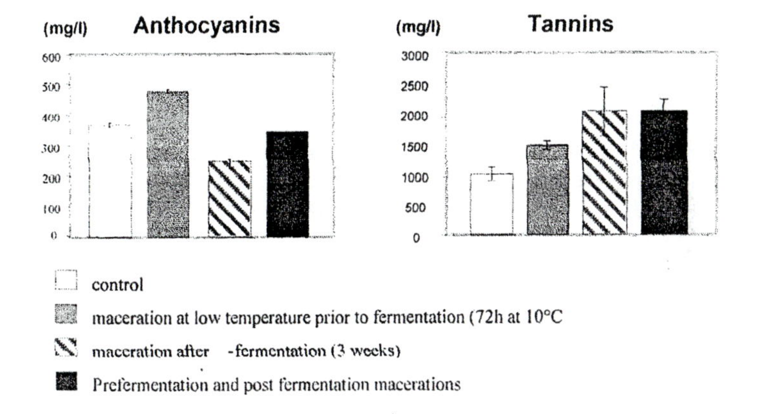

Figure 5. Effect of maceration on anthocyanin and tannin extraction

Enzymatic oxidation reactions are limited in red wine-making since the fermenting yeast consume most of the available oxygen (*36*). Chemical oxidation is usually restricted but can be much increased by microoxygenation performed in order to enhance color intensity and stabilize wine pigments. This technique allows continuous oxygen supply in quantities small enough to ensure that no accumulation takes place and that the exchange occurs in a very small volume zone so as to mimic oxidation conditions encountered in barrel ageing. Thus, the influence of controlled oxygenation on the phenolic composition and color characteristics of the red wine was monitored over a seven-month storage period by comparison with a control (same wine saturated with N_2) (*37*). The set of analytical data recorded along the seven-month period was processed by principal component analysis. Distribution of the wine samples along the first and second principal components accounting, respectively, for 69% and 16% of the variance showed that the first axis corresponds to an "ageing" axis whereas the second one can be interpreted as an "oxygenation" axis. High levels of free anthocyanins and native tannins as well as pigments showing the same properties as free anthocyanins with respect to sulfite bleaching and copigmentation or self-association characterized the young wines. In contrast, wine ageing was associated with larger amounts of pyranoanthocyanins and direct flavanol anthocyanin adducts and increased resistance to sulfite bleaching. The major variables contributing to the oxygenation axis are concentration of ethyl-linked anthocyanin-flavanol pigments and color intensity (*i.e.* sum of absorbances at 420, 520 and 620 nm).

Thus, it seems that the major changes induced by micro-oxygenation are due to acetaldehyde reactions (figure 6). In a first stage, acetaldehyde-induced reaction between anthocyanins and flavanols leads to purple ethyl-linked pigments. These intermediate pigments have been shown to be rather unstable (*38*). They may then proceed to flavanyl-pyranoanthocyanin (*12*) that accumulate with time since analogous pigments were found to be general markers of wine ageing. As well, oxygen may favor formation of pyranoanthocyanins and flavanyl-pyranoanthocyanins through acetaldehyde-anthocyanin cyloadditions.

Conclusion

Owing to the high reactivity of flavonoids and the numerous reaction pathways, a huge variety of pigments can be generated in the course of winemaking and ageing. Their quantitative importance depends on many factors. Firstly, pH determines the ratio between flavylium cations and hydrated forms of anthocyanins and controls the extent of some reaction pathways (interflavanic bond and ethyl-bond cleavage as well acetaldehyde-induced polymerizations).

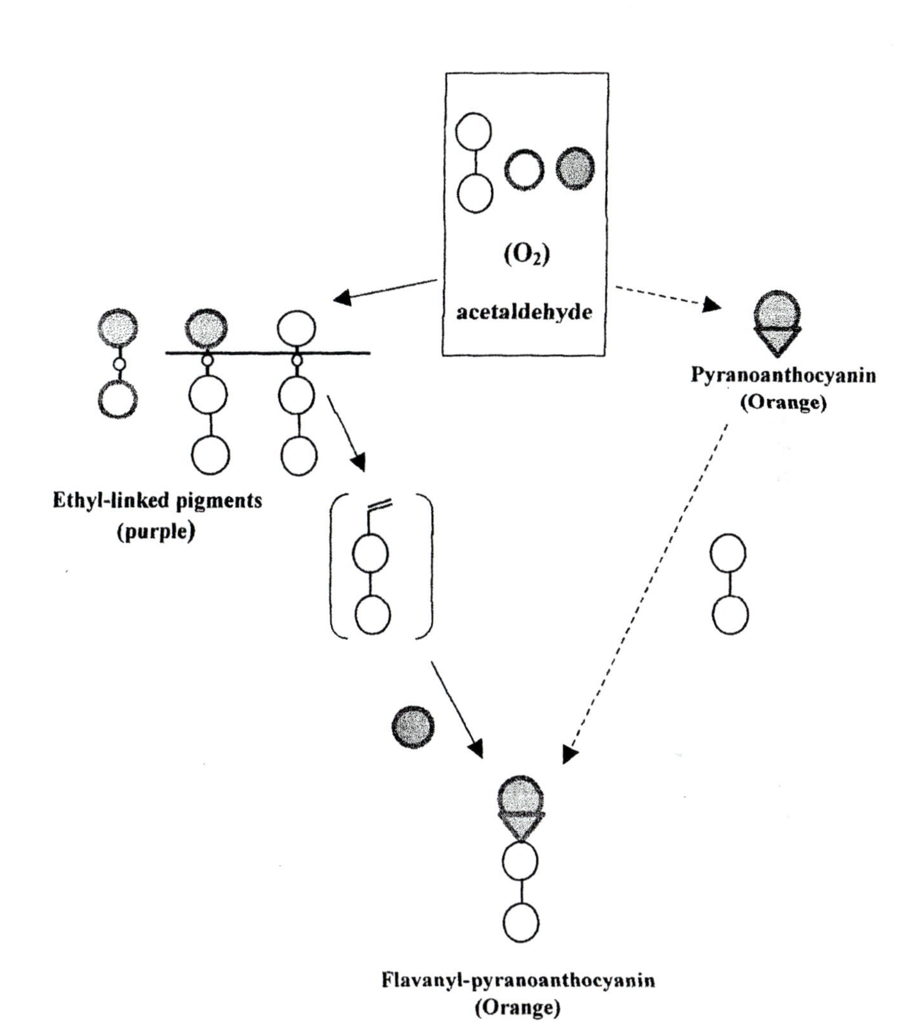

Figure 6. Oxidation pigment marker

As a result, pyranoanthocyanins, F-A^+, F-Et-A^+ and A^+-Et-AOH pigments are more specifically formed at lower pH values. Moreover, the relative proportions of anthocyanins and flavanols affect the nature and size of wine pigments. Excess of anthocyanins is expected to significantly increase anthocyanins polymers (A^+-Et-AOH, A^+-AOH), more stable than their precursors. The presence of large amounts of tannins favours conversion of anthocyanins to stable anthocyanin-tannin pigments. However, many unstable products may arise from tannin tannin reactions. The competition issue between polymerization processes, formation of tannin-anthocyanin adducts and hydration of genuine anthocyanins followed by subsequent degradations is not currently established nor, consequently, the impact on wine color stability. Molecular oxygen is another important factor that mainly promotes acetaldehyde reactions between anthocyanins and flavanols. Therefore, purple ethyl-linked pigments are short-term oxidation markers while orange pyranoanthocyanins are the feature of long-term wine ageing. Their occurrence in wines should significantly increase color stability.

Although the taste of anthocyanin polymers is still unknown, the average size of these products is expected to have a significant impact on wine astringency, as astringency increases with tannin chain length (*39*). Reactions of anthocyanins with tanins are usually believed to yield higher molecular weight species, through acetaldehyde-induced polymerisation and formation of AF adducts. However, interflavanic bond cleavage of tannins also occurs under mild acidic conditions such as encountered in wine (*40*) (*32*). These reactions and subsequent nucleophilic addition of other tannins or anthocyanins onto the resulting carbocations may yield products of lower molecular weight, depending on the nucleophile structure. In fact, the presence of large amounts of monomeric nucleophiles, including hydrated anthocyanins as well as flavanol monomers, leads to a decrease of the mean degree of polymerization of wine tannins whereas predominance of polymeric species results in an increase of their average chain length. Therefore, the loss of astringency associated with wine ageing should be regarded as an average size reduction of wine molecules rather than a precipitation phenomenon provoked by the polymerization of wine constituents as is commonly admitted. Whether incorporation of anthocyanin units within the polymeric chain of tannins affects interactions with other macromolecules such as polysaccharides and proteins and further reduces astringency is still unknown and needs to be investigated.

References

1. Brouillard, R.; Delaporte, B. Chemistry of Anthocyanin Pigments. 2. Kinetic and Thermodynamic Study of Proton Transfer, Hydration, and Tautomeric

Reactions of Malvidin 3-Glucoside. *J. Amer. Chem. Soc.* **1977**, *99*, 8461-8468.

2. Goto, T.; Kondo, T. Structure and molecular stacking of anthocyanins. Flower color variation. *Angew. Chem. Int. Ed. Engl.* **1991**, *30*, 17-33.
3. Brouillard, R.; Dangles, O. Flavonoids and flower colour. In *The flavonoids. Advances in research since 1986*; Harborne, J. B., Ed.; Chapman and Hall: London, 1993; pp 565-588.
4. Haslam, E. Anthocyanin copigmentation - fruit and floral pigments. In *Practical polyphenolics. from structure to molecular recognition and physiological action*; Haslam, E., Ed.; Cambridge University Press: Cambridge, 1998; pp 265-297.
5. Brouillard, R.; Dangles, O. Anthocyanin molecular interactions : the first step in the formation of new pigments during wine aging. *Food Chem.* **1994**, *51*, 365-371.
6. Fulcrand, H.; Benabdeljalil, C.; Rigaud, J.; Cheynier, V.; Moutounet, M. A new class of wine pigments yielded by reactions between pyruvic acid and grape anthocyanins. *Phytochemistry* **1998**, *47*, 1401-1407.
7. Prieur, C.; Rigaud, J.; Cheynier, V.; Moutounet, M. Oligomeric and polymeric procyanidins from grape seeds (*Vitis vinifera*). *Phytochemistry* **1994**, *36*, 781-784.
8. Souquet, J.-M.; Cheynier, V.; Brossaud, F.; Moutounet, M. Polymeric proanthocyanidins from grape skins. *Phytochemistry* **1996**, *43*, 509-512.
9. Fulcrand, H.; Guyot, S.; Roux, E. L.; Remy, S.; Souquet, J.-M.; Doco, T.; Cheynier, V. Electrospray contribution to structural analysis of condensed tannin oligomers and polymers. In *Plant polyphenols 2. Chemistry, biology, pharmacology, ecology*; Gross, G. G.,Hemingway, R. W.,Yoshida, T.,Branham, S. J., Eds.; Kluwer Academic/Plenum Publisher: New York, Boston, Dordrecht, London, Moscow, 1999; pp 223-244.
10. Kennedy, J. A.; Hayasaka, Y.; Vidal, S.; Waters, E. J.; Jones, G. P. Composition of grape skin proanthocyanidins at different stages of berry development. *Journal of Agricultural and Food Chemistry* **2001**, *49*, 5348-5355.
11. Remy, S.; Fulcrand, H.; Labarbe, B.; Cheynier, V.; Moutounet, M. First confirmation in red wine of products resulting from direct anthocyanin-tannin reactions. *Journal of the Science of Food and Agriculture* **2000**, *80*, 745-751.
12. Atanasova, V. Réactions des composés phénoliques induites dans les vins rouges par la technique de micro-oxygénation. caractérisation d enouveaux produits de condensation des anthocyanes avec l'acétaldéhyde, ENSAM, 2003.

13. Wildenradt, H. L.; Singleton, V. L. The production of acetaldehyde as a result of oxidation of phenolic compounds and its relation to wine aging. *American Journal of Enology and Viticulture* **1974**, *25*, 119-126.
14. Fulcrand, H.; Doco, T.; Es-Safi, N.; Cheynier, V.; Moutounet, M. Study of the acetaldehyde induced polymerisation of flavan-3-ols by liquid chromatography ion spray mass spectrometry. *J. Chromatogr. A* **1996**, *752*, 85-91.
15. Saucier, C.; Little, D.; Glories, Y. First evidence of acetaldehyde-flavanol condensation products in red wine. *American Journal of Enology and Viticulture* **1997**, *48*, 370-373.
16. Cheynier, V.; Fulcrand, H.; Sarni, P.; Moutounet, M. Application des techniques analytiques à l'étude des composés phénoliques et de leurs réactions au cours de la vinification. *Analusis* **1997**, *25*, M14-M21.
17. Atanasova, V.; Fulcrand, H.; Le Guerneve, C.; Cheynier, V.; Moutounet, M. Structure of a new dimeric acetaldehyde malvidin 3-glucoside condensation product. *Tetrahedron Letters* **2002**, *43*, 6151-6153.
18. Atanosova, V.; Fulcrand, H.; Le Guernevé, C.; Dangles, O.; Cheynier, V. *Polyphenol Communications 2002, Marrakech*; Vol. 2, p 417-418.
19. Fulcrand, H.; Cheynier, V.; Oszmianski, J.; Moutounet, M. An oxidized tartaric acid residue as a new bridge potentially competing with acetaldehyde in flavan-3-ol condensation. *Phytochemistry* **1997**, *46*, 223-227.
20. Es-Safi, N. E.; Cheynier, V.; Moutounet, M. Study of the reactions between (+)-catechin and furfural derivatives in the presence or absence of anthocyanins and their implication in food color change. *Journal of Agricultural and Food Chemistry* **2000**, *48*, 5946-5954.
21. Es-Safi, N. E.; Guerneve, C. L.; Fulcrand, H.; Cheynier, V.; Moutounet, M. Xanthylium salts formation involved in wine colour changes. *International Journal of Food Science and Technology* **2000**, *35*, 63-74.
22. BenAbdeljalil, C.; Cheynier, V.; Fulcrand, H.; Hakiki, A.; Mosaddak, M.; Moutounet, M. Mise en évidence de nouveaux pigments formés par réaction des anthocyanes avec des métabolites de levures. *Sciences des Aliments* **2000**, *20*, 203-220.
23. Bakker, J.; Bridle, P.; Honda, T.; Kuwano, H.; Saito, N.; Terahara, N.; Timberlake, C. F. Identification of an Anthocyanin Occuring in Some Red Wines. *Phytochemistry* **1997**, *44*, 1375-1382.
24. Francia-Aricha, E. M.; Guerra, M. T.; Rivas-Gonzalo, J. C.; Santos-Buelga, C. New anthocyanin pigments formed after condensation with flavanols. *Journal of Agricultural and Food Chemistry* **1997**, *45*, 2262-2266.

25. Mateus, N.; Silva, A. M. S.; Vercauteren, J.; Freitas, V. d. Occurrence of anthocyanin-derived pigments in red wines. *Journal of Agricultural and Food Chemistry* **2001**, *49*, 4836-4840.
26. Asentorfer, R. E.; Hayasaka , Y.; Jones, G. P. Isolation and structures of oligomeric wine pigments by bisulfite-mediated Ion-exchange Chromatography. *Journal of Agricultural and Food Chemistry* **2001**, *49*, 5957-5963.
27. Mateus, N.; Silva, A. M. S.; Rivas-Gonzalo, J. C.; Santos-Buelga, C.; De Freitas, V. A new class of blue anthocyanin-derived pigments isolated from red wines. *Journal of Agricultural And Food Chemistry* **2003**, *51*, 1919-1923.
28. Es-Safi, N.; Fulcrand, H.; Cheynier, V.; Moutounet, M. Competition between (+)-catechin and (-)-epicatechin in acetaldehyde-induced polymerization of flavanols. *Journal of Agricultural and Food Chemistry* **1999**, *47*, 2088-2095.
29. Es-Safi, N.; Fulcrand, H.; Cheynier, V.; Moutounet, M.; Hmamouchi, M.; Essassi, E. M. Kinetic studies of acetaldehyde-induced condensation of flavan-3-ols and malvidin-3-glucoside in model solution systems. *JIEP 96* **1996**, 279-280.
30. Cheynier, V.; Es-Safi, N.-E.; Fulcrand, H. *International Congress on Pigments in Food and Technology, Sevilla (Spain)*, p 23-35.
31. Salas, E.; Fulcrand, H.; Meudec, E.; Cheynier, V. Reactions of Anthocyanins and tannins in model solutions. *Journal of Agricultural and Food Chemistry* **2003**, *in press*.
32. Vidal, S.; Cartalade, D.; Souquet, J.; Fulcrand, H.; Cheynier, V. Changes in proanthocyanidin chain-length in wine-like model solutions. *Journal of Agricultural and Food Chemistry* **2002**, *50*, 2261-2266.
33. Oszmianski, J.; Cheynier, C.; Moutounet, M. Iron-catalyzed oxidation of (+)-catechin in wine-like model solutions. *Journal of Agricultural and Food Chemistry* **1996**, *44*, 1972-1975.
34. Es-Safi, N. E.; Guerneve, C. L.; Cheynier, V.; Moutounet, M. New phenolic compounds formed by evolution of (+)-catechin and glyoxylic acid in hydroalcoholic solution and their implication in color changes of grape-derived food. *Journal of Agricultural and Food Chemistry* **2000**, *48*, 4233-4240.
35. Remy-Tanneau, S.; Guerneve, C. L.; Meudec, E.; Cheynier, V. Characterization of a colorless anthocyanin-flavan-3-ol dimer containing both carbon-carbon and ether interflavanoid linkages by NMR and mass spectrometries. *Journal of Agricultural and Food Chemistry* **2003,** in press.
36. Cheynier, V.; Hidalgo-Arellano, I.; Souquet, J.-M.; Moutounet, M. Estimation of the oxidative changes in phenolic compounds of Carignane

during wine making. *American Journal of Enology and Viticulture* **1997**, *48*, 225-228.

37. Atanasova, V.; Fulcrand, H.; Cheynier, V.; Moutounet, M. Effect of oxygenation on polyphenol changes occurring in the course of wine making. *Analytica Chimica Acta* **2002**, *458*, 15-27.
38. Escribano-Bailon, T.; Alvarez-Garcia, M.; Rivas-Gonzalo, J. C.; Heredia, F. J.; Santos-Buelga, C. Color and stability of pigments derived from the acetaldehyde-mediated condensation between malvidin-3-O-glucoside and (+)-catechin. *Journal of Agricultural and Food Chemistry* **2001**, *49*, 1213-1217.
39. Vidal, S.; Francis, L.; Guyot, S.; Marnet, N.; Kwiatkowski, M.; Gawel, R.; Cheynier, V.; Waters, E. J. The mouth-feel properties of grape and apple proanthocyanidins in a wine-like medium. *Journal of the Science of Food and Agriculture* **2003**, *83*, 564-573.
40. Haslam, E. In vino veritas : oligomeric procyanidins and the ageing of red wines. *Phytochemistry* **1980**, *19*, 2577-2582.

Chapter 7

New Pigments Produced in Red Wines via Different Enological Processes

C. Gómez-Cordovés

Instituto de Fermentaciones Industriales (C.S.I.C), Juan de Laeierva, 3 Madrid, 28006 Spain

This paper summarizes the results obtained by our group, and those of others with whom we collaborate, with respect to new pigments formed by different enological processes. The reaction mechanisms that form these new pigments, their structures, and some of their properties are discussed.

Introduction

Red wines owe their color to anthocyanins extracted from the grape during the winemaking process. In young wines, several types of anthocyanin contribute to the initial color, including free anthocyanins (mostly flavylium [red] and quinoid [blue] cations), auto-associated anthocyanins, and anthocyanins co-pigmented with other phenolic wine compounds such as hydroxycinnamic esters and catechins. Together, these render young wines a maximum absorption of between 520 and 530 nm.

With the conservation, maturation and ageing of wine, the anthocyanins undergo changes, becoming more complex and of greater molecular weight. This can occur through condensation, polymerization and oxidation reactions and causes the maximum absorbance of the wine to shift to shorter wavelengths, the wines taking on more orange tones.

Since Somers published his first observations on anthocyanin-tannin condensations, many advances have been made in the identification of new pigments thanks to high performance liquid chromatography-electrospray mass spectrometry (HPLC-ESI-MS). Some of these pigments form as the must turns into wine via the condensation of grape anthocyanins with products formed during alcoholic fermentation, such as acetaldehyde, pyruvic acid and vinylphenols, etc. Compounds formed from the union between anthocyanins and flavanols via ethyl or vinyl bridges (derived from acetaldehyde) also appear during maturation and ageing.

Anthocyanin-tannin condensation via acetaldehyde

Some authors have shown that the acetaldehyde present in wine has two origins: the metabolism of yeasts during alcoholic fermentation (Romano *et al.*, 1994) and the oxidation of ethanol in the presence of polyphenols during ageing (Wildenradt and Singleton , 1974).

The reaction mechanism proposed by Timberlake and Bridle (1976), suggests that acetaldehyde, in the form of a carbo-cation, reacts with flavanol (tannin) at position C-6 or C-8 (Figure 1 (1)). This, via several condensation reactions, gives rise to tannin-ethyl-anthocyanin derivatives (Figure 1 (2)) such as malvidin-3-glucoside-ethyl-(epi)catechin and malvidin-3-(6-*p*-coumaroyl)-glucoside-ethyl-(epi)catechin.

The presence of the dimers malvidin-3-glucoside-ethyl-catechin (*m/z* 809, $[M]^+$) and malvidin-3-(6-*p*-coumaroyl)-glucoside-ethyl-catechin (*m/z* 955, $[M]^+$) has been confirmed in wine and wine fractions with the help of ESI-MS. Other pigments (see below), in which the anthocyanin is joined to a flavanol via a vinyl fragment, can also be produced in the presence of acetaldehyde.

Several authors (Bakker *et al.,* 1993; García-Viguera *et al.*, 1994; Escribano-Bailón *et al.*, 1996; Es-Safi *et al.*, 1999) have noted that the reaction via acetaldehyde is quicker than the direct condensation observed by Somers (1971). The initial products become new pigments with a high degree of polymerization, and they eventually precipitate.

The red-violet color of malvidin-3-glucoside - (+)-catechin condensation products (via acetaldehyde), which some authors have suggested is an intramolecular copigmentation effect (Escribano-Bailón *et al.*, 1996), is more pH-stable and SO_2 discoloration-stable than are free anthocyanins (Escribano-Bailón *et al.*, 2001). However, these adducts are also less stable in aqueous solution because of the breakage of the ethyl bridge that joins the anthocyanin to the catechin (Escribano-Bailón *et al.*, 2001).

Anthocyanin-pyruvate condensation

Pyruvic acid (pyruvate) is a product of yeast glycolysis. The decarboxylation of this compound leads to the formation of acetaldehyde, which in turn is reduced to ethanol. In the fermentation of wine, the maximum concentration of pyruvic acid occurs approximately when half of the sugars have been fermented (Whiting and Coggins, 1960). These pyruvate molecules are later metabolized by the yeast.

The compound formed from the union of malvidin-3-glucoside with pyruvate, vitisin A, was first detected in alcoholically strengthened red wines (Bakker and Timberlake, 1997) and later isolated from grape pressings by Fulcrand *et al.* (1998). These authors determined its origin, proposed a mechanism for its formation, and provided its structure (deduced from studies of model solutions). According to these studies, the reaction between pyruvic acid and grape anthocyanins involves a series of steps (Figure 2) similar to those required for the formation of anthocyanin-4-vinylphenol adducts.

The structures initially proposed by Bakker and by Fulcrand were slightly different. Later work involving nuclear magnetic resonance (Mateus *et al.*, 2001) and mass spectrometry (Asenstorfer *et al.*, 2001; Hayasaka y Asenstorfer, 2002) confirmed the structure proposed by Fulcrand *et al.* (1998).

Adducts of malvidin-3-glucoside and of its *p*-coumaryl and acetyl forms acylated with pyruvate, also known as vinylformic adducts, have been identified in red wines (Revilla *et al.*, 1999; Vivar-Quintana *et al.*, 1999; Asenstorfer *et al.*, 2001; Revilla and González-San José, 2001; Hayasaka and Asenstorfer, 2002) and Port by several groups (Bakker *et al.*, 1997a; Romero and Bakker, 2000b; Mateus *et al.*, 2002a). Similarly, pyruvic derivatives of delphinidin, petunidin and peonidin-3-glucosides have been reported, as have derivatives of delphinidin and petunidin-3-(6-acetyl)-glucoside, and petunidin and peonidin-3-(6-*p*-coumaroyl)-glucoside.

C-4/C-5 cycloaddition can also involve other secondary metabolites originating in the activities of yeasts. Acetaldehyde, acetone, acetoine (3-hydroxybutan-2-one), oxaloacetic and diacetylic acid etc., can all be made from pyruvate. All these compounds show keto-enol tautomerism and can therefore react, in their enol form, with anthocyanin pigments via a mechanism similar to that of pyruvate and 4-vinyphenol. Pigments derived from anthocyanins which have a CH=CH group between the C-4 and the hydroxyl group of C-5 have been identified. These are formed through the condensation of anthocyanins with acetaldehyde (vinyl adduct) (Fig. 1(3)). These adducts of malvidin-3-glucoside and its acetylated ester (also known as vitisin B and acetylvitisin B) were first

+ CH_3-CHO

(3)

vitisin B

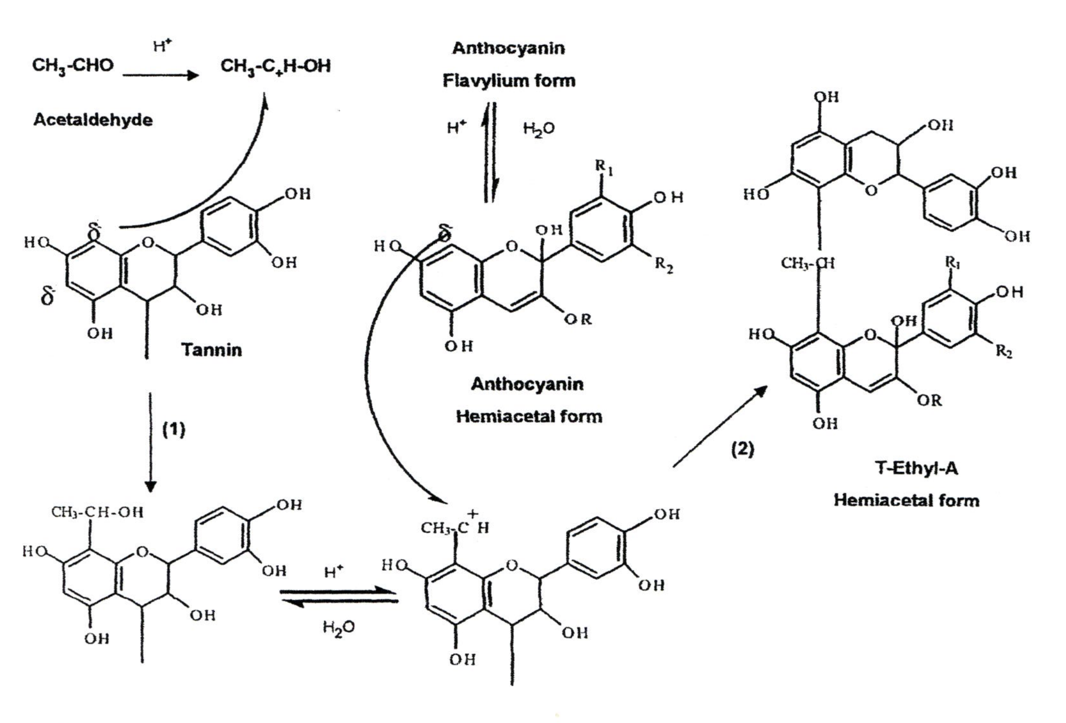

Figure 1. Tannin-Anthocyanin reactions mediated by acetaldehyde (Cheynier, 2001).

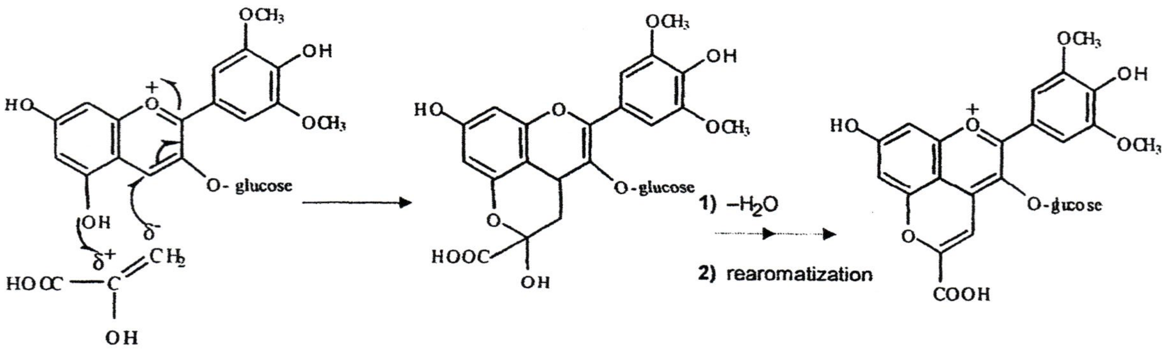

Figure 2. Reaction between Piruvic acid and Malvidin-3-glucoside (Fulcrand *et al.*, 1998).

isolated from Port (Bakker *et al.*, 1997) and later from red wines (Revilla *et al.*, 1999; Vivar-Quintana *et al.*, 1999, 2002; Atanasova *et al.*, 2002; Hayasaka and Asenstorfer, 2002). The corresponding vinyl derivative of malvidin-3-(6-*p*-coumaroyl)-glucoside (*p*-coumarylvitisin B) has recently been identified in wine fractions (Vivar-Quintana *et al.*, 2002).

Anthocyanin-4-vinylphenol and anthocyanin-vinylflavanol condensations

The 4-vinylphenol in wine comes from the decarboxylation of *p*-coumaric acid via the yeast enzyme cinnamate decarboxylase (CD) (Chantonnet *et al.*, 1993). Other cinnamic acids, such as caffeic, ferulic and sinapic acid suffer this same process, generating 4-vinylcatechol, 4-vinylguaiacol and 4-vinylsyringol, respectively. Like 4-vinylphenol, these volatile compounds are associated with aromas that give unpleasant hints to the wine (Etiévant, 1981).

The products of condensation between anthocyanins and 4-vinylphenol, first detected and isolated from polymer membranes used in the microfiltration of red wine (Cameira dos Santos *et al.*, 1996), have been studied by UV-visible, mass, and H^+ NMR spectrometry. This has confirmed the presence of adducts formed between the most abundant anthocyanins (malvidin-3-glucoside and malvidin-3-(6-*p*-coumaroyl)-glucoside) and 4-vinylphenol (Fulcrand *et al.*, 1996). The proposed structures are shown in Fig 3 (a).

The pigments thus formed are red-orange and have two mesomeric forms, one corresponding to the malvidin type (λ_{max} 534 nm) (Fig. 4 (a)) and the other (more common) to the pelargonidin type (λ_{max} 506 nm) (Fig. 4 (b)), (Fulcrand et al., 1996). These develop at the pH of wine (pH 3-4) and could justify – at least partially – the gradual changes in color from red to slate tones that occur during ageing since they are more stable and more resistant to discoloration by SO_2 than the anthocyanidin-3-glucosides (Sarni-Manchado et al., 1996). Their formation also reduces the volatile phenol content of red wines.

Using mass spectrometry (nano-ESI-MS/MS), Hayasaka and Asenstorfer (2002) characterized other new vinyl anthocyanins in red wines with substitutions at C-4, such as adducts of malvidin-3-glucoside with 4-vinylcatechol, 4-vinylguaiacol and 4-vinylsyringol (Figure 3 (a)), as well as adducts of peonidín and petunidin-3-glucosides with 4-vinylphenol. The pigments resulting from the condensation of malvidin-3-(6-acetyl)-glucoside, malvidin-3-(6-caffeyl)-glucoside and peonidin-3-(6-*p*-coumaroyl)-glucoside with 4-vinylphenol have also been described (Asenstorfer *et al.*, 2001; Mateus *et al.*, 2002a).

(a)

(b)

(a) Adducts of Malvidin-3-glucoside and 4-vinyilphenols

R_1: H; R_2: H = Malvidn -3-glucoside-4-vinylphenol (Fulcrand *et al*., 1996a).

R_1: H; R_2: OH = Malvidin-3-glucoside-4-vinylcatechol

R_1: H; R_2: OCH_3 = Malvidin-3-glucoside-4-vinylguaiacol

R_1: OCH_3; R_2: OCH_3 = Malvidín-3-glucoside-4-vinylsyringol

(Hayasaka y Asenstorfer, 2002)

(b) Adducts of Malvidin- 3-glucoside and vinylflavanols (Francia- Arichae*t al*., 1997).

R: H = Malvidin-3-glucoside-vinylcatechin

R: (+)-catechin = Malvidin-3-glucoside-vinyldicatechin

Figure 3. 4-Vinylphenols and vinylflavanols mavidin-3-glucoside derivatives.

Adducts of malvidin-3-glucoside with vinylcatechin, vinylepicatechin or vinyldiepicatechin (procyanidin B2), reported in model solutions containing malvidin-3-glucoside, acetaldehyde and the respective flavan-3-ol (Francia-Aricha *et al.*, 1997) (Fig. 3, (b)), have also been detected in wine fractions (Mateus *et al.*, 2002a, b; Atanasova *et al.*, 2002). In the latter work, the authors showed, via thiolysis, that both proanthocyanidins and monomeric flavanols are involved in the formation of anthocyanin-vinylflavanol and anthocyanin-ethyl-flavanol pigments induced by acetaldehyde.

The mechanism by which these vinylflavanol pigments form is not entirely clear. It is suggested that they might come from the reaction between the flavylium cation of the anthocyanin and a flavanol molecule containing a vinyl residue at C-8, following the mechanism proposed by Fulcrand *et al.* (1996) for the formation of 4-vinylphenol derivatives (Figures 3 and 4) (Francia-Aricha *et al.*, 1997; Mateus *et al.*, 2002 a,b). The last step in the mechanism would be cycloaddition, via which the vinylflavanol residue joins with the flavylium unit. This oxidative addition would cause the new pigments formed to be extensively conjugated, allowing the aromatization of the new pyrane ring (ring D) (Fig. 5).

Like the 4-vinylphenol derivatives, vinylflavanol derivatives are orange red and have a maximum absorption of between 500 and 511 nm. Because of the substitution at C-4, they are also more resistant to discoloration by changes in pH and SO_2 than are pigments with an ethyl bridge (Vivar-Quintana *et al.*, 2002). This greater stability may also explain why they tend not to suffer polycondensation (Francia-Aricha *et al.*, 1997).

Compounds with substitutions at C-4, including anthocyanins derived through reactions with pyruvate and other metabolites of fermentation, have hypsochromatically displaced spectra with respect to that of malvidin 3-glucoside (18-19 nm for the vinylformic adduct and 36-39 nm for the vinyl adduct, depending on the solvent). This represents a change towards yellow-orange - exactly that which happens during the maturation of wine. Their spectra also show an absorption peak in the UV region (352-370 nm) characteristic of anthocyanins with a substitution at C-4. As mentioned above, this substitution confers total or partial resistance to SO_2 discoloration, as well as greater color intensity in the pH interval 3.5 - 7 compared to malvidin 3-glucoside (Bakker and Timberlake, 1997). Therefore, even in low concentration, greater color intensity is provided in the conditions reigning in wine.

Vivar-Quintana *et al.* (2002) indicate that the true importance of these pigments to the color of wine cannot be judged given their scant contribution to wine chromatographic profiles. Under the acidic conditions this type of analysis requires, the anthocyanins are in their red, flavylium cationic state, while at the pH of wine ($\cong$ 3.5) they exist mainly in a weakly-colored pseudo-base state. Small quantities of vinyl derivatives could therefore contribute significantly to the color.

Anthocyanidin derivatives formed by different enological processes

This section provides results recently obtained – the fruit of collaboration between different work groups. The pigments discussed are those identified in processes such as:

*The retention of anthocyanins and pyruvic anthocyanin derivatives by the cell walls of yeasts used in the fermentation of red wines.

*The synthesis of pigments derived from anthocyanins in wines made from the Graciano, Tempranillo and Cabernet Sauvignon varieties.

*Variations in the new anthocyanin derivatives in wines of different vintages during ageing in bottles.

*Formation of new pigments during the manufacture of sparkling wines (*cavas*).

The retention of anthocyanins and pyruvic anthocyanin derivatives by the cell walls of yeasts used in the fermentation of red wines

The making of red wine involves maceration of the grape skins during alcoholic fermentation. The aim is that pigments in the skins spread throughout the must. However, some of the anthocyanins released are retained by the yeast cell walls and are therefore lost to the wine when the lees is removed.

The cell wall of *Saccharomyces cerevisiae* is made of mannoproteins bound to oligopolysaccharides which remain exposed on the outside of the cell. These mannoproteins are also bound to glucanos and chitin (Pretorius, 2000; Suárez, 1997). The different polarities and the hydrophilic or hydrophobic nature of these wall polymers define the capacity of yeast to retain or adsorb different wine molecules such as volatile compounds, fatty acids or pigments (Vasserot *et al.*, 1997).

Given the importance of colour to red wines, and the quantity of pigments lost when the lees is removed, was important to investigate the adsorption of anthocyanins by yeast cell walls during wine-making: 1) to determine whether there are any differences between yeasts strains, and 2) to determine the degree of retention of different anthocyanins.

Ten small scale fermentations were established with 5 L of crushed grapes from *Vitis vinifera L.* cv. Cabernet Sauvignon. Each must was inoculated with a population of 10^8 cfu/mL of one of the ten yeast strains, all these belonged to *Saccharomyces bayanus* or *cerevisiae.* Three yeast strains of each regions were isolated from grapes collected in the Spanish *apellation controlée* regions of La

oxidation

a

4-vinylphenol

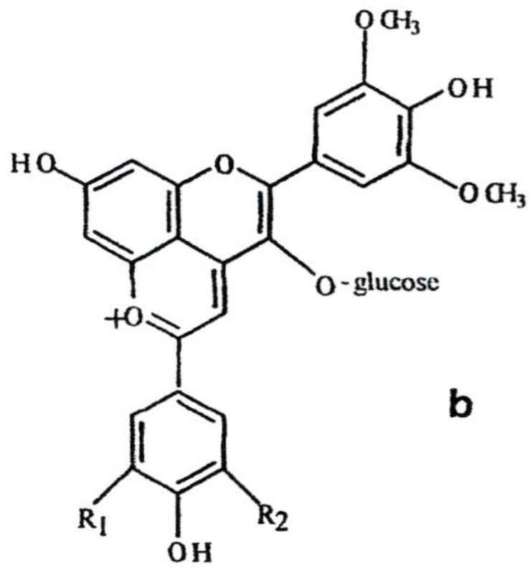

b

Figure 4. Proposed reaction mechanims between malvidin-3-glucoside and 4-vinylphenol (Fulcrand *et al.*, 1996).

oxidation

vinylflavanol

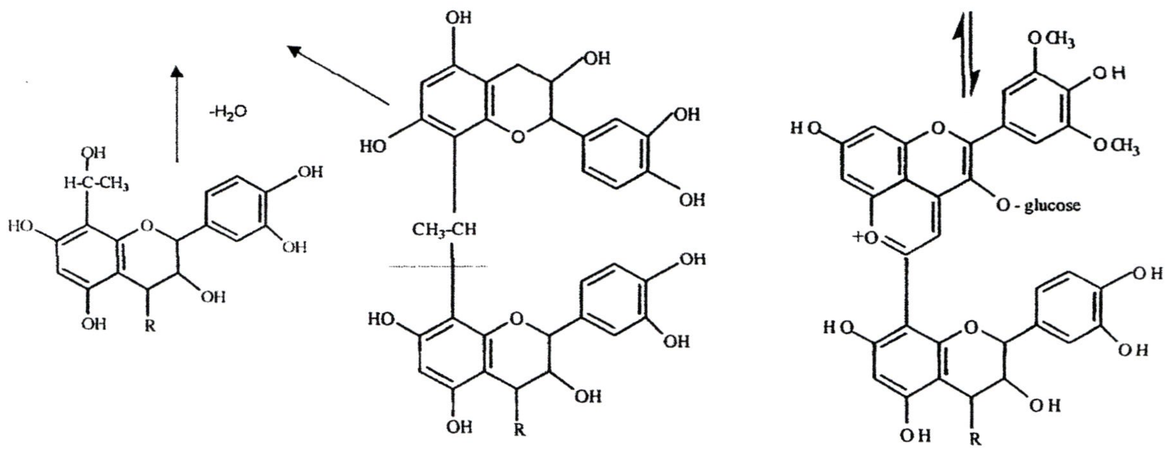

Figure 5. Proposed mechanism reaction between malvidin-3-glucoside and vinylflavnols (Francia-Aricha *et al.*, 1997; Mateus *et al.*, 2002 a,b)

Rioja, Navarra and Ribera del Duero . The commercial yeast S6U (*Saccharomyces uvarum*) (Lallemand Inc., Canada) was also used.

The anthocyanins adsorbed onto the yeast cell walls were recovered from the lees of each wine in preparation. These were washed with distilled water and then extracted with formic acid:methanol, as well as samples of the finished wines were analysed spectrophotometrically to determine colour, and by HPLC-ESI-MS/HPLC-DAD to confirm and evaluate their anthocyanin content. The following anthocyanins were identified in both wines and cell wall adsorbates: delphinidin (D), cyanidin (Cy), petunidin (Pt), peonidin (Pn) and malvidin (M), as their 3-glucoside (3G), pyruvic (Py), acetyl glucoside (6Ac), cinnamoyl (*p*-coumaryl (6Cm) and caffeyl (6Caf)) glucoside derivates.

The mean total anthocyanin content of the wines fermented with the ten yeast strains was 430 mg/L. The mean total adsorbate anthocyanin content was 74 mg/L (Table 1). The distribution of anthocyanins was very different between the wines and the cells walls. In general, in the wines (Fig. 6 (a)), anthocyanins glycosylated at position 3 (3G) were the most common (59 %), followed by acetyl derivatives (6Ac) (33 %), then cinnamoyl derivatives at 5.4 %, and pyruvic derivatives at 2.9 % (Table 1). This is the normal distribution for Cabernet-Sauvignon wines. The glycosylated, acetyl and pyruvic derivatives were slightly less common in the adsorbates than in the wines. However, the quantities of *p*-coumaryl and caffeyl derivatives were very much greater (Fig. 6 (b)).

Table 1 shows the ratio of each anthocyanin family adsorbed onto the cells walls/ in the wine (expressed as a percentage of adsorption). Glycosylated anthocyanins are the most common both in grapes and the wine. This is logical since glycosylation is a necessary step for the stability and mobility of anthocyanins. Later, some of the glycosyl derivatives are acylated by acyltransferases. However, the percentage adsorption of the acylated derivatives (acetyl, caffeyl and *p*-coumaryl) is much greater than that of the non-acylated derivates (3G).

The elution order of the reverse phase HPLC column was in decreasing order of anthocyanins polarity. This is similar to the adsorption profile of the cell walls *p*-coumaryl derivatives followed by caffeyl and then acetyl derivatives: i.e., the same order of polarity.

This greater affinity for more apolar acyl derivatives, which are more retained in the reverse phase column, indicates that, with respect to anthocyanins, the cell wall is an apolar and hydrophobic fashion than the solvent (wine). This agrees with the results obtained by Lubbers *et al* (1994), who indicate a greater fixing capacity of volatile hydrophobic compounds of yeast cell walls in their study on the adsorption of aromatic compounds. However, it disagrees with the results of Vasserot *et al.* (1997) who, in a study of the five monoglycoside anthocyanins, report a greater percentage adsorption of hydrophilic anthocyanins (delphinidin and petunidin).

Table 1. Mean total anthocyanin content and derivatives in wines and cell wall adsorbates (mg/L).

Sample	STRAIN	3G	PY	6AC	6CAF	6CM	Totals
Wines	Mean	254.15	12.56	140.49	2.38	20.89	430.47
	sd	19.19	2.26	10.80	0.30	3.35	32.33
	%	59.04	2.92	32.64	0.55	4.85	100.00
Adsorbates	Mean	33.04	0.31	19.90	1.40	18.85	73.50
	sd	7.01	0.14	4.55	0.52	3.99	15.56
	%	44.96	0.42	27.08	1.91	25.64	100.00
	%Adsorption	13.00	2.45	14.16	58.81	90.21	17.07

The anthocyanin profiles (Fig 6 a, b) show that malvidin-3G and its acylated derivatives are those most adsorbed and delphinidin-3G (and their acetylderivate) are the least. The non-adsorption of cyanidin glycoside might be explained by its low initial concentration in the wine. It is also the starting anthocyanin for the formation of all others through the action of flavonyl-3-hydroxylase and methyl-transferase. Further, because of its highly hydrophilic nature, what is availiable is more likely to remain in the wine than to enter the yeast cell walls.

There was very little adsorption of pyruvic derivatives, possibly because of their low content in wine. Further, pyruvic derivatives are mostly formed from the pyruvate made during alcoholic fermentation; their presence increases after removal of the lees.

The different CV_s for pyruvic derivatives in wines (CV 18) and cell walls (CV 33) calculated from Table 1 (Morata *et al.*, 2003), appear to indicate that the yeast have different fermentative behavior and a different structure cell wall also. This appear to confirm, especially when the values for caffeylderivates (CV 36) are compared with those coumaroylderivates (CV 21).

The influence of anthocyanin adsorption by cell walls on wine colour

The colour of the cell wall adsorbates is much less intense than that of the wines themselves, although the former do have greater tonality. This reduced colour intensity is a consequence of the lower total adsorbed anthocyanin content of the walls (Table 1). The greater tonality is due to an increase in yellow tones and an important reduction in blue (Morata *et al.*, 2003). The cell wall adsorbates showed a greater percentage yellow and a fall in blue colour. This statistically correlated with high acetyl derivative contents - especially of the most apolar (petunidin, peonidin and malvidin) (Morata *et al.*, 2003).

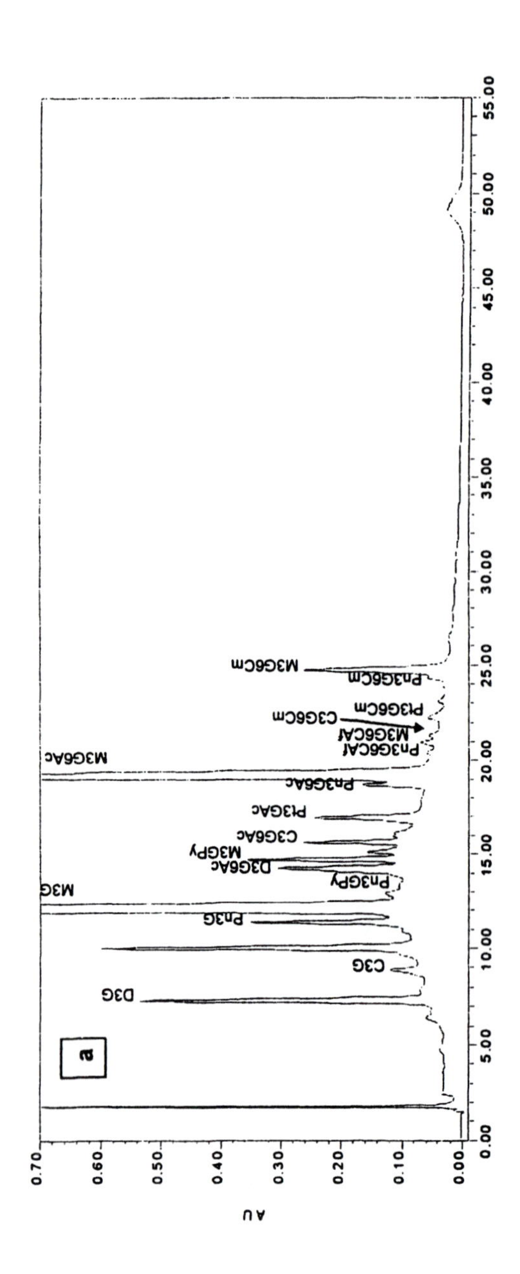
a
D3G
C3G
Pn3G
M3G
Pn3GPy
D3G6Ac
M3GPy
C3G6Ac
Pt3GAc
Pn3G6Ac
M3G6Ac
Pn3G6Caf
M3G6Caf
C3G6Cm
Pt3G6Cm
Pn3G6Cm
M3G6Cm
AU
0.00
0.10
0.20
0.30
0.40
0.50
0.60
0.70
0.00
5.00
10.00
15.00
20.00
25.00
30.00
35.00
40.00
45.00
50.00
55.00

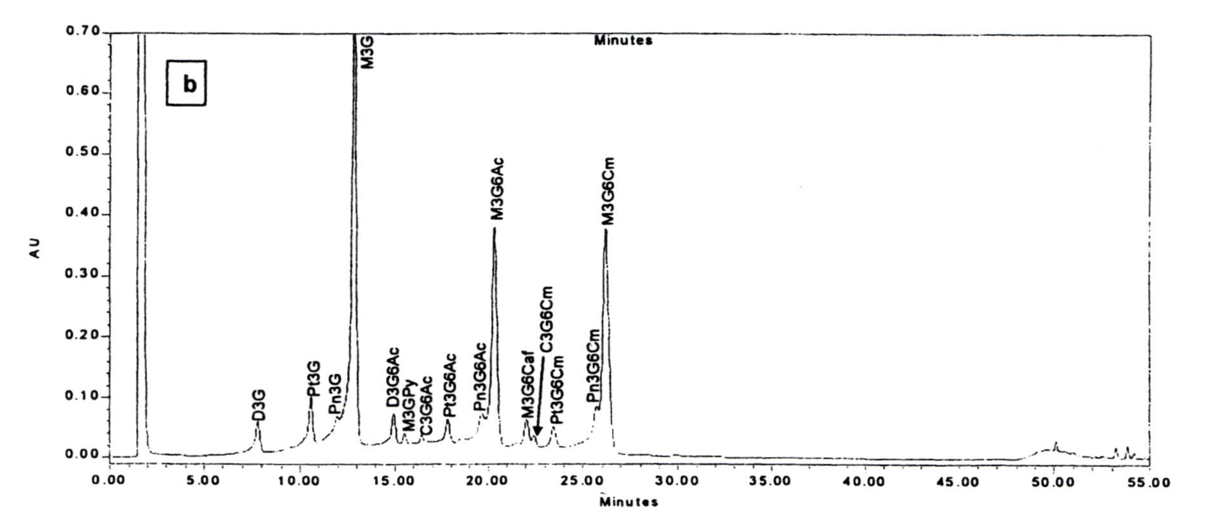

Figure 6. HPLC-DAD chromatograms for strain yeast from A. O. Ribera del Duero. a. Anthocyanins in wine. b. Anthocyanins adsorbed by the cell wall.

Pigments derived from anthocyanin in wines made from the Graciano, Tempranillo and Cabernet-Sauvignon varieties

The anthocyanin profile of young wines made from the Spanish *Vitis vinifera* cv. Graciano, Tempranillo, and and French cv. Cabernet Sauvignon, were determined by HPLC-ESI-MS. All grape varieties grown in the same geographical area and elaborated at the Viticulture and Enology Station of Navarra (EVENA), situated in the northeast of Spain (vintage 2000).

The anthocyanin HPLC chromatograms corresponding to Graciano, Tempranillo and Cabernet Sauvignon wines after 12 months in bottle, are illustrated in Figure 7 (a, b and c, respectively).

The 3-glucosides (peaks 2, 3, 4, 6 and 7) as well as the 3-(6"-acetylglucosides) (peaks 9, 12, 14, 16 and 18), the 3-(6"-*p*-coumaroyl)glucosides (peaks 17, 21, 23, 27 and 28), and the 3-(6"-caffeoyl)glucosides (peaks 19 and 20) were identified in the three varieties studied. The molecular ions $[M]^+$ as well as the fragments corresponding to the anthocyanidin after cleavage of the (esterified) glucose moiety were detected (Table 2) (Monagas *et al.*, 2003).

Of particular interest was the presence of peak **22**, which had a mass spectra identical to malvidin-3-(6"-*p*-coumaroyl)glucoside **(28)** (*m/z* 639, 331), but with slight differences in the UV spectra: malvidin-3-(6"-*p*-coumaroyl)glucoside spectra showed an extra band at 320 nm. This peak was proposed to be the *cis* isomer of malvidin-3-(6"-*p*-coumaroy)glucoside, peak **28** being the *trans* isomer. Further confirmation was achieved by exposing a wine sample under to UV light (254 nm, 6h). A significant increase of peak **22**, accompanied by a decrease of the rest of the other anthocyanins was observed. Even more, mass signals corresponding to the possible presence of peonidin-3-(6"-*p*-coumaroyl)glucoside *cis* isomer were also detected at a retention time of 21.1 min coeluting with cyanidin-3-(6"-*p*-coumaroyl)glucoside **(21)**. Although *trans-cis* photoisomeration have been reported to occur *in vivo* in flower anthocyanins acylated with hydroxycinnamic acids (George et al. 2001), always predominating the *trans* isomer, to our knowledge this is the first time that a *cis* isomer of a *p*-coumaric acid acylated anthocyanin is reported in a fermentation-derived product of the *Vitis vinifera* spp.

A first family of new pigments (peaks 8, 10, 11 and 15), all of them corresponding to the product resulting from the C-4/C-5 cycloaddition of pyruvic acid, eluted soon after malvidin-3-glucoside and had a polarity very similar to that of the anthocyanidin-3-(6"-acetyl)glucosides (Table 2). Peak 10 corresponded to malvidin-3-glucoside pyruvate (Bakker et al., 1997, Fulcrand et al., 1998). Mass signals of *m/z* 517, 355 were found to coelute with malvidin-3-glucoside pyruvate, corresponding to the malvidin-vinyl adduct resulting from the reaction of malvidin-3-glucoside with acetaldehyde (Benabdeljalil et al., 2000), or vitisin B. Adducts such as malvidin-3-(6"-acetyl)-glucoside pyruvate (11), and malvidin-3-(6"-*p*-coumaroyl)-glucoside pyruvate (15) were also detected. Peak 8, corresponded to peonidin-3-glucoside pyruvate and it was only present in the Graciano wines studied.

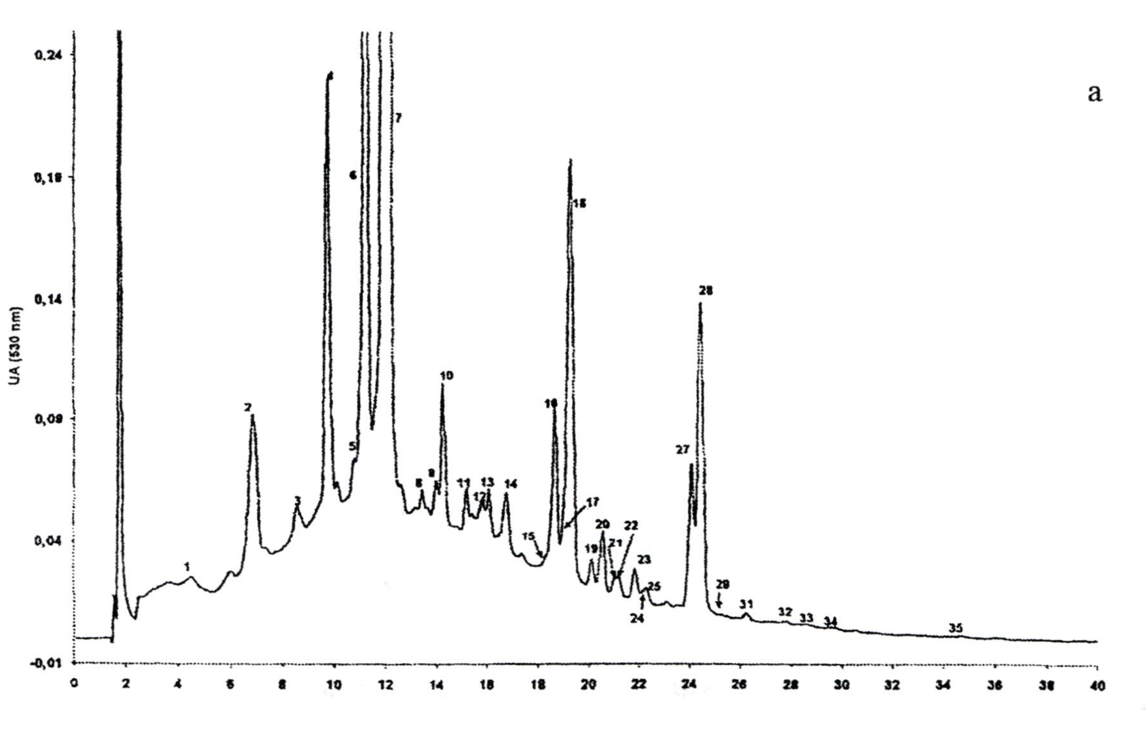

Figure 7. HPLC chromatograms corresponding to Graciano, Tempranillo and Cabernet Sauvignon wines after 12 months in bottle (a, b and c, respectively).

Continued on next page.

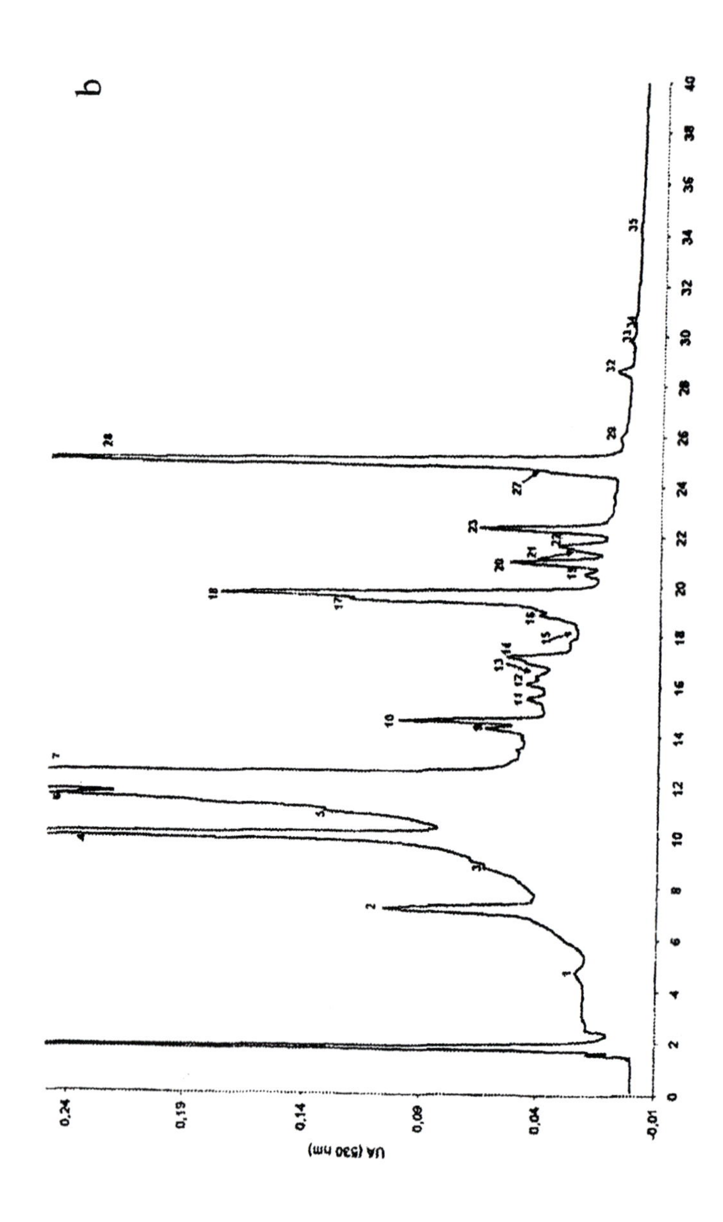
b
UA (530 nm)
0,24
0,19
0,14
0,09
0,04
-0,01

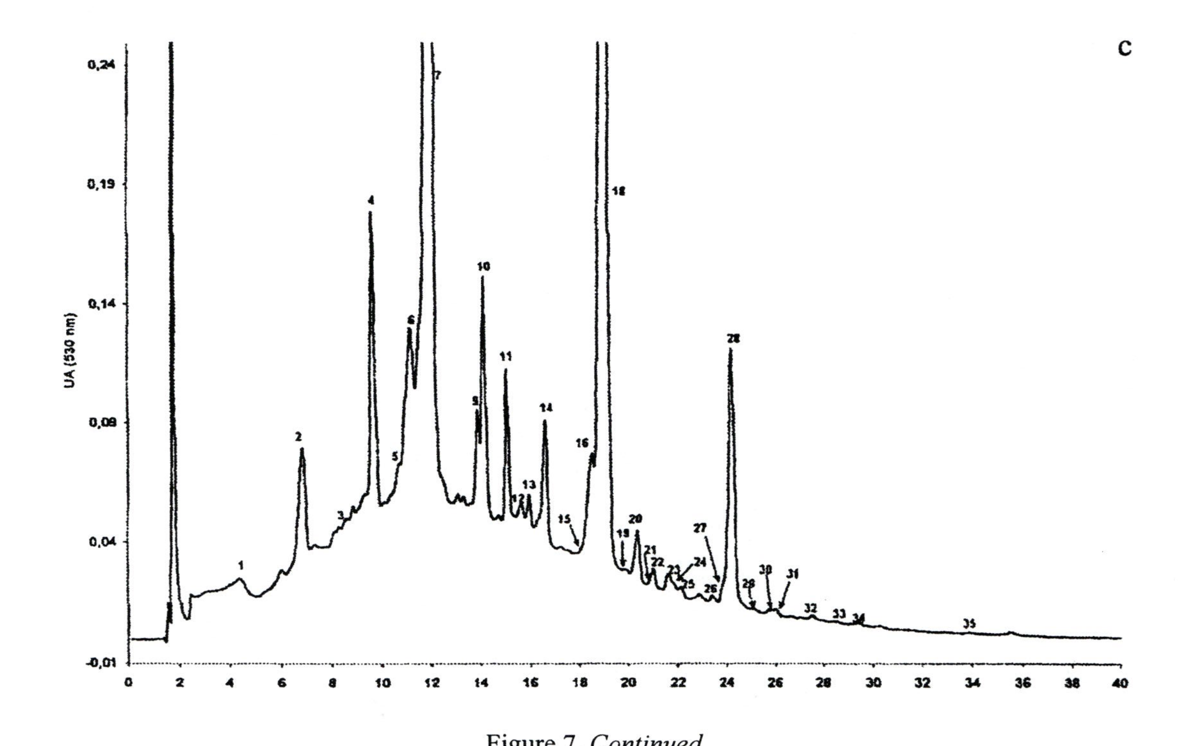

Figure 7. *Continued.*

A second family of anthocyanin-derived pigments, present only at very low levels and exhibiting very low polarity (peaks 24, 26, 29-35), was also constituted by C4-substituted structurally related anthocyanins (Table 2). They all exhibit a malvidin-based skeleton and lower absorption maxima (504-514 nm) than the normal glucoside and acyl glucoside anthocyanins. Peak 32, having the same molecular mass (*m/z* 609) of peonidin-3-(6"-*p*-coumaroyl)glucoside (27), but a different ion fragment (*m/z* 447), corresponds to malvidin-3-glucoside-4-vinylphenol. The equivalent 4-vinylphenol adducts of malvidin-3-(6"-acetyl)-glucoside (34) and malvidin-3-(6"-*p*-coumaroyl)-glucoside (35) were also identified. Other cinnamic acids such as caffeic, ferulic, and sinapic acid can undergo the same decarboxylation reaction of *p*-coumaric acid to respectively form 4-vinylcatechol, 4-vinylguaiacol, and 4-vinylsyringol. Peak 29 could correspond to malvidin-3-glucoside-4-vinylcatechol, and peak 33 could be the pigment malvidin-3-glucoside-4-vinylguaiacol. The formation of anthocyanin-4-vinylderivates reduces the volatile 4-vinylphenols content associated with off-flavor notes of red wines.

Additional anthocyanin-derived pigments (peaks 24, 26, 30, 31) of oligomeric nature were also detected in Graciano and/or Cabernet Sauvignon wines. Two peaks (24, 31), presenting a $[M]^+$ of *m/z* 805, were identified as malvidin-3-glucoside-vinyl-catechin (or epicatechin), while the corresponding vinyl-catechin (or epicatechin) derived pigments of malvidin-3-(6"-acetylglucoside), with a $[M]^+$ of *m/z* 847 (26, 30) were only identified in the Cabernet Sauvignon wines corresponding to the vintage year, production area, and fermentation conditions of this study (Monagas *et al.,* 2003). Both compounds have been previously reported in grape marc (Asenstorfer *et al.*, 2001) and wine fractions (Asenstorfer *et al.*, 2001, Atanasova *et al.*, 2002, Hayasaka and Asenstorfer, 2002), although the corresponding ion fragments were not observed in our study due to the different fragmentation conditions employed.

A third family of anthocyanin-derived pigments (peaks **1**, **13** and **25**) also included oligomeric condensed-pigments (Table 2). Peak **1**, corresponded to the dimer resulting from the direct condensation reaction between malvidin-3-glucoside and (epi)catechin, previously reported in wines (Remy *et al.*, 2000, Vivar-Quintana *et al.*, 1999, 2002). Peak **13**, detected in the same region as the anthocyanidin-3-(6"-acetyl) glucosides, was assigned to a malvidin-(epi)catechin dimer linked by an ethyl brigde (acetaldehyde-mediated condensation) first reported in model solutions by FAB-MS (Bakker *et al.,* 1993) and further detected in wines by ESI-MS by several authors. As reported by other authors (Atanasova *et al.*, 2002, Vivar-Quintana *et al.*, 2002), several peaks were observed in the mass chromatograms of this ethyl-linked dimer corresponding to the possible stereo isomers (*R, S*), that result from the presence of an asymmetric carbon within the ethyl bridge, and regio isomers (C8-C8, C8-C6) resulting from the different attachment positions of the flavanol unit (catechin and epicatechin) to the anthocyanin. Another dimer, malvidin-3-(6"-*p*-coumaroyl)-glucoside-ethyl-(epi)catechin **(25)**, previously reported in one year-old Port wine fractions (Mateus *et al.*, 2002a), was identified only in the Graciano and Cabernet Sauvignon wines studied. As for vinyl-catechin condensed pigments, no ion fragments were found for any of these pigments.

Varietal influence in the occurrence of different wine anthocyanin-derived pigments

The occurrence of anthocyanin-derived pigments appeared to be related to the concentration of the corresponding anthocyanin precursors in wine. It was found that the wines presenting the highest levels of a given anthocyanin-derived pigment, were also the ones containing the largest concentration of its corresponding anthocyanin precursor, observation also reported by Romero and Bakker (2000a). For instance, Graciano wines, containing high levels of peonidins, were the only ones with the condensed pigment peonidin-3-glucoside pyruvate **(8)**. In similar way, the high concentration of acetyl-glucosides, specially malvidin-3-(6"-acetyl)-glucoside, in Cabernet Sauvignon wines resulted in the formation of high levels of malvidin-3-(6"-acetyl)-glucoside pyruvate **(11)**. Wines made from both, Graciano and Cabernet Sauvignon grapes also presented oligomeric-type pigments such as malvidin-3-glucoside-vinyl-catechin (or epicatechin) **(24, 31)** and malvidin-3-(6"-*p*-coumaroyl)-glucoside-ethyl-(epi)catechin **(25)**, while only in Cabernet Sauvignon wines the presence of malvidin-3-(6"-acetyl)-glucoside-vinyl-catechin (or epicatechin) **(26, 30)** was detected, suggesting a higher degree of anthocyanin condensation in wines from these two varieties (Monagas *et al.*, 2003).

Variations in the new anthocyanin derivatives in wines of different vintages during ageing in bottles

Wines made from *Vitis vinifera* L. cv. Merlot grapes (clone 343 from France, vintages 1997 and 2001) were analyzed by HPLC-DAD and HPLC-ESI-MS to establish their pyruvic and vinyl derivative contents and to study their behavior during ageing in bottles (reducing medium).

The wines were supplied by the *Estación de Viticultura y Enología de Navarra* (EVENA). Although different vintages were used, both the vineyards where the grapes were grown and the ageing process (performed in bottles stored in cellars at a temperature of 13-17 °C) were the same. Figure 8 (Suárez *et al.*, 2003) shows the anthocyanin derivatives HPLC chromatogram corresponding to 1997 vintage .

Table 2. Anthocyanins and anthocyanin-derived pigments identified in Graciano, Tempranillo and Cabernet Sauvignon wines by HPLC/ESI-MS.

No.	t_R (min)	λ_{max} (nm)	$[M]^+$ (*m/z*)	Fragments (*m/z*)	Compound	GRA	TEM	CS
1	4.5	530	781		Malvidin-3-glucoside-(epi)catechin	*	*	*
2	6.9	524	465	303	Delphinidin-3-glucoside	*	*	*
3	8.6	515	449	287	Cyanidin-3-glucoside	*	*	*
4	9.8	526	479	317	Petunidin-3-glucoside	*	*	*
5	10.8	536	657	331	Unknown	*	*	*
6	11.2	516	463	301	Peonidin-3-glucoside	*	*	*
7	12.0	520	493	331	Malvidin-3-glucoside	*	*	*
8	13.4	509	531	369	Peonidin-3-glucoside pyruvate	*	-	-
9	14.0	533	507	303	Delphinidin-3-(6”-acetylglucoside)	*	*	*
10	14.3	513	561	399	Malvidin-3-glucoside pyruvate	*	*	*
11	15.2	518	603	399	Malvidin-3-(6”-acetylglucoside) pyruvate	*	*	*
12	15.8	516	491	287	Cyanidin-3-(6”-acetylglucoside)	*	*	*
13	16.1	543	809		Malvidin-3-glucoside-ethyl-(epi)catechin	*	*	*
14	16.2	532	521	317	Petunidin-3-(6”-acetylglucoside)	*	*	*
15	18.0	513	707	399	Malvidin-3-(6”-*p*-coumaroylglucoside) pyruvate	*	*	*
16	18.7	520	505	301	Peonidin-3-(6”-acetylglucoside)	*	*	*
17	19.0	532	611	303	Delphinidin-3-(6”- *p*-coumaroylglucoside)	*	*	-
18	19.3	530	535	331	Malvidin-3-(6”-acetylglucoside)	*	*	*
19	20.1	524	625	301	Peonidin-3-(6”-caffeoylglucoside)	*	*	*
20	20.6	536	655	331	Malvidin-3-(6”-caffeoylglucoside)	*	*	*
21	21.1	527	595	287	Cyanidin-3-(6”-*p*-coumaroylglucoside)	*	*	*
22	21.2	537	639	331	Malvidin-3-(6”-*p*-coumaroylglucoside)- *cis* isomer	*	*	*
23	21.9	532	625	317	Petunidin-3-(6”-*p*-coumaroylglucoside)	*	*	*
24	22.2	503	805		Malvidin-3-glucoside-vinyl-catechin	*	-	*
25	22.3	540	955		Malvidin-3-(6”-p-coumaroylglucoside)-ethyl-(epi)catechin	*	-	*

Table 2. *Continued.*

No	t_R (min)	λ_{max} (nm)	$[M]^+$ (*m/z*)	Fragments (*m/z*)	Compound	GRA	TEM	CS
26	23.6	508	847		Malvidin-3-(6”-acetylglucoside)-vinyl-catechin	-	-	*
27	24.1	524	609	301	Peonidin-3-(6”-*p*-coumaroylglucoside)	*	*	*
28	24.4	535	639	331	Malvidin-3-(6”-*p*-coumaroylglucoside)- *trans* isomer	*	*	*
29	25.3	514	625	463	Malvidin-3-glucoside-4-vinylcatechol	*	*	*
30	26.0	514	847		Malvidin-3-(6”-acetylglucoside)-vinyl-epicatechin	-	-	*
31	26.3	508	805		Malvidin-3-glucoside-vinyl-epicatechin	*	-	*
32	27.8	504	609	447	Malvidin-3-glucoside-4-vinylphenol	*	*	*
33	28.6	504	639	477	Malvidin-3-glucoside-4-vinylguaiacol	*	*	*
34	29.7	509	651	447	Malvidin-3-(6”-acetylglucoside)-4-vinylphenol	*	*	*
35	34.7	504	755	447	Malvidin-3-(6”- *p*-coumaroylglucoside)-4-vinylphenol	*	*	*

GRA= Graciano; TEM= Tempranillo; CS= Cabernet Sauvignon. t_R = retention time. * detected. - not detected. (Monagas *et al.*,2003).

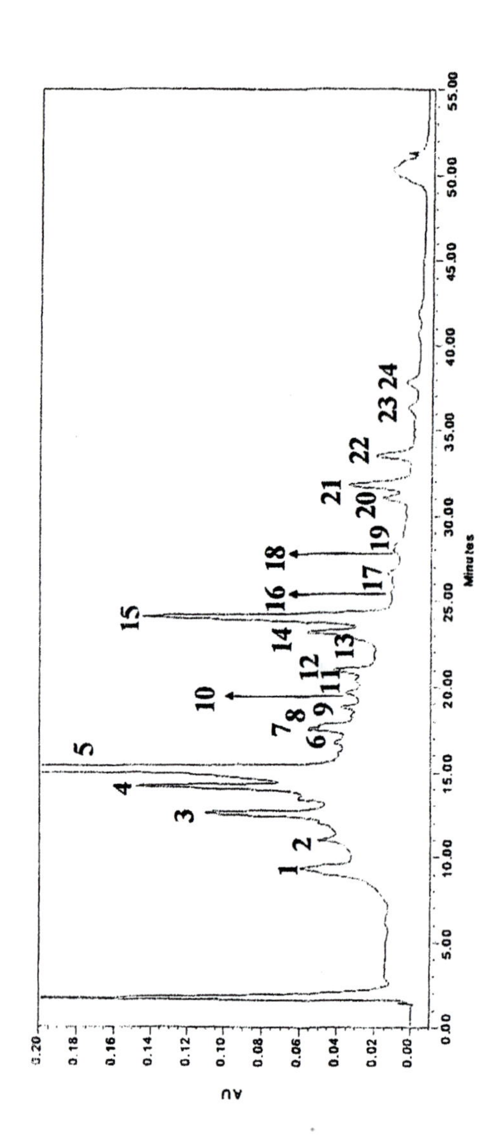
AU
Minutes
0.00
0.02
0.04
0.06
0.08
0.10
0.12
0.14
0.16
0.18
0.20
0.00
5.00
10.00
15.00
20.00
25.00
30.00
35.00
40.00
45.00
50.00
55.00
1
2
3
4
5
6
7
8
9
10
11
12
13
14
15
16
17
18
19
20
21
22
23
24

1. Delphinidin-3-glucoside 2. Cyanidin-3-glucoside 3. Petunidin-3-glucoside
4. Peonidin-3-glucoside 5. Malvidin-3-glucoside 6. Peonidin-3-glucoside pyruvate
7. Delphinidin-3-(6-acetylglucoside) 8. Malvidin-3-glucoside pyruvate
9. Malvidin-3-(6-acetylglucoside) pyruvate 10. Cyanidin-3-(6-acetylglucoside)
11. Malvidin-3-glucoside-ethyl-(epi)catechin 12. Petunidin-3-(6-acetylglucoside)
13. Malvidin-3-(6-*p*-coumaroylglucoside) pyruvate 14. Peonidin-3-(6-acetylglucoside)
15. Malvidin-3-(6-acetylglucoside) 16. Peonidin-3-(6-caffeoylglucoside)
17. Malvidin-3-(6-caffeoylglucoside) 18. Malvidin-3-(6-coumaroylglucoside) -*cis* isomer
19.Petunidin-3-(6-*p*-coumaroylglucoside) 20. Peonidin-3-(6-*p*-coumaroylglucoside)
21. Malvidin-3-(6-*p*-coumaroylglucoside) -*trans* isomer
22. Malvidin-3-(6-acetylglucoside)-vinyl-epicatechin
23. Malvidin-3-glucoside-4-vinylphenol 24. Malvidin-3-(6-acetylglucoside)-4-vinylphenol

Figure 8. HPLC chromatogram of anthocyanin derivatives of Merlot wine corresponding to 1997 vintage.

Figure 9 shows an important reduction in glucosyl, acetyl and pyruvic derivatives after the first two years of ageing. The profiles of the three oldest vintages (wines 3-5 years old) were similar, although differences can be seen due to variations in the grapes at each harvest and to differences during manufacture (even though the same process was used for all).

Though the percentages of each of the derivatives with respect to their precursor are not the same for every vintage, probably due to the quantities of pyruvic acid and acetaldehyde formed during fermentation, these percentages increase as the wines age.

Malvidin-3-glucoside-4vinylphenol and malvidin-3(6 acetyl) glucoside-4 vinylphenol form from the second year in the bottle, and increase slightly during ageing (Suárez *et al.*, 2003).

Formation of new pigments during the making of sparkling *rosé* wines

The formation of anthocyanin pigments was analyzed by HPLC-ESI-MS and monitored by HPLC-DAD during the second fermentation of a base *Vitis vinifera* L cv Garnacha wine (following the *champenoise* method to obtain a *cava* or sparkling wine).

Figures 10 show the chromatograms of the base wine (Fig. 10 (B W)) (made in the same way as any rosé, i.e., a brief period of maceration with the skins to adequately extract the anthocyanins) and of the final wine after the second fermentation in the bottle (Fig. 10 (9m W)). Degorging was performed after nine months fermentation and ageing with the yeast. This is the minimum time required by Spanish legislation for a wine to be classified as a *cava*.

Comparison of the percentages of each derivative in both types of wine with respect to the own precursor malvidin-3-glucoside shows the most abundant in the base wine to be malvidin-3-glucoside-pyruvate (vitisin A) (23.7% of the concentration of the glucoside). The lowest concentration was seen for malvidin-3(6 acetyl)glucoside–4 vinylphenol, with 5.3%. In the *cava*, the most abundant derivative with respect to malvidin-3-glucoside was malividin-3 glucoside-4-vinylguaiacol (14.6%), although vitisin A showed a very similar value (12.7%) and the lowest is malvidin-3-glucoside-4 vinylcatechol (1.6%) (Table 3). From these percentages, it can be seen that malvidin-3-glucoside-4 vinylcatechol and vitisin B are the derivatives that suffers the greatest reduction nine months into the second fermentation and ageing process. The percentages of vitisin A and malvidin-3-glucoside-4-vinylphenol were reduced to one half their values with respect to the base wine ones while those of the acetylated vinylphenol derivatives - malvidin-3(6 acetyl) glucoside-4-vinylcatechol and vinylphenol - suffered no changes. The percentage of malvidin-3-glucoside-4-vinylguiacol with respect to the glucoside increased by 40%. Malvidin-3 (6-coumaroyl) glucoside-4 vinylphenol and malvidin3 (6-acetyl) glucoside-4

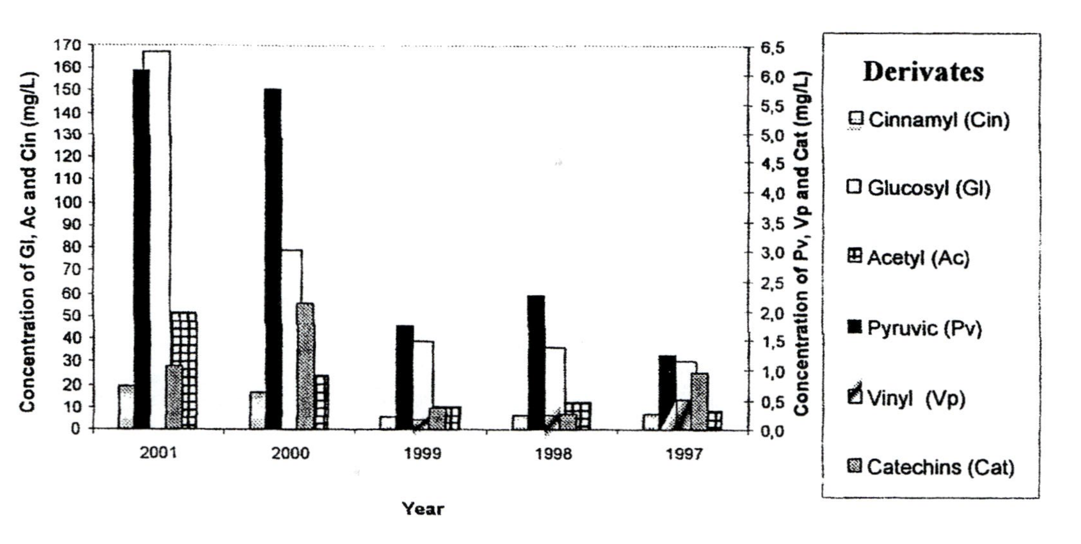

Figure 9. Profiles of anthocyanin derivatives of Merlot wines from different vintages ageing in bottles.

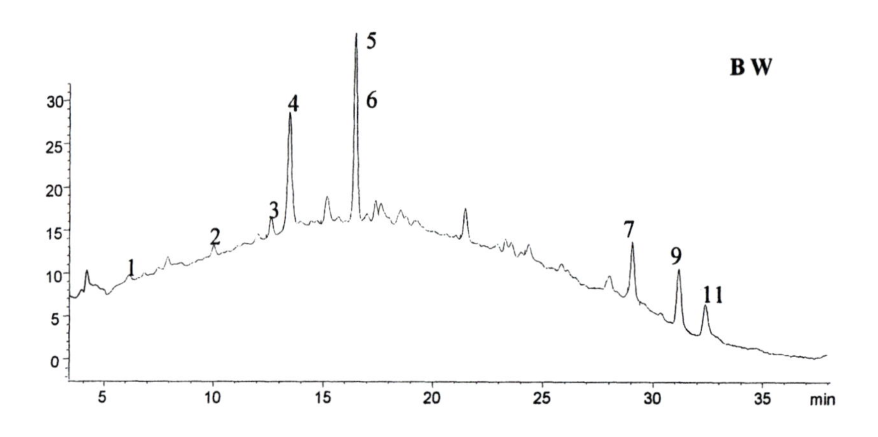

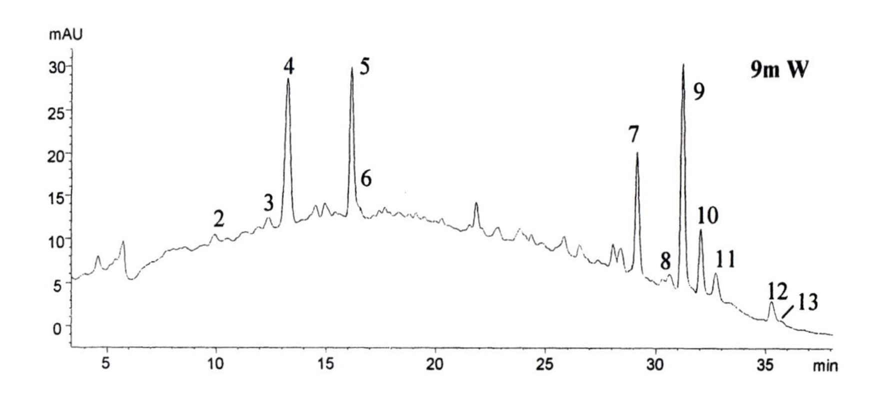

1. Delfinidin-3-glucoside
2. Petunidin-3-glucoside
3. Peonidin-3-glucoside
4. Malvidin-3-glucoside
5. Malvidin-3-glucoside pyruvate (vitisin A)
6. Malvidin-3-glucoside 4 vinyl (vitisin B)
7. Malvidin-3-glucoside-4 vinylcatechol
8. Malvidin-3 (acetyl)-glucoside 4 vinylcatechol
9. Malvidin-3-glucoside 4 vinylphenol
10. Malvidin-3 (acetyl)-glucoside 4 vinylguaiacol
11. Malvidin-3 (acetyl)-glucoside 4 vinylphenol
12. Malvidin-3 (coumaroyl)-glucoside 4 vinylphenol
13. Malvidin-3 (acetyl)-glucoside 4 vinylsyringol

Figure 10. HPLC chromatogram of anthocyanin derivatives of Garnacha sparkling wines: B W- base wine. 9m, W- sparkling (*cava*) wine.

Table 3 .Variations on the percentages over Malvidin-3-glucoside of anthocyanin derivated pigments in sparkling wines after 9 months of ageing on yeasts.

	M-G	Vit B	Vit A	MG-vy-catechol	MAcG-vy-catechol	MG-vy-phenol	MG-vy-guaiacol	MAcG-vy-phenol
Base wine	100 (3.4)*	14.2	23.7	6.5	7.4	6.2	8.3	5.3
Sparkling wine	100 (6.9)*	3.9	12.7	1.6	7.9	3.6	14.6	4.5

*concentration (ppm). G: glucoside; Ac: acetyl; v: vinyl

vinylsyringol (*m/z* 711, $[M]^+$) form during the second fermentation and ageing (Pozo-Bayón *et al.* 2003)

The stability of the 4-vinyl derivatives is therefore greater than the new pigments of the *cava*, and they form during the nine months during which the base wine is transformed into a sparkling wine.

References

Asenstorfer, R.E., Y. Hayasaka, and G.P. Jones. 2001. Isolation and structures of oligomeric wine pigments by bisulfite-mediated ion-exchange chromatography. J. Agric. Food Chem. 49, 5957-5963.

Atanasova, V., H. Fulcrand, V. Cheynier, and M. Moutounet. 2002. Effect of oxygenation on polyphenol changes occuring in the course of wine-making. Anal. Chim. Acta. 458, 15-27.

Bakker, J., A. Picinelli, and P. Bridle. 1993. Model wine solutions: colour and composition changes during ageing. Vitis. 32, 111-118.

Bakker, J., and C.F. Timberlake. 1997. Isolation, identification and characterization of new color-stable anthocyanins occuring in some red wines. J. Agric. Food Chem. 45, 35-43.

Bakker, J., P. Bridle, T. Honda, H. Kuwano, N. Saito, N. Terahara, and C. Timberlake. 1997. Identification of an anthocyanin occuring in some red wines. Phytochem. 44, 1375-1382.

Benabdeljalil, C., V. Cheynier, H. Fulcrand, A. Hakiki, M. Mosaddak, and M. Moutounet. 2000. Mise en évidence de nouveaux pigments formés par réaction des anthocyanes avec des métabolites de levure. Sci. Alim. 20, 203-220.

Cameira dos Santos, P.J., J.Y. Briollouet, V. Cheynier, and M. Moutounet. 1996. Detection and partial characterization of new anthocyanin derived pigments in wines. J. Sci. Food Agric. 70, 204-208.

Chantonnet, P., D. Dubourdieu, J-N. Boidron, and V. Lavigne. 1993. Synthesis of volatile phenols by *Saccharomyces cerevisiae* in wines. J. Sci. Food Agric. 62, 191-202.

Cheynier, V. 2001. Grape polyphenols and their reactions in wine. *Polyphénols Actualités*. No.21 (Décembre 2001).

Escribano Bailón, T.; Dangles, O.; Brouillard, R. 1996. Coupling reactions between flavylium ions and catechin. Phytochem. 41, 1583-1592.

Escribano-Bailón, T.; Alvarez-García, M.; Rivas-Gonzalo, J.C.; Heredia, F.J.; Santos-Buelga, C. 2001. Color and stability of pigments derived from the acetaldehyde-mediated condensation between malvidin-3-*O*-glucoside and (+)-catechin. J Agric Food Chem. 49, 1213-1217.

Es-Safi, N.; Fulcrand, H.; Cheynier, V.; Moutounet, M. 1999. Studies on the acetaldehyde-induced condensation of (-)-epicatechin and malvidin 3-*O*-glucoside in a model solution system. J Agric Food Chem. 47, 2096-2102.

Etiévant, P.X. 1981. Volatil phenols determination in wines. *J Agric Food Chem*. 29, 65-67.

Francia-Aricha, E.M.; Guerra, M.T.; Rivas-Gonzalo, J.C.; Santos-Buelga, V. 1997. New anthocyanin pigments formed after condensation with flavanols. J Agric Food Chem. 45, 2262-2266

Fulcrand, H; Cameira Dos Santos, P.J.; Sarni-Manchado, P.; Cheynier, V.; FabreBonvin, J. 1996. Structure of new anthocyanin-derived wine pigments. J Chem Soc, Perkin Trans *1,* 735-739.

Fulcrand, H., C. Benabdeljalil, J. Rigaud, V. Cheynier, and M. Moutounet. 1998. A new class of wine pigments generated by reaction between pyruvic acid and grape anthocyanins. Phytochem. 47, 1401-1407.

George, F., P. Figueiredo, K. Toki, F. Tatsuzawa, N. Saito, and R. Brouillard. 2001. Influence of *trans-cis* isomerisation of coumaric acid substituents on colour variance and stabilisation in anthocyanins. Phytochem. 57, 791-795.

García-Viguera, C.; Bridle, P.; Bakker, J. 1994. The effect of pH on the formation of coloured compounds in model solutions containing anthocyanins, catechin and acetaldehyde. *Vitis*. 33, 37-40.

Hayasaka, Y., and R.E. Asenstorfer. 2002. Screening for potencial pigments derived from anthocyanins in red wine using nanoelectrospray tandem mass spectrometry. J. Agric. Food Chem. 50, 756-761.

Lubbers, S., Charpentier, C., Feuillat, M., and Voilley, A. 1994. Influence of yeast walls on the behavior of aroma compounds in a model wine. Am. J. Enol. Vitic., 45, 29-33.

Mateus, N.; Silva M.S., A.; Vercauteren, J.; De Freitas, V. 2001. Occurrence of anthocyanin derived pigments in red wines. J Agric Food Chem. 49, 4836-4840.

Mateus, N., S. De Pascual-Teresa, J.C. Rivas-Gonzalo, C. Santos-Buelga, and V. De Freitas. 2002a. Structural diversity of anthocyanin-derived pigments in port wines. Food Chem. 76, 335-342.

Mateus, N.; Silva M.S., A..; Santos-Buelga, C.; Rivas-Gonzalo, J.C.; De Freitas, V. 2002b. Identification of anthocyanin-flavanol pigments in red wines by NMR and mass spectrometry. J Agric Food Chem. 50, 2110-2116.

Morata, A., Gómez-Cordovés, C., Suberviola, J., Bartolomé, B., Colomo, B. and Suarez, J.A. 2003. Adsorption of anthocyanins by yeast cell wall during the fermentation of red wines. J.Agric. Food Chem. 51, (14), 4084-4088.

Monagas, M., Nuñez, V., Bartolomé, B. and Gómez-Cordovés, C. 2003. Anthocyanin-derived pigments in Graciano, Tempranillo and Cabernet Sauvignon wines produced in Spain. Am. J. Enol. Vitic. 54, 163-169.

Pozo-Bayón, M.A., Gómez-Cordovés, C. and Polo, M.C. 2003. Estudio de la composición antociánica de vinos espumosos elaborados con variedades tintas y de los cambios que se producen durante el envejecimiento. "Study of anthocyanin composition of red sparkling wines and their changes during the ageing". Jornadas Científicas de los grupos de Investigación Enológica. Published by Gobierno de la Rioja. 173-174

Pretorius., I. S. 2000. Tailoring wine yeast for the new millennium: novel approaches to the ancient art of winemaking. Yeast 16, 675-729.

Remy, S., H. Fulcrand, B. Labarbe, V. Cheynier, and M. Moutounet. 2000. First confirmation in red wine of products resulting from direct anthocyanin-tannin reactions. J. Sci. Food Agric. 80,745-751

Revilla, I.; Pérez-Magariño, S.; González-San José, M.L.; Beltrán, S. 1999. Identification of anthocyanin derivaties in grape skin extracts and red wines by liquid chromatography with diode array and mass spectrometric detection. *J Chrom A*. 847, 83-90.

Revilla, I.; San-José, M.L. 2001. Effect of different oak woods on aged wine color and anthocyanin composition. *Eur Food Res Technol*. 213, 281-285.

Romano, P.; Sizzi, G.; Turbanti, L.; Polsinelli, M. 1994. Acetaldehyde production in *Saccharomyces cerevisiae* wine yeasts. FEMS Microbiol. Lett. 118, 213-218.

Romero, C., Bakker, J. 2000a. Effect of acetaldehyde and several acids on the formation of vitisin A in model wine anthocyanin and colour evolution. Int J Food Sci Tech. 35, 129-140.

Romero, C., and J. Bakker. 2000b. Anthocyanin and colour evolution during maturation of four port wines: effect of pyruvic acid addition. J. Sci. Food Agric. 81, 252-260.

Sarni-Manchado, P.; Fulcrand, H.; Souquet, J-M.; Cheynier, V.; Moutounet, M. 1996. Stability and color of unreported wine anthocyanin-derived pigments. J. Food Sci. 61, 938-941.

Somers, T.C. 1971. The phenolic nature of wine pigments. Phytochem. **10**, 2175-2186.

Suarez, J. A. 1997 Levaduras vínicas. Funcionalidad y uso en bodega. *Ed. MundiPrensa S. A.*, 57-60.

Suárez, R., Bartolomé, B. and Gómez-Cordovés, C. 2003. Estudio de la composición fenólica, familias y antocianos, y colorimétrica de vinos Merlot en función de la añada y el envejecimiento. "Phenolic composition and color of wines cv. Merlot in relation to vintage and ageing". Jornadas Científicas de los grupos de Investigación Enológica. Published by Gobierno de la Rioja. 161-163.

Timberlake, C.F. and Bridle, P. 1976. Interactions between anthocyanins, phenolic compounds, and acetaldehyde and their significance in red wines. *Am J Enol Vitic*: 27, 97-105.

Vasserot, Y, Caillet, S. and Maujean, A. 1997. Study of anthocyanin adsorption by yeast lees. Effect of some physicochemical parameters. Am. J. Enol. Vitic., 48, 433-437.

Vivar-Quintana, A.M., C. Santos-Buelga, E. Francia-Aricha, and J.C. Rivas-Gonzalo. 1999. Formation of anthocyanin-derived pigments in experimental red wines. Food Sci. Tech. Int. 5,347-352.

Vivar-Quintana, A.M., C. Santos-Buelga, and J.C. Rivas-Gonzalo. 2002. Anthocyanin-derived pigments and colour of red wines. Anal. Chim. Acta. 458, 147-155.

Whiting, G.C.; Coggins, P.A. 1960. Organic acid metabolism in cider and perry fermentations. III. Keto-acids in cider-apple juices and ciders. J Sci Food Agric. 11: 705-709.

Wildenradt, H.L.; Singleton, V.L. 1974. The production of aldehydes as a result of oxidation of polyphenolic compounds and its relation to wine ageing. Am J Enol Vitic. 25: 119-126.

Chapter 8

Factors Affecting the Formation of Red Wine Pigments

David F. Lee[1,2], Ewald E. Swinny[1,2], Robert E. Asenstorfer[1], and Graham P. Jones[1,2,*]

[1]School of Agriculture and Wine, University of Adelaide, Waite Campus, PMB 1, Glen Osmond, South Australia 5064, Australia
[2]Cooperative Research Centre for Viticulture, P.O. Box 154, Glen Osmond, South Australia 5064, Australia

The development of pigments during red wine fermentation has been shown to proceed by the conversion of the relatively unstable grape-derived anthocyanins to more complex and color stable pigments through the formation of both ethyl-linked and vitisin-like pigments. These ethyl-linked pigments are more resistant to hydration and bisulfite bleaching than the parent anthocyanins, a phenomena explained by steric hindrance to nucleophilic attack and by alteration of the electronic properties of the anthocyanin entity. Furthermore, they have been shown to be formed rapidly during fermentation and likely to consume much of the anthocyanins extracted from the grapes. Because they are purple in color it is suggested that the purple hues of young red wines, particularly immediately after the completion of fermention, are a result of the high concentrations of these pigments. The ethyl-linked pigments are degraded quickly with time, however, and may provide a pool of reactive intermediates in the formation of more stable red wine pigments. The vitisin-like compounds, on the other hand, are long lived and have been shown to persist in red wine over many decades. Whilst the reaction to produce the ethyl-linked pigments requires the loss of a proton in the reaction sequence, the production of the vitisin-type compounds depends on an oxidation step. The presence of atmospheric oxygen has been shown to be necessary for the formation of vitisin A in model systems but in red wine ferments the rate of vitisin A synthesis may be correlated with active oxidants rather than just oxygen concentration.

Abreviations: Mv (malvidin-3-glucoside), HVPE (high voltage paper electrophoresis)

Anthocyanin pigments extracted from grape skins during vinification are responsible for the color of red wines. Malvidin-3-glucoside (Mv) is the most common anthocyanin found in *Vitis vinifera* and, similar to many other anthocyanins, it forms colorless hydrated species at wine pH, it is bleached by sulphur dioxide and also readily oxidises. Somers *(1)* proposed that during wine ageing anthocyanins are converted to more stable, brick-red colored polymeric pigments formed through reactions with other grape-derived phenolics, principally flavan-3-ol monomers and polymers. The stability of these compounds was attributed to the blocking of the 4-position to nucleophilic attack (by bisulphite for example) and a reduction of the reactivity of the C-2 position such that the polymeric pigments are less susceptible to pH-dependent hydration reactions (*2-5*). The synthesis and characterisation of a number of color stable, 4-substituted anthocyanin pigments by Timberlake and Bridle (*6*) supported these ideas. The identification and characterisation of Somer's-type polymeric pigments in wine has proved difficult even using modern liquid chromatographic techniques and electrospray mass spectrometry *(7)*. However, a large group of anthocyanin-derived wine pigments, many of which are appreciably more stable than the parent anthocyanin, have been recently found. They may be generated from reaction between anthocyanins and fermentation-derived carbonyl compounds (e.g. vitisin A (**1**, Figure 1) and B (**2**), formed from pyruvic acid and acetaldehyde respectively) (*8-10*) or wine-derived compounds containing a vinyl group (e.g. pigment A (**3**) *(11,12)* and pigment B2-III (**4**) *(13)* formed from vinylphenol and vinylcatechin respectively). These types of pigments have been shown to occur widely in red wines, particularly in fortified Port wines *(8, 14-15)* and Australian Shiraz wines *(16-17*). They are orange-red pigments with absorbance maxima around 500nm at wine pH.

Another group of compounds identified in wine arise from the condensation of anthocyanins with flavanols by a carbonyl compound such as acetaldehyde to form ethyl-bridged or related dimers and multimers. These compounds have been extensively studied *(18-26)*. A representative compound of this group is Mv-ethyl-catechin **(5)** a compound identified from mass spectral data of model systems and subsequently observed in wines *(15)*. These pigments have absorption maxima around 544 nm giving them purple color at wine pH. The 4-position of the anthocyanin is not substituted in these compounds and they would therefore be expected to have similar stabilities to the parent anthocyanins. However, much evidence has been obtained to show that both diastereoisomers of Mv-ethyl-catechin are resistant to bisulfite bleaching and hydration *(21)*. On the other hand they are not long-lived. Garcia-Viguera et al. *(19)* along with others *(22)* have shown that their concentrations decline rapidly in model wines

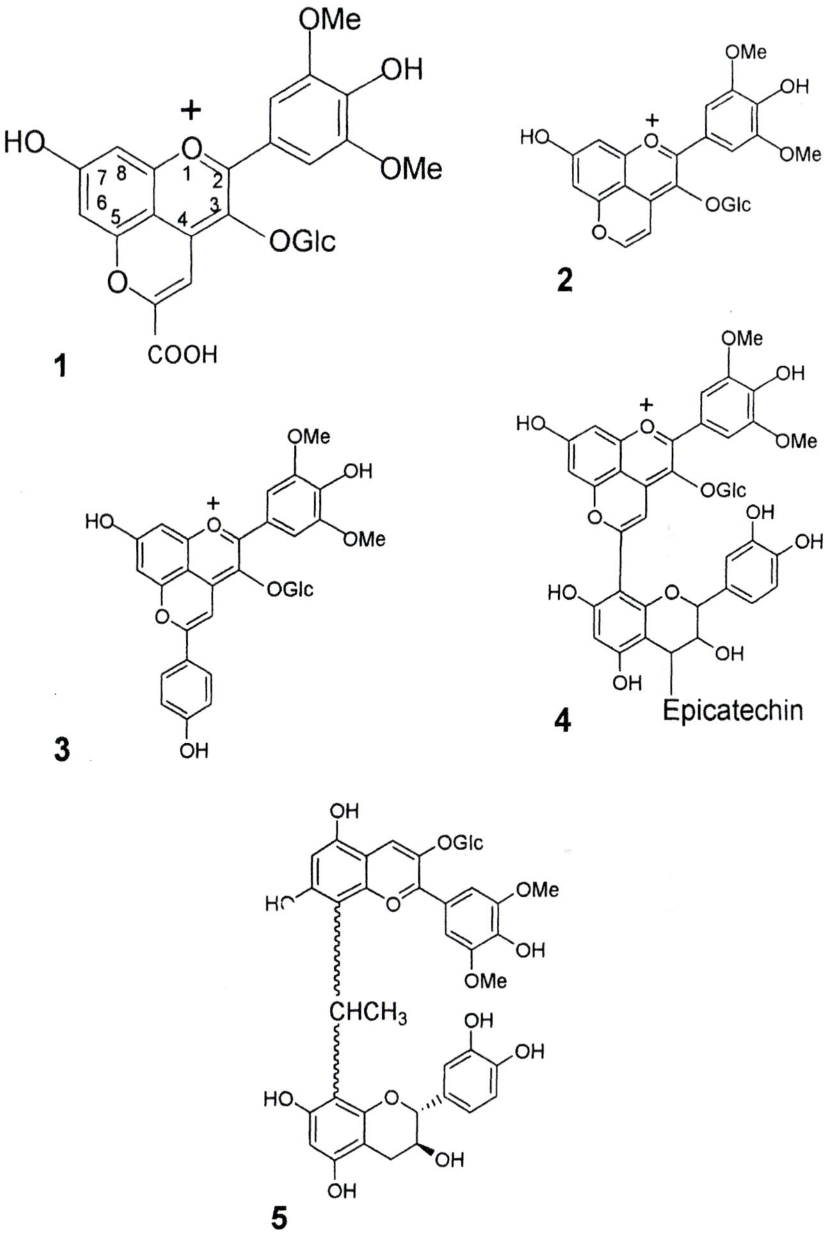

Figure 1. Structures of some oligomeric pigments identified in wine, grape marc and model wine solutions. ***(1)*** *vitisin A,* ***(2)*** *vitisin B,* ***(3)*** *pigment A,* ***(4)*** *pigment B2-III ,* ***(5)*** *Mv-ethyl-(epi)catechin.*

as a function of increased temperature and lower pH with the appearance of more complex pigments.

A knowledge of the factors controlling the formation of the various types of pigments is important in winemaking. The formation of ethyl-linked pigments may provide a mechanism for the extraction of anthocyanins from grape skins. They may also be a source of intermediates that are involved in the formation of more long-lived, more complex pigments. Similarly, the vitisin-like pigments provide a pool of stable pigments which are long-lived in wines. Their production may be influenced during fermentation, maturation and ageing by factors such as oxygen ingress, yeast type and other winemaking parameters.

Isolation and preparation of pigments

Mv was isolated from Shiraz grape skins as previously reported *(26)*. After re-crystallisation, Mv was obtained as 94% pure by HPLC. Vitisin A was extracted from a four year old Shiraz wine as reported previously *(27)*. After a final clean-up on a C_{18}-bonded reverse phase column vitisin A was obtained in 97% purity. Ethyl-linked dimers of malvidin-3-glucose and either catechin or epi-catechin were prepared by reacting malvidin-3-glucose (300mg/L) with the respective flavanol (200mg/L) in the presence of acetaldehyde (100mg/L) in a model wine system (pH 2.0) at room temperature (20°C) for 2 days. Progress of the reaction was monitored by HPLC. The pigments were isolated on a LH-20 column (34 x 2 cm) equilibrated with 50% methanol (0.1% TFA). Elution (1 mL/min) with 50% methanol (0.1% TFA) followed by 60% methanol (0.1% TFA) separated the ethyl-linked compounds from unreacted Mv and catechin / epicatechin. Purification was performed on a Waters Preparative HPLC with a C18 column (20 x 2.5 cm) equilibrated with 0.01M HCl using a solvent flow rate of 2 mL / min. Pigment I was eluted with 30% methanol (0.01M HCl) and pigment II with 50% methanol (0.01M HCl) with purities in the range 90 – 95% by HPLC. Pigment I was found to be sufficiently stable to afford isolation and purification. Nomenclature of the pigments is defined by the flavanol from which the pigment was formed and its elution profile. Pigments C-I and C-II are the first and second eluting (by HPLC) ethyl-linked pigments formed between catechin and Mv. Pigments EC-I and EC-II are the first and second eluting ethyl-linked pigments formed between epicatechin and Mv.

Determination of ionisation constants using HVPE

Ionisation constants of vitisin A and the various ethyl-linked malvidin-3-glucose / catechin / epicatechin dimers were determined by high voltage paper electrophoresis (HVPE) using chromatography paper (Chr. 1, Whatman, England) as previously reported *(28)*. All relative mobilities (Rm) were compared to anionic standards orange G and xylene cyanol. Fructose was used as the neutral standard and glucosamine as the single positively charged standard. Additionally, open-strip paper electrophoresis was carried out using glass fibre paper (GF/A, Whatman). A voltage gradient of 7.5 V/cm was used and the electrophoresis unit cooled to 15°C to help dissipate heat and limit evaporation of the buffer *(28)*.

Relative oxidative stability and longevity of vitisin A in wines

Mv and vitsin A were made up in model wine solution consisting of saturated potassium hydrogen tartrate in 10% (v/v) ethanol adjusted to pH 3.6 using tartaric acid at appropriate concentrations to give absorbances of approximately 1.0 at 520nm. At this pH the concentration of Mv required was 260mg/L compared to a concentration of 29 mg/L for vitisin A. 25 mL portions of each solution, in duplicate, were placed in open containers and allowed to undergo oxidation at room temperature (21°C). The concentrations were measured at regular intervals using HPLC. The resulting data were fitted to first order reaction kinetics and half lives and rate constants calculated using MacCurvefit version 1.4 (Kevin Raner Software, Australia).

The long term stability of vitisin A in wine was also estimated. A series of premium wines from a commercial winery in the Barossa Valley, South Australia made from 1958 to 1998 inclusive were analysed by HPLC for their Mv and vitisin A concentrations. These wines were made from Shiraz grapes sourced from a single vineyard, vinified using the same traditional methods over the span of vintages and stored under optimal conditions. Whilst vintage to vintage concentrations of pigments may have varied they are deemed to be small in comparison to the large changes observed with age. As such, the wines provided an indication of the persistence of Mv and vitisin A in wine over an extended period.

Fermentations involving malvidin-3-glucose and catechin

The fate of Mv and catechin in model fermentations was investigated. To flasks containing 100ml chemically defined grape juice media (CDJM) *(29)* was added malvidin-3-glucose (500 mg/L) (a), catechin (300 mg/L) (b) and malvidin-3-glucose (500 mg/L) plus catechin (300 mg/L) (c). A control with no additions was also run. The flasks were then inoculated with yeast (*Saccharomyces cerevisiae* EC1118) to a concentration of $1x10^6$ cells/mL. Each treatment was carried out in triplicate. The fermentations were conducted at 30°C in the dark and the flasks containing malvidin-3-glucose were additionally protected from light during transfers using aluminium foil. The progress of fermentation was monitored by measuring sugar content by refractive index. Once an indication that the sugar content was below 6.5 g/L a

Clinitest (Bayer Corporation, UK) was used to verify that the ferment had gone to completion. The ferments were centrifuged and the wine decanted from the lees. A sample (20mL) of each wine was stored in the dark at room temperature for aging studies. The lees (approx. 5mL) were extracted with 2x10mL methanol/formic acid (90:10) with agitation. The methanol was removed by rotary evaporation, the remaining solution was made up to 5mL with water and pH adjusted to approx 3.5 with sodium hydroxide *(29)*. Samples were analysed immediately after the completion of fermentation and then on a weekly basis by HPLC. The clarified wines were also monitored by UV-Vis spectrophotometry.

Synthesis of vitisin A during winemaking

Small lot wines were made from Shiraz grapes as previously reported *(27)*. During winemaking 10 mL samples were taken after each plunging (every 8 hr) for the analysis of Mv, vitisin A, pyruvate and glucose by HPLC and apparent dissolved oxygen concentration using a calibrated oxygen meter (Digital Oxygen System Model 10, Rank Bros. Ltd, Cambridge, UK).

HPLC analyses

HPLC analyses were carried out on a Waters system (Waters 501 pump, Wisp auto-sampler 710B; Waters Corporation, Milford, Massachusetts, USA) equipped with a diode array detector (Waters 996). Two systems were employed. The first used a reverse phase C18 column (250 x 4 mm Licrosphere 100; Merck, Darmstadt, Germany) protected by a C18 (NovaPak; Waters) guard column. A binary solvent system consisting of solvent A (dilute HCl , pH 2.4) and solvent B (80% acetonitrile solution acidified using concentrated HCl to pH 2.1) with a flow rate of 0.6 mL/min and column temperature of 30 ^{0}C were used. The linear gradient consisted of 0% to 100% solvent B over 50 min. The second system used a Platinum EPS C18 Rocket (Alltech) column (53x7 mm) protected by a guard column of the same material. A binary solvent system consisting of (A) 2% v/v aqueous formic acid and (B) 100% acetonitrile was used. The elution profile consisted 10% B for 2 minutes, a linear gradient of 10% B to 60% B over 8 minutes, 60% B to 100% B over 2 minutes and 100% B for 2 minutes. A flow rate was 1.5 mL/min and column temperature of 30^{0}C were used. Detection was by a photodiode array detector in the range 250 – 600 nm. All samples were filtered using a 0.45 μm syringe filter (Schleicher und Schuell GmbH, Dassel, Germany). The data were analysed using Waters Millennium software Version 3.05.

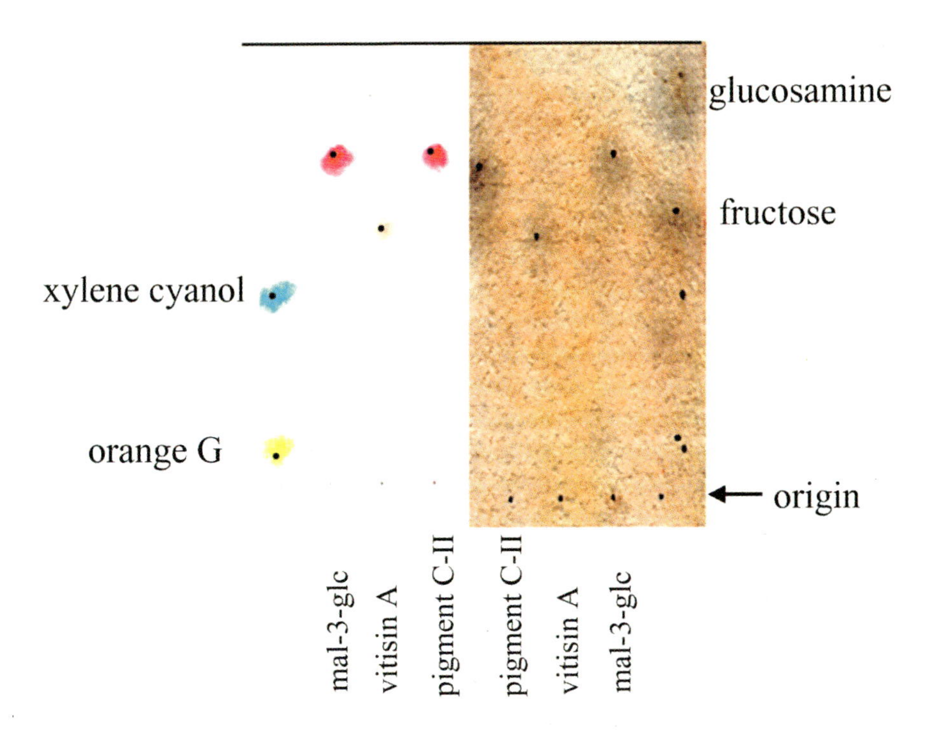

Figure 2. Glass fibre open-strip electrophoresis of pigment C-II, Mv and vitisin A at pH 2.2. The right hand side of each electrophoretogram is stained with silver nitrate to reveal the migration of the non-colored standards, neutral fructose and single positively charged glucosamine. Mv and pigment C-II appear as red spots whereas vitisin A is coloured orange at this pH.

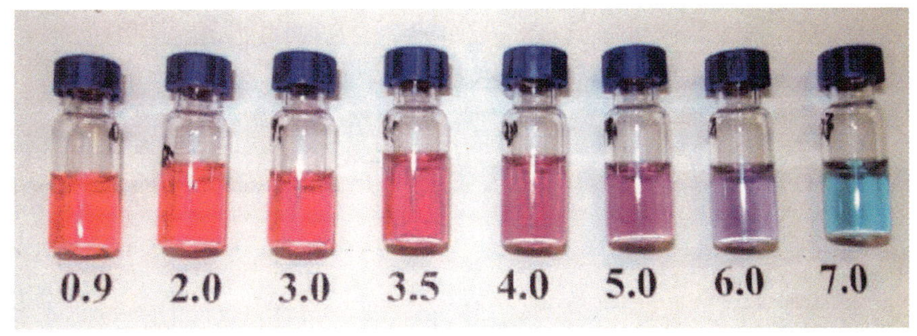

Figure 3. Effect of pH on the color and colour expression of pigment C-II. Concentration of pigment in each case is 2×10^{-5} M.

Table I. pKa_1 and pKa_2 Values for Ethyl-linked Dimers of Mv and Catechin / Epicatechin as Determined by HVPE.

	Mv-et-C-I	Mv-et-C-II	Mv-et-EC-I	Mv-et-EC- II
pKa_1	2.11±0.08	1.96±0.08	2.34±0.08	2.17±0.08
pKa_2	4.61±0.06	4.77±0.04	4.53±0.04	4.70±0.04

UV-vis spectroscopy

UV-vis spectroscopy was carried out at room temperature on a Helios α (Thermo Spectronic) spectrophotometer using a 2 mm path length quartz cell. Spectra were recorded from 250 to 700 nm.

Results and Discussion

The resistance of pigment C-II to bisulfite bleaching and hydration has been well documented *(4)*. Using the same techniques it has been shown that pigments C-I, EC-I and EC-II have similar properties to pigment C-II (data not shown). Further, the first two ionisation constants, pKa_1 and pKa_2, for all four pigments, as determined by HVPE, vary only slightly (Table 1). pKa_1 and pKa_2 values are in the ranges 2.0 – 2.3 and 4.5 to 4.7 respectively. These values compare with those for Mv of 1.76 (±0.07) and 5.36 (±0.04) and pKa_1 = 0.95 (± 0.10), pKa_2 = 3.56 (± 0.06), pKa_3 = 5.38 (± 0.07) for vitisin A *(31)*. Figure 2 shows a glass fibre open-strip electrophoretogram at pH 2.2 of pigment C-II (mv-et-cat II), Mv and vitisin A. Mv and pigment C-II have similar relative electrophoretic mobilities, half way between the neutral standard fructose and the mono cationic standard glucosamine and consistent with their respective measured pKa_1s. Both appear as reddish-purple spots. It is noteworthy that vitisin A gives an orange spot on the electrophoretogram and moves as an uncharged molecule at pH 2.2. This is in agreement with the absorbance maximum of the neutral quinonoidal base of vitisin A at 499nm.

pKa_1 and pKa_2 represent the protonation and de-protonation of the neutral quinonoidal base respectively. Introduction of an electron donating substituent at the nucleophilic C-4 position is expected to stabilise the quinonoidal base against gaining a proton, particularly on the C-ring, thus lowering pKa_1. On the other hand substitution of an *p*-hydroxyphenylmethine substituent at the electrophilic C-8 site, as in pigment C-II, is expected to facilitate the loss of a proton, particularly on the A-ring, thereby lowering pKa_2 *(32)*. The slightly higher pKa1 of the ethyl-linked pigments and the lower pKa1 value of vitisin A combined with an 0.8 unit lower value of pKa_2 for the ethyl-linked pigments compared to Mv support this proposal.

The 4-position in the ethyl-linked pigments is not substituted and yet these pigments are quite resistant to bisulfite bleaching and hydration. The impact of pH on stability and color change of pigment C-II is shown in Figure 3. The stability of these pigments has been attributed to conformational effects where the flavanol B-ring is folded under and blocks the anthocyanin C-ring to nucleophilic attack. This is illustrated in Figure 4 where pigments C-I (a) and C-II (b) have been subjected to conformational energy minimisation using MM2 *(33)*. NMR data, not presented, suggests that the stereochemistry at the ethyl group of pigment C-II has the S configuration.

Fermentations involving malvidin-3-glucose and catechin

Ethyl-linked pigments are generated rapidly during fermentation. Figure 5 shows the decreases in malvidin-3-glucose concentration and the development of other pigments during yeast fermentation in chemically defined grape juice medium with Mv and catechin added. Substantial amounts of pigments C-I and C-II (mvcat-ethyl) as well as low levels of other pigments (mvcat-other) are seen to form during fermentation. These other pigments may include trimers and higher oligomers of catechin linked to Mv as well as vitisin-like compounds which are observed to form in fermentations with added Mv but without added catechin (mv-other). A significant proportion of the Mv is lost during these fermentations as a result of binding to the yeast lees (mv-mv, Figure 5) *(34)*. The yeast lees also bind some of the more complex pigments (the broad peak centred at 7.4min) as well as the ethyl-linked pigments (Figure 6(a) & (b)). Whilst these reactions have been observed in model systems where acetaldehyde has been added, and related compounds have been identified in wines and rose ciders *(35)*, the data shows that the ethyl-linked pigments formed in wine are generated as a result of fermentation and it is proposed that the acetaldehyde exported by the yeast cell during fermentation is responsible for the coupling reactions.

Synthesis of vitisin A during winemaking

The reaction sequence to produce the ethyl-linked pigments involves a Baeyer-type condensation where the oxidation state of the anthocyanin entity

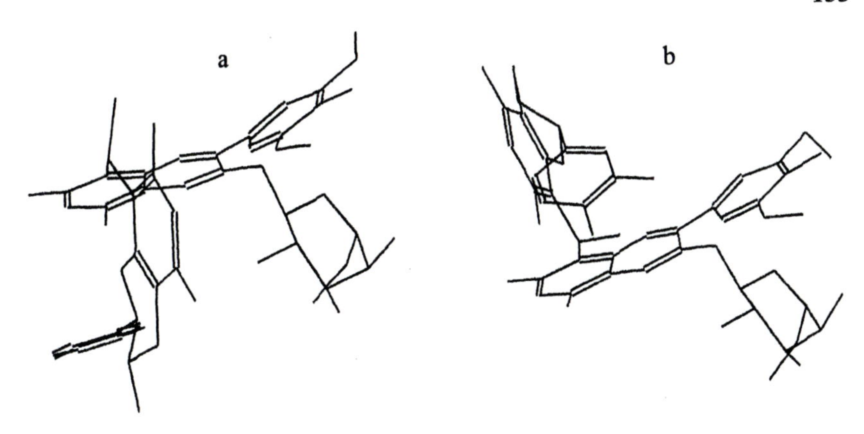

Figure 4. Minimum energy conformations of pigments C-I (a) and C-II (b) determined by MM2 molecular mechanics calculation. The B ring of the catechin moiety is folded under the malvidin C-ring and it is suggested that this protects the latter from attack by nucleophiles at the 2 or 4 positions.

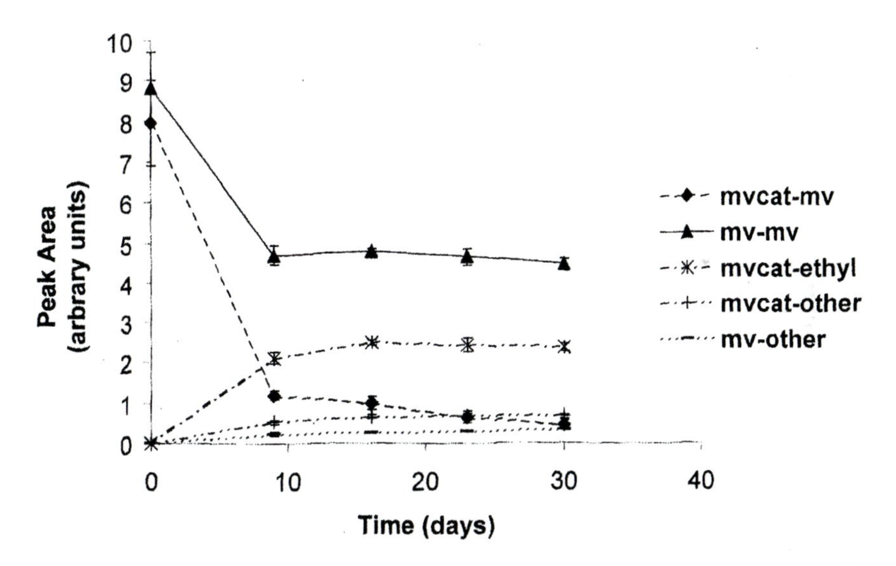

Figure 5. Development of pigments, monitored by HPLC at 520nm, in model fermentations to which either Mv (500mg/L) or Mv (500mg/L) and catechin (300mg/L) had been added prior to the start of fermentation. The decline in Mv concentration (mv-mv) during fermentation in those ferments to which only Mv had been added is mostly attributed to the binding of Mv to the yeast lees. The decline in Mv concentration (mvcat-mv) during fermentation in those ferments to which Mv and catechin had been added is attributed to the formation of wine pigments (mvcat-ethyl, mvcat-other) together with loss due to binding to yeast lees.

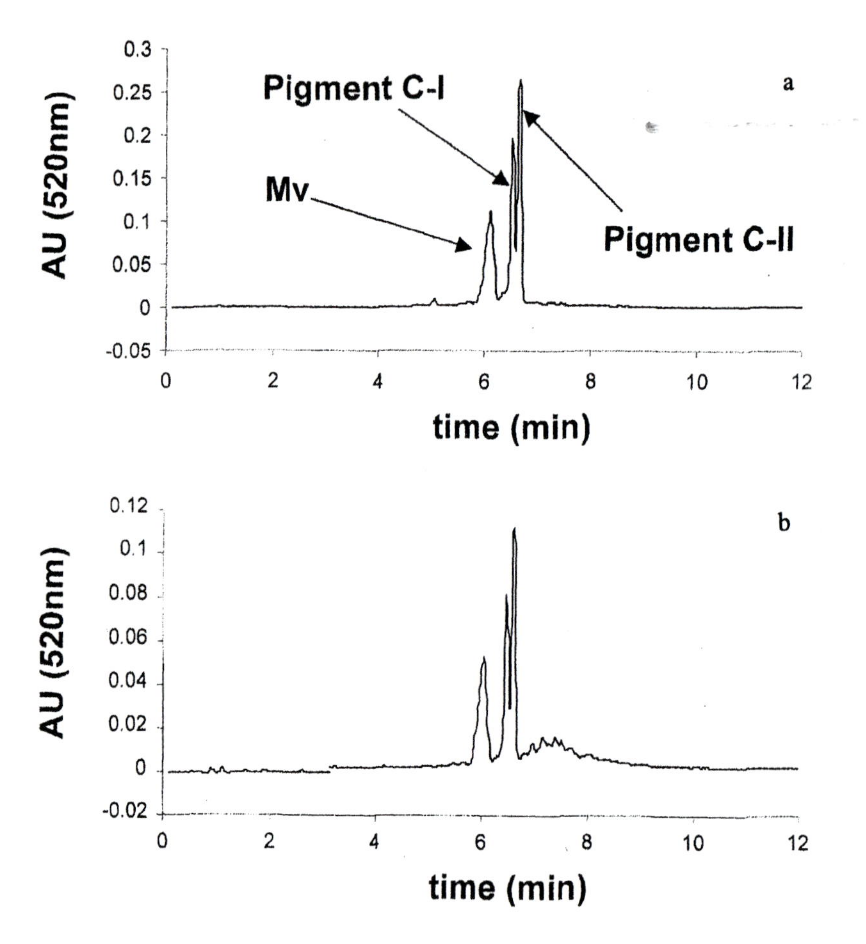

Figure 6. HPLC chromatograms using the rocket column, measured at 520nm, of (a) pigments present at the end (9 days) of a model ferment to which Mv (500mg/L) and catechin (300mg/L) had been added prior to the start of fermentation; (b) pigments extracted from the yeast lees after completion of fermentation (9 days).

remains the same. The production of the vitisin-type compounds depends, however, on a two electron oxidation step to reform the anthocyanin entity. The presence of atmospheric oxygen has been shown to be necessary for the formation of vitisin A in model systems *(20)* but in red wine ferments the rate of vitisin A synthesis is not correlated with molecular oxygen concentration. In replicated small lot fermentations of Shiraz grapes, where malvidin-3-glucose, pyruvate, glucose and dissolved oxygen concentrations were measured, it was shown that the rate of vitisin A formation was not limited by substrate concentrations *(27)*. Most of the vitisin A formation occurred during fermentation when yeast activity was the highest and oxygen concentration was low (see Figure 7). At the end of fermentation, oxygen concentration increased towards its saturation level (8 – 10mg/L) and yet there was no sudden increase in the rate of vitisin A formation even though substrate concentrations remained non-limiting. It is suggested that rather than oxygen itself is a reactive oxygen species that is responsible for the formation of vitisin A. This reactive oxygen species may be generated from molecular oxygen by the yeast during active fermentation or may be generated by other components of the wine such as polyphenols. Therefore, appropriate cap management techniques and / or the presence of components in the wine which promote the production of active oxygen species, such as ellagitannin, may be used to enhance the formation of stable wine pigments such as vitisin A. Introduction of controlled amounts of oxygen into wine during maturation (micro oxygenation) may be a key factor in the formation of the vitisin-like pigments *(36)*.

Stability and longevity of wine pigments

Loss of a particular pigment in wine occurs as a result of transformation into different pigments, or to degradation, or to precipitation, or to a combination of these processes. Unlike the reversible effects of pH and bisulfite bleaching, oxidation is one of the key factors affecting pigment longevity. It has responsibility for the generation of more complex pigments from grape anthocyanins as well as oxidative degradation of pigments. The resistance to oxidation of malvidin-3-glucose and vitisin A are shown in Figure 8. It is pointed out that whilst the solution of vitisin A retained its original orange-red color, that of the malvidin3-glucoside turned brown. The data gave rate constants and half lives of 0.808 month^{-1} and 25.8 days for Mv and 0.016 month^{-1} and 50.0 months for vitisin A in a container open to air at 21°C. Ethyl-linked pigments behave in a similar manner to Mv. They quickly loose color when exposed to atmospheric oxygen and turn brown (data not shown).

Ethyl-linked pigments in wine-like media are also short lived. Francia-Arichia et al. *(13)* showed that ethyl-linked procyanidin B2-Mv dimers at pH 3.2 in a model solution reached a maximum after 10 days but then declined to negligible levels after 120 days. The rate constants for the formation and disappearance of ethyl-linked pigments formed from other procyanidins show

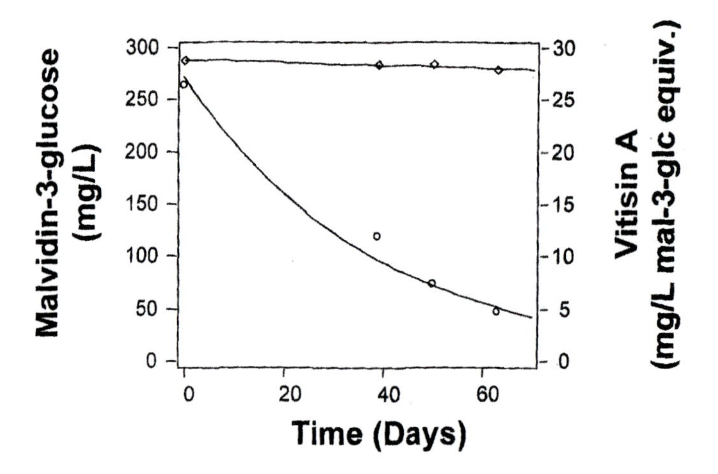

Figure 7. Stability of Mv (circles) and vitisin A (diamonds) to atmospheric oxidation as measured by the time-dependant loss of absorbance at 520nm.

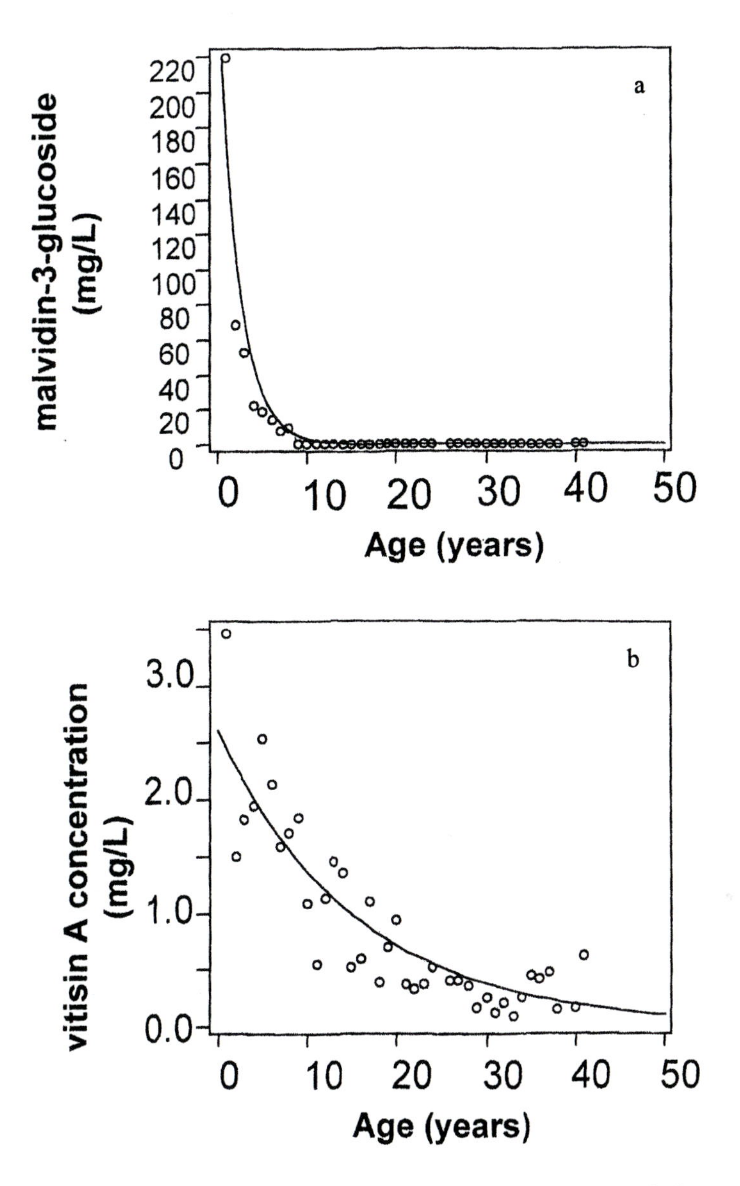

Figure 8. Concentration of (a) Mv and (b) vitisin A in a commercial Shiraz wine made from grapes sourced from a single vineyard and made by the same vinification process over 40 vintages.

similar patterns to that for procyanidin B2 *(13,23)*. This decrease was accompanied by the formation of vitisin-like pigments. Similar formation and loss profiles have been observed with catechin *(22)* in model systems with added malvidin-3-glucose and acetaldehyde although only trace quantities of vitisin-like compounds were reported in either systems.

Unlike the ethyl-linked pigments the vitisin-like pigments are appreciably more temporally stable. The longevity of vitisin A in wines compared to that of malvidin-3-glucose is shown in Figures 9a and 9b for wines spanning 41 vintages. Mv can be measured in wines up to 8 years old whereas vitisin A can be measured even after 41 years of maturation. The concentration data were fitted to first order kinetics and pseudo stability parameters of 0.443 $year^{-1}$ and $t_{1/2}$ of 1.6 years obtained for Mv and 0.065 $year^{-1}$ and $t_{1/2}$ of 10.7 years for vitisin A. Whilst there may be a contribution from vitisin A synthesis during ageing the data clearly indicate the longevity of vitisin A in wine.

Concluding remarks

Many of the observed oligomeric pigments are present in wine in low concentrations. Whilst these wine pigments provide a pool of color-stable compounds in wine, a question to be asked is 'how important are they'? In aged wine, where the concentration of grape-derived anthocyanins is negligible, the stable C4-substituted wine pigments may contribute significantly towards the color of wine. The color of vitisin A at wine pH is brick red (λ_{max} = 501 nm *(31)*), and therefore it is possible that the tawny color of aged wine is due to the presence of these vitisin-like wine pigments. In younger wines where there are higher concentrations of grape-derived anthocyanins and ethyl-linked pigments, these C4-substituted pigments will still contribute to the total pool of pigments and therefore may be considered as important pigments in the overall expression of wine color. Currently, it is not possible to estimate the relative contribution of the oligomeric pigments to wine color because molar absorbances for the individual compounds are not available and methods need to be developed for the routine quantification of these compounds. In addition, whilst ever increasing numbers of wine pigments are being identified, it is important to recognise that a large proportion of the color of older wines is not accounted for. The HPLC chromatograms of pigments isolated from aging red wines usually show an increasing proportion of the color in the amorphous baseline 'hump' in proportion to the age of the wine (see Figure 10). Separation and detection techniques to better define the components in the 'hump' need to be developed (see Waterhouse et al, Kennedy and Hayasaka, this issue).

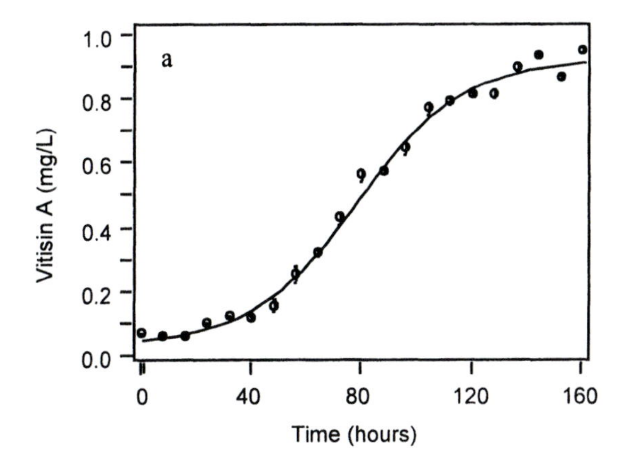

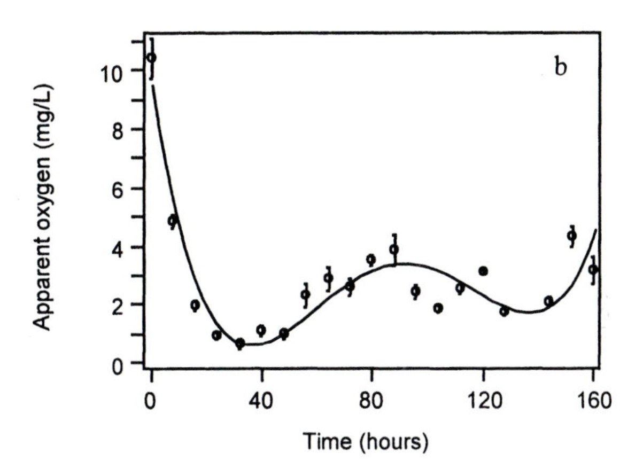

Figure 9. The change in concentration of (a) vitisin A and (b) concentration of dissolved oxygen as a function of time during fermentation. Each point represents the mean of three replicates and the vertical bars represent the standard error of the mean. The solid lines represent the lines of fit for (a) a logistic function or (b) a four factor polynomial (27).

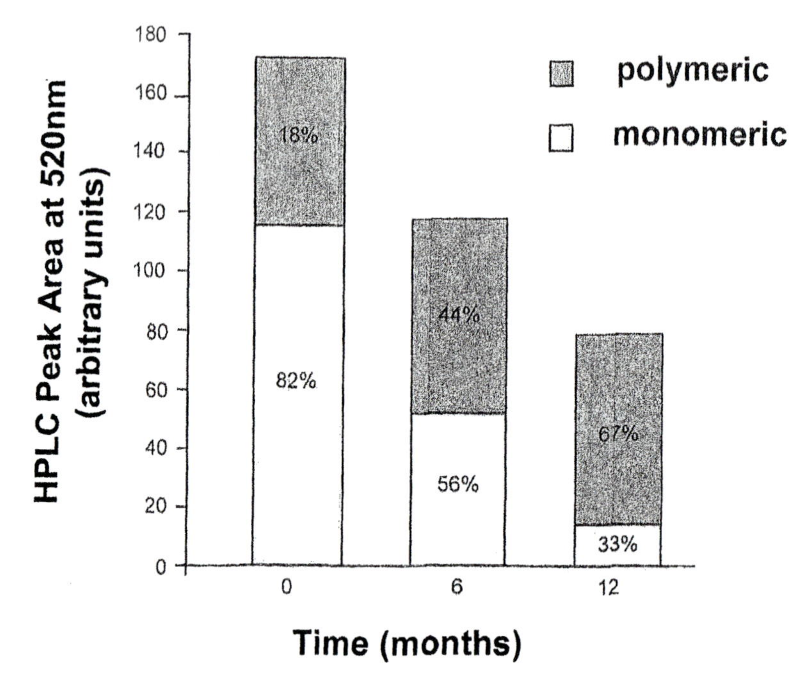

Figure 10. Proportion of monomeric anthocyanin pigment (Mv, Mv-3-(acetyl)glucoside, Mv-3-(p-coumaryl)-glucoside, vitisin A, acetyl-vitisin A, p-coumaryl-vitisin A) to polymeric pigment (unidentified pigments in the 'hump') as defined by the area under the HPLC chromatogram at 520 nm. The wine was made from Shiraz grapes and was measured directly after fermentation and then at 6 months and at 12 months of age.

Acknowledgments

P. Iland and E. Waters are thanked for their helpful comments. This project was supported by Australia's grapegrowers and winemakers through their investment body, the Grape and Wine Research and Development Corporation, with matching funds from the Federal Government and by the Commonwealth Cooperative Research Centres Program, conducted by the CRC for Viticulture and by the Australian Research Council (ARC).

References

1 Somers, T. C. *Phytochemistry,* **1971**, *10*, 2175-2186.
2 Brouillard, R.; Dubois, J-E. *J. Amer. Chem. Soc.* **1977**, *99*, 1359-1364.

3 Brouillard, R.; Delaport, B. *J. Amer. Chem. Soc.* **1977**, *99*, 8461-8468.
4 Cheminat, A.; Brouillard, R. *Tetrahedron Lett.* **1986**, *27*, 4457-4460.
5 Santos, H.; Turner, D.L.; Lima, J. C.; Figueiredo, P.; Pina, F.; Macanita, A. L. *Phytochemistry* **1993**, *33*, 1227-1232.
6 Timberlake, C.F.; Bridle, P. *Chem Ind.* October 1968, p 1489.
7 Remy, S.; Fulcrand, H.; Labarbe, B.; Cheynier, V.; Moutounet, M. *J. Agric. Food Chem.* **2000**, *80*, 745-751.
8 Bakker, J.; Timberlake, C.F. *J. Agric. Food Chem.* **1997**, *45*, 35-43.
9 Bakker, J.; Bridle, P.; Honda, T.; Kuwano, H.; Saito, N.; Terahara, N.; Timberlake, C. F. *Phytochemistry* **1997**, *44*, 1375-1382.
10 Cameira-dos-Santos, P-J.; Brillouet, J-M.; Cheynier, V.; Moutounet, M. *J. Sci. Food Agric.* **2000**, *70*, 204-208.
11 Fulcrand, H.; Benabdeljalil, C.; Rigaud, J.; Cheynier, V.; Moutounet, M. *Phytochemistry* **1998**, *47*, 1401-1407.
12 Fulcrand, H.; Dos Santos, P-J., C.; Sarni-Manchado, P.; Cheynier, V.; Favre-Bonvin, J. *J. Chem. Soc. Perkin Trans. 1*, **1996**, 735-739.
13 Francia-Aricha, E. M.; Guerra, M. T.; Rivas-Gonzalo, J. C.; Santos-Buelga, C. *J. Agric. Food Chem.* **1997**, *45*, 2262-2266.
14 Bakker, J.; Timberlake, C.F. *J. Sci. Food Agric.* **1985**, *36*, 1325-1333.
15 Mateus, N.; de Pascual-Teresa, S.; Rivas-Gonzalo, J.C.; Santos-Buelga, C.; de Freitas, V. *Food Chemistry* **2002**, *76*, 335-343.
16 Jones, G.P.; Asenstorfer, R.E. Australian Society of Oenology and Viticulture Seminar on Phenolics and Extraction, Adelaide, South Australia, **1998,** 33-37.
17 Asenstorfer, R.E.; Hayasaka, Y.; Jones, G.P. *J. Food Agric. Chem.* **2001**, *49*, 5957 – 5963.
18 Timberlake, C.F.; Bridle, P. *J. Sci. Food Agric.* **1977**, *28*, 539-544.
19 Garcia-Viguera, C.; Bridle, P.; Bakker, J. *Vitis* **1994**, *33*, 37-40.
20 Bakker, J.; Picinelli, A.; Brindle, P. *Vitis* **1993,** *32*, 111-118.
21 Escribano-Bailon, T.; Dangles, O.; Brouillard, R. *Phytochemistry* **1996** *41*, 1583-1592.
22 Rivas-Gonzalo, J.C.; Bravo-Haro, S.; Santos-Buelga, C. *J. Agric. Food Chem.* **1995**, *43*, 1444-1449.
23 Dallas, C.; Ricardo-da-Silva, J.M.; Laureanto, O. *J. Agric. Food Chem.* **1996**, *44*, 2402-2407.
24 Dallas, C.; Ricardo-da-Silva, J.M.; Laureano. O. *J. Sci. Food Agric.* **1996**, *70*, 493-500.
25 El-Safi, N.-E.; Fulcrand, H.; Cheynier, V.; Moutounet, M. *J. Agric. Food Chem.* **1999**, *47*, 2096-2102.
26 Vivar-Quintana, A.M.; Santos-Buelga, C.; Francia-Aricha, E.; Rivas-Gonzalo, J.C. *Food Sci. Tech. Int.* **1999**, *5*, 347-352.
27 Asenstorfer, R.E.; Iland, R.G.; Markides, A.J.; Jones, G.P. *Aust. J. Grape and Wine Research* **2003**, *9*, 40 – 46.

28 Asenstorfer, R.E.; Iland, P.G.; Tate, M.E.; Jones, G.P. *Analytical Biochemistry* **2003**, *318*, 291- 299.

29 Jiranek, V.; Langridge, P.; Henschke, P.A. *Am. J. Enol. Vitic.* **1995**, *46*, 75-83.

30 Morata, A.; Gomez-Cordoves, M. C.; Suberviola, J.; Bartolome, B.; Colomo, B.; Suarez, J. A. *J. Agric. Food Chem.* **2003**, *51*, 4084-4088.

31 Asenstorfer R. E. PhD thesis, University of Adelaide, Australia, **2001.**

32 Perrin, D.D.; Dempsey, B.; Serjeant, E.P. p*Ka* Prediction for Organic Acids and Bases, Chapman and Hall, London, **1981**.

33 Burkert, U.; N. L. Allinger, N.L. Molecular Mechanics ACS Press, Washington, D.C., USA, **1982**.

34 Vasserot, Y.; Caillet, S.; Maujean, A. *Am. J. Enol. Vitic.* **1997**, *48*, 433-437.

35 Shoji, T.; Goda, Y.; Toyoda, M.; Yanagida, A.; Kanda, T. *Phytochemistry* **2002**, *59*, 183-189.

36 Atanasova, V.; Fulcrand, H.; Cheynier, V.; Moutounet, M. *Analytica Chim. Acta* **2002**, *458*, 15-27.

Chapter 9

Flavanols and Anthocyanins as Potent Compounds in the Formation of New Pigments during Storage and Aging of Red Wine

Nour-Eddine Es-Safi[1,2] and Véronique Cheynier[1]

[1]UMR Sciences Pour l'Œnologie, INRA, 2 place Viala, 34060 Montpellier, France
[2]École Normale Supérieure, Laboratoire de Chimie Organique et d'Etudes Physico-chimiques, B.P. 5118, Takaddoum Rabat, Morocco

During storage and ageing of red wine, the pigment composition becomes progressively more complex. There is a steady shift from the original red color due to the anthocyanins extracted from grapes to a more yellow tint. This is explained by the progressive and irreversible conversion of anthocyanins and of flavanols to more stable pigments, including purple flavylium salts and yellow xanthylium derivatives, in particular through reactions with aldehydes. The new pigments formed are also much more resistant than oligomeric proanthocyanidins towards acid-catalyzed cleavage. In this paper, the structures of these compounds, some of which have been newly established, as well as the mechanisms of their formation in wine are discussed.

Anthocyanins, flavanols and phenolic acids are the main polyphenols of grapes and grape-derived foods, responsible for their major organoleptic properties. Anthocyanins are directly involved in the color of most red fruits including grapes (*1*) while flavanols are thought to make an important contribution to the astringency and bitterness of plant derived foods (*2,3*).

The anthocyanins of most important grapes cultivars have been identified and the levels in which they are found have also been studied (*4*). Naturally occurring grape anthocyanins are located in grape skins and their structures have at least one sugar. In *Vitis vinifera* varieties, the only anthocyanins present are derivatives of malvidin, delphinidin, petunidin, peonidin and cyanidin, with a glucose in the 3 position. Non *vinifera* grapes also contain 3,5 diglucoside forms of these five anthocyanidins. The sugar group may be further acylated with an acetic, p-coumaric, or caffeic acid group. Quantitatively, malvidin 3-O-glucoside is the predominant grape anthocyanin, and can account for more than 90 % of the anthocyanins.

In grapes, flavan-3-ol monomers and proanthocyanidins, which are flavan-3-ol oligomers and polymers, are located in the skins and seeds (*5-10*). The major flavan-3-ols identified in grapes include (+)-catechin, (-)-epicatechin, and (-)-epicatechin 3-O-gallate as well as dimeric and trimeric procyanidins based on these units (*9*). However, concentration of oligomers is relatively low compared to that of larger molecular weight flavanols (*5,8,10*).

During red wine making, significant proportions of grape polyphenols are extracted (*11,12*). Their concentrations in red wine are affected by wine production practices such as maceration time, pressing, maturation and fining in addition to that of the grape variety used as the raw material for the preparation of wine. Monitoring of phenolic compounds throughout red-wine making indicated that phenolic acids and anthocyanins diffuse faster than flavanols (*13*).

Once extracted, all phenolic compounds undergo various types of reactions, themselves depending on the presence of other wine components as well as on the storage conditions (*14*).

Reactions of the extracted grape phenolics begin with the crushing of the berry and the breakdown of its cellular compartmentalisation as shown by appearance of polymeric pigments from the very earliest stages (*15, 16*). They continue throughout fermentation and ageing and lead to a great diversity of products, thus adding to the complexity of wine phenolic composition. The new compounds formed often show specific organoleptic properties, distinct from those of their precursors, which explains why, as the wine ages, its sensory characteristics are modified.

The color of red wines which is intimately related to their polyphenolic content is obviously affected by these transformations. The initial color of red wine is fundamentally due to the extraction of anthocyanin pigments from the black grape skins during vinification. During conservation, the contribution of anthocyanins to wine color gradually decreases as they proceed to new, stable oligomeric/polymeric pigments (*15-21*). Concomitantly, red wine color changes from a bright red to a red brown tint.

Various mechanisms have been proposed to explain the formation of these pigments, resulting from the reaction between free anthocyanins extracted from grapes and other phenolic compounds, particularly flavanols (*18*). Processes involving either direct condensation between flavanols and anthocyanins giving rise to xanthylium salts with a yellow brown hue or reactions mediated by acetaldehyde with the formation of violet pigments (*18*, *22-32*) have been studied in model solution systems. It was also shown that other aldehydes such as glyoxylic acid, furfural or HMF, can replace acetaldehyde in the latter mechanism (*33-36*) where, in addition to the red pigments, yellow xanthylium salts were obtained. The presence of some of the obtained compounds has actually been observed in grape derived foods (*33*, *37*).

The present paper summarizes and gathers our findings about aldehyde-induced reactions involving or not anthocyanins and giving various purple and yellow adducts, as well as the structures of the new formed pigments. The implication and impact of these reactions and these new pigments on red wine color is also discussed.

Material and Methods

Reactions

Solutions containing malvidin 3-O-glucoside or cyanidin 3-O-glucoside and (+)-catechin or (-)-epicatechin (20 mM) were prepared in 0.5 mL of solutions containing a mixture of $CH_3COOH/EtOH/H_20$ (17/50/433, V/V/V) and giving a pH of 2.2. The aldehydes (acetaldehyde, glyoxylic acid, furfuraldehyde or HMF) were added in order to give a final concentration of 20 mM.

The prepared solutions were monitored by liquid chromatography coupled with a diode array detector (DAD) and with an electrospray mass spectrometry (ESI-MS) detector. Isolation of the formed pigments was achieved by HPLC at the semipreparative scale.

Analytical HPLC/DAD analyses

HPLC/DAD analyses were performed by means of a Waters 2690 Separation Module system including a solvent and a sample management system, a Waters 996 photodiode array detector and a Millenium 32 chromatography manager software. UV-visible spectra were recorded from 250 to 600 nm. The column was a reversed-phase Lichrospher 100-RP18 (5 µm packing, 250 x 4 mm i.d.) protected with a guard column of the same material. Elution conditions were as follows: 1 mL/min flow rate; temperature 30 °C; solvent A, water-formic acid (98:2,v/v); solvent B, acetonitrile-water-formic acid (80:18:2, v/v); elution from 5 to 30 % B in 40 min, from 30 to 40 % B in 10 min and from 40 to 100 % B in 5 min, followed by washing and re-equilibrating the column.

MS Apparatus and LC/ESI-MS analyses

MS measurements were performed on a Sciex API I Plus simple quadruple mass spectrometer equipped with an electrospray ionisation source. The mass spectrometer was operated in negative or positive-ion mode. Ion spray voltage/orifice voltages were selected at - 4 KV/- 70 V and + 5 KV/+ 60 V, respectively.

HPLC separations were carried out on a narrowbore reversed-phase column with an ABI 140 B solvent delivery system (Applied Biosystems, Weiterstadt, Germany). The column was connected with the ion spray interface via a fused-silica capillary (length 100 cm, 100 µm i.d.). The reaction mixture was injected with a rotary valve (Rheodyne Model 8125) fitted with a 20 µL sample loop. The separation was achieved on a Lichrospher 100-RP18 column (5 µm packing, 250x4 mm i.d., Merck, Darmstadt, Germany), with a flow rate of 280 µL/min. The elution was done with solvents A and B used in HPLC/DAD analysis and the conditions adapted as follows: isocratic elution with 10 % B for 4 min, linear gradients from 10 to 15 % B in 11 min, from 15 to 50 % B in 25 min, and from 50 to 100 % B in 5 min, followed by washing and reconditioning of the column. The absorbance at 280 nm was monitored by an ABI 785A programmable absorbance detector and by a Waters 990 diode-array detector linked to a 990 system manager software.

Results and Discussion

Reactions of anthocyanins and flavanols in the presence of aldehydes, namely acetaldehyde, glyoxylic acid, furfural, and hydroxymethylfurfural, have been studied in wine like model solution systems, enabling to demonstrate the formation of purple and yellow pigments, as discussed below. Identification of the formed compounds was achieved on the basis of their UV-visible, 1D and 2D NMR and mass spectra (*38-42*).

Formation of purple pigments from colored precursors:

Reactions taking place in a model solution containing malvidin 3-O-glucoside, (-)-epicatechin and acetaldehyde was explored by HPLC/DAD and HPLC/ESI-MS analysis (*29*). Acetaldehyde is a product of yeast metabolism but may also result from ethanol oxidation or from decarboxylation of pyruvic acid during fermentation of grapes (*43*).

HPLC-ESI-MS analysis of the mixture conducted in the positive ion mode showed the presence of both ethyl-linked flavanol dimers, detected at *m/z*: 607 amu, and ethyl-linked adducts involving both (-)-epicatechin and the anthocyanin, detected as the flavylium cations or protonated quinonoidal bases at *m/z*: 809 amu (Figure 1, R= CH_3, R_1= H, R_2= OH). The latter (Figure 2, **a**, **b**) were characterized by absorbance maxima bathochromically shifted (around 540 nm) compared to that of malvidin 3-O-glucoside (525 nm) as shown in Figure 2. This bathochromic shift was more accentuated for compounds eluted at the end of the chromatogram where absorbance maxima attained 555 nm (Figure 2, compounds **1-11**). Although it was not possible to determine individually the mass of each of these pigments, series of mass signals corresponding to ethyl-linked trimers and tetramers containing two epicatechin and one malvidin 3-O-glucoside (*m/z*: 1125 amu), one epicatechin and two malvidin 3-O-glucoside (*m/z*: 664 amu), three epicatechin and one malvidin 3-O-glucoside (*m/z*: 1441 amu), two epicatechin and two malvidin 3-O-glucoside (*m/z*: 822 amu), were detected in this region of the HPLC profile, suggesting that the bathochromic shift was associated with further polymerisation.

Monitoring of the reaction also indicated that ethyl-linked flavanol-anthocyanin adducts are more stable than ethyl-linked flavanol oligomers.

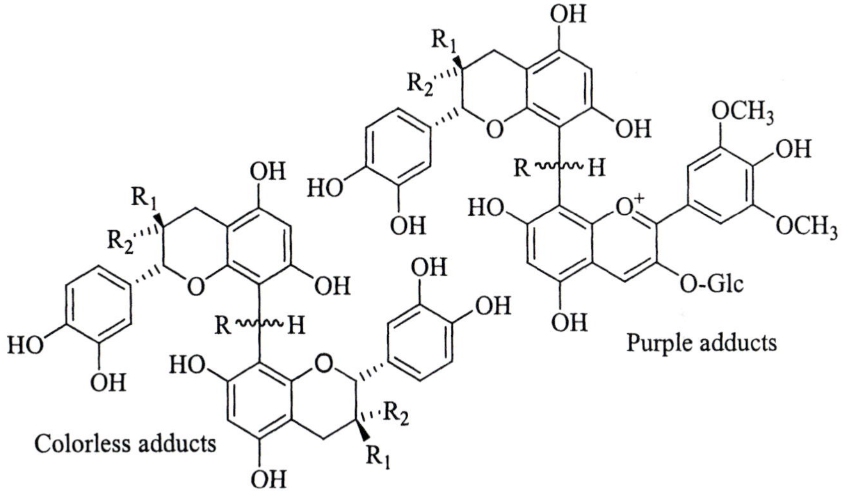

Figure 1: Example of colorless adducts and colored purple pigments detected in model solution containing (-)-epicatechin (R_1= H, R_2= OH) or (+)-catechin (R_1= OH, R_2= H), malvidin 3-O-glucoside and acetaldehyde (R= CH_3), glyoxylic acid (R= COOH), furfural (R= furyl group), HMF (R= hydroxymethylfuryl group).

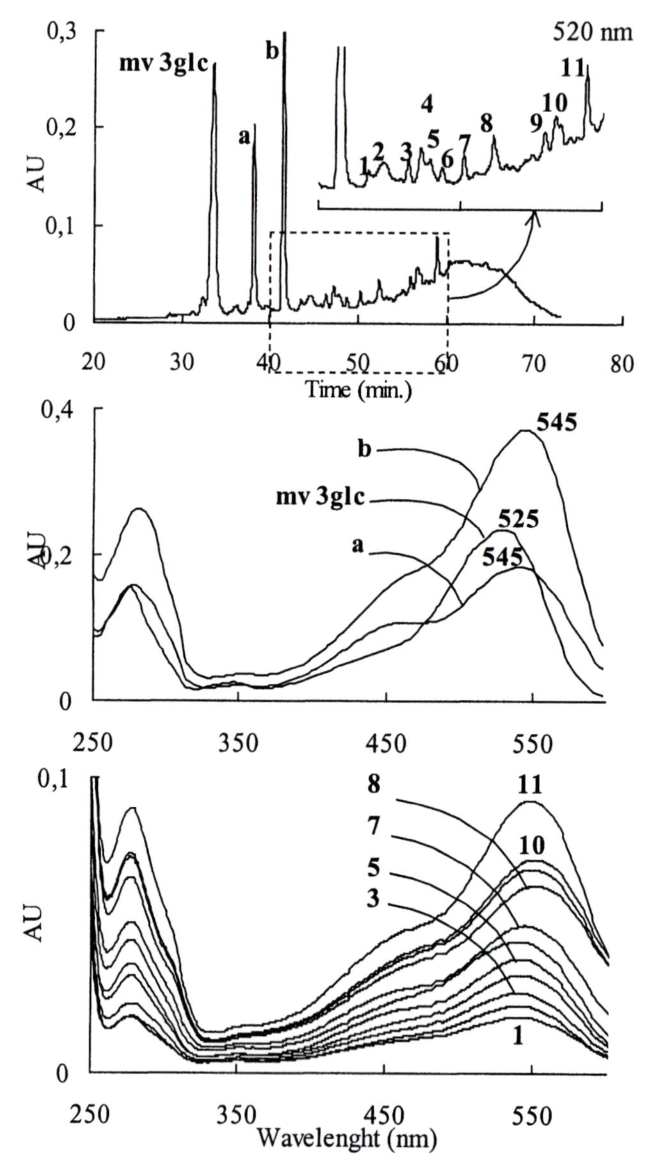

Figure 2: HPLC chromatogram profile of a solution containing (-)-epicatechin, malvidin 3-O-glucoside and acetaldehyde recorded after 24 hours of reaction (up) and UV-visible spectra of the major pigments detected in the solution (middle and bottom), (mv 3glc: malvidin 3-O-glucoside). (Adapted with permission from reference 31. Copyright 2002 American Chemical Society.)

From a mechanistic point of view, the reaction starts with protonation of acetaldehyde, followed by addition onto a nucleophilic position of the flavanol (C6 or C8) and dehydration of the resulting protonated adduct to yield a new carbocation, which suffers nucleophilic attack either by another flavanol or by the anthocyanin (*44*) (Figure 3).

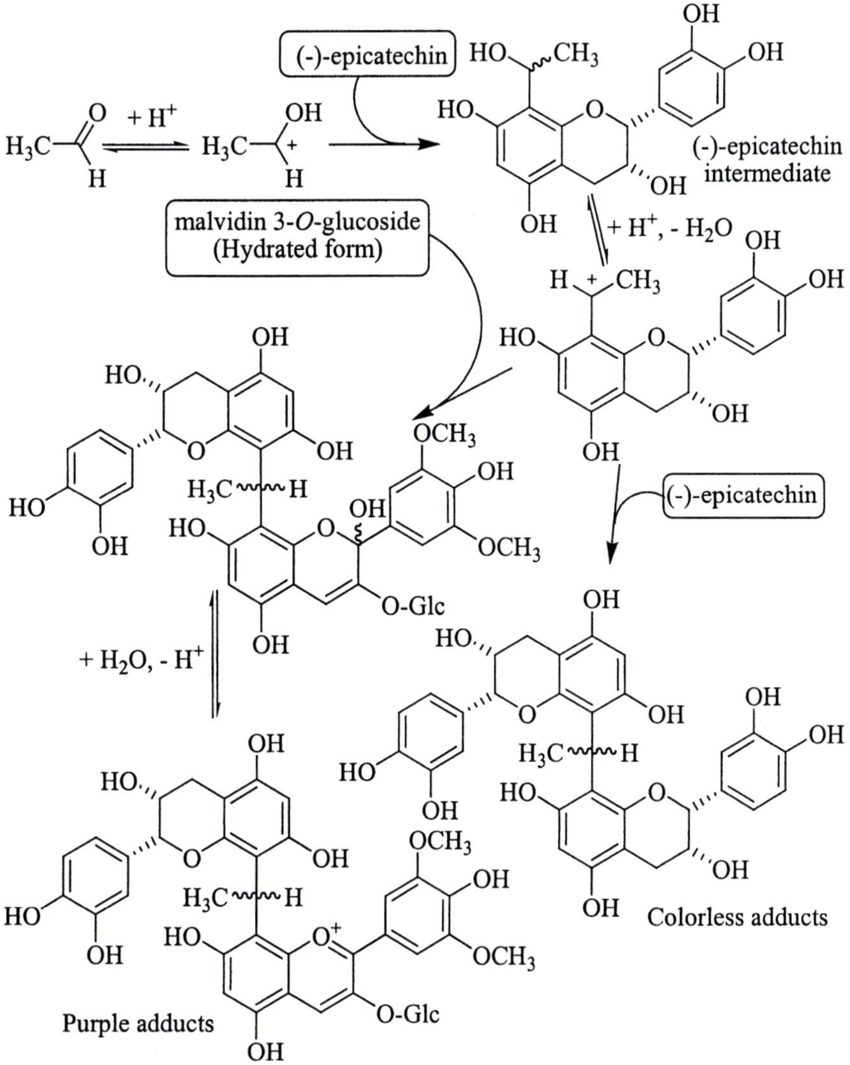

Figure 3: General mechanism of the acetaldehyde-induced polymerization of flavan-3-ols and anthocyanins. (Adapted with permission from reference 31. Copyright 2002 American Chemical Society.)

In this reaction, both the flavanol and the anthocyanin act as nucleophiles while the protonated form of the aldehyde is the electrophilic species. It must be noted that the flavylium form of the anthocyanin is an electrophilic entity while the hydrated hemiketal form is a nucleophilic entity. During the addition process, the anthocyanin acts thus as a nucleophile in its hydrated hemiketal form and the obtained adduct is then dehydrated to the corresponding flavylium (Figure 3).

It must also be noted that the anthocyanin and the flavanol compete in the different steps of the reaction. Nevertheless, and due to the low proportion of the hydrated form of the anthocyanin at the pH value of the reaction medium (2.2) which considerably decreases its ability to act as a nucleophile, this competition is in favor of the flavanol which is predominantly attacked first. This was shown by detection by LC/ESI-MS of the flavanol intermediate at *m/z*: 333 amu in the negative ion mode, while the corresponding anthocyanin intermediate was not detected. Isolation of the flavanol intermediate was achieved through HPLC at the semi-preparative scale and further reaction with malvidin 3-O-glucoside was shown to afford the two ethyl-bridged anthocyanin-flavanol adducts.

The anthocyanin-ethyl-flavanol adducts thus formed are purple pigments (λ_{max} 540 nm), and much more resistant towards acid-catalyzed cleavage than the equivalent colorless ethyl-linked flavanol dimers (*29*). This presumably results from the displacement of the hydration equilibrium toward the flavylium cation form, due to self association, since similar synthetic pigments have been shown to be greatly stabilized by sandwich-type stacking (*28*).

In flavanol, both the C6 and C8 tops are reactive positions so that successive condensations lead to numerous oligomers and polymers (*44*). In contrast, anthocyanins seem to react predominantly through one position only. In fact, ethyl-linked malvidin polymers were also reported (*32*), indicating that the anthocyanin can also react through the two positions 6 and 8 of the hemiketal form, the position 6 being less reactive than the 8 position. In the case of the ethyl-bridged malvidin dimer, it was demonstrated (*32*) that one of the two anthocyanin units is in the hemiketal form (Mv-Et-MvOH) in a rather wide pH range. This may explain the ability of such compounds to give more polymerized adducts (*45*). As indicated above, when the anthocyanin is linked to a flavanol through an ethyl bridge and at the pH value used for this experiment (2.2), it is mostly under the flavylium form which cannot act as nucleophile for further condensation with acetaldehyde. As a consequence, in acetaldehyde-induced flavanol-anthocyanin reactions, anthocyanins act as terminal units ending off the polycondensation reaction while flavanols are propagation units maintaining the polymerization process.

The formation of trimeric and tetrameric colored derivatives could be envisaged from the colorless or the colored dimers through successive oligomerization through ethyl bridges following the mechanism discussed above. It may also proceed through depolymerization and recombination processes in which cleavage of the ethyl bridge is followed by reaction of the obtained intermediate with a flavanol or an anthocyanin moiety (*31*).

Since the linkage between (-)-epicatechin and the bridge is more sensitive to acid conditions than that between the ethyl group and malvidin 3-O-glucoside (*29*), colorless ethyl-linked flavanol polymers are gradually replaced with ethyl-linked pigments.

When acetaldehyde was replaced with glyoxylic acid, similar colorless and colored compounds were detected. Glyoxylic acid results from metal-ion catalysed oxidation of tartaric acid (*46, 47*). LC/DAD analysis of a solution containing malvidin 3-O-glucoside, (+)-catechin and glyoxylic acid enabled the detection of colorless species and pigments with λ_{max} around 280 nm and 545 nm, respectively. Upon LC-MS analysis conducted in the positive ion mode, molecular ions located at *m/z:* 637 amu for four colorless products and 839 amu for two pigments were observed. This is consistent with structures consisting of a (+)-catechin unit bridged through a carboxy-methine group to another flavanol or flavylium moiety, respectively (Figure 1, R= COOH, R_1= OH, R_2= H), formed by the aldehyde-induced condensation mechanism described earlier with acetaldehyde (*23, 29, 44*).

In addition, other compounds exhibiting UV-visible spectra with absorption maxima at 295 nm and a shoulder around 340 nm were also identified in the model solution containing (+)-catechin and glyoxylic acid. These compounds were isolated and analyzed by ESI-MS and NMR spectroscopy. Their structures were shown to consist of a (+)-catechin unit substituted by aldehydic groups through the C-6 and/or C-8 carbon atoms (*40, 42*) (Figure 4). Their formation may involve the loss of a formic acid molecule of the intermediate adduct, resulting from addition of protonated glyoxylic acid onto the flavanol (Figure 4).

Reaction was finally conducted in presence of furfural or HMF. These aldehydes are sugar dehydration products which may be formed during toasting of oak and then extracted during barrel ageing of red wine. LC/DAD and LC/ESI-MS analysis of the incubated model solutions allowed to demonstrate the formation of four colorless adducts (λ_{max} = 280 nm) detected at *m/z*: 657 and 687 amu, respectively, for furfural and HMF, and two pigments (λ_{max} = 545 nm) detected at *m/z*: 859 and 889 amu, respectively, in the case of furfural or HMF. This is consistent with one (+)-catechin moiety linked to one (+)-catechin or flavylium moiety through a furfuryl bridge for colorless or colored adducts (Figure 2, R= furyl or hydroxymethylfuryl, R_1= OH, R_2 =H). Similar compounds with similar UV-visible characteristics and similar structures were finally observed when cyanidin 3-O-glucoside was used instead of malvidin 3-O-glucoside.

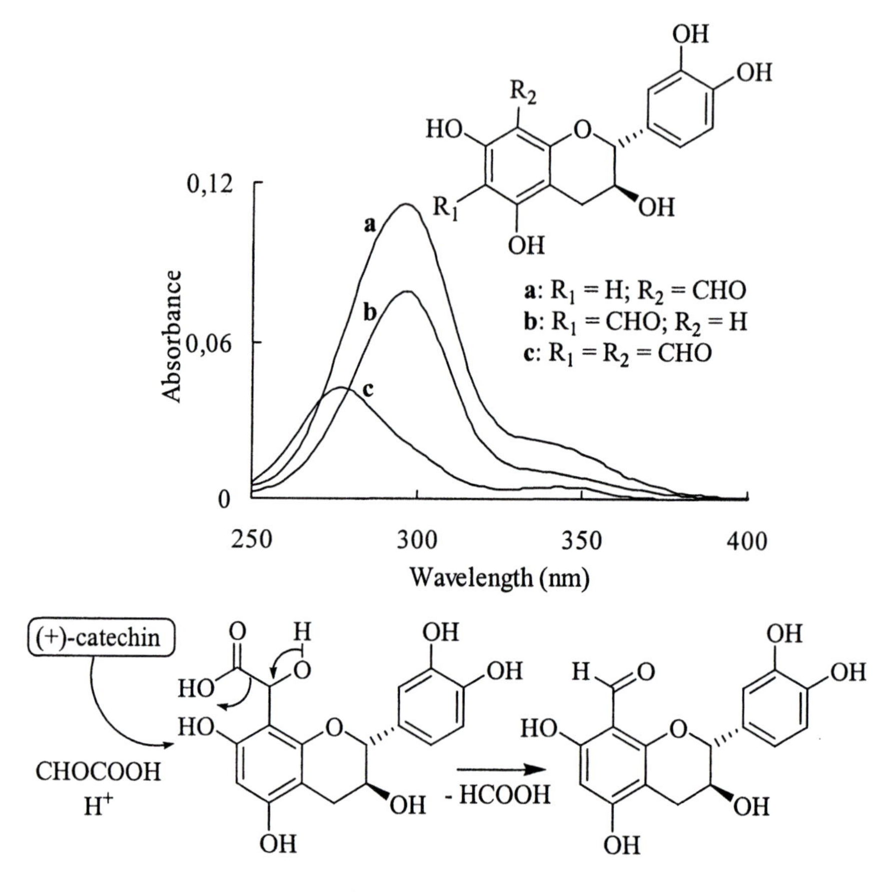

Figure 4: Structures, UV-visible spectra and mechanism of formation of the isolated (+)-catechin formylated derivatives. (Reproduced with permission from reference 40. Copyright 2000 Elsevier.)

Formation of yellow pigments from non colored precursors

The probable involvement of xanthylium salts in red wine color change is supported by the fact that their absorption maxima match well with the increase in yellow color around 420 nm observed during ageing of red wine. Thus, Liao et al. (*22*) demonstrated in model systems that reactions between anthocyanins and flavan-3-ols give rise to orange-yellow (440 nm) pigmented products. Based on the earlier observations of Jurd & Somers (24) and Hrazdina & Borzell (*48*), such products have been postulated to be xanthylium salts proceeding from direct anthocyanin-flavanol adducts.

The formation of xanthylium pigments may also result from non enzymatic reactions of colorless precursors such as flavanols as previously reported (*18*, *24*, *49*). Thus, catechin monomer has been shown to yield yellow derivatives by various oxidation processes. In the presence of tartaric acid and iron or copper ions, yellow pigments with absorption maxima at 440 and 460 nm were formed along with the colorless (+)-catechin carboxymethine bridged dimers (*46*, *47*). These pigments were analysed by 1D and 2D NMR and identified as non esterified and esterified xanthylium salts respectively (*33*, *38*, *39*, *42*) (Figure 5).

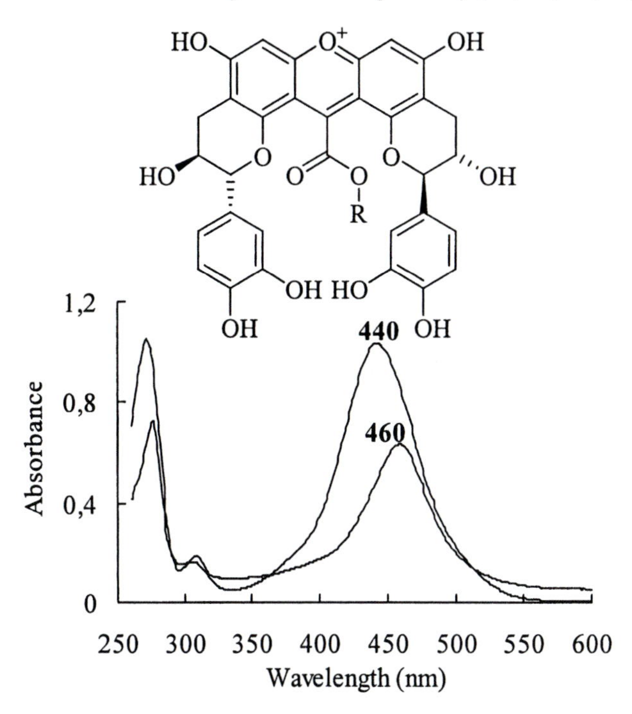

Figure 5: Structures and UV-visible spectra of the isolated esterified (R= Et, λ_{max}= 460 nm) and non esterified (R= H, λ_{max}= 440 nm) xanthylium salts derivatives. (Reproduced with permission from reference 38. Copyright 1999 Elsevier.)

From a mechanistic point of view, the formation of the identified xanthylium salts involve dehydration of the first colorless dimers giving a xanthene derivative followed by an oxidation process. Detection of the xanthene adducts was also achieved through LC/MS analysis. The esterification was shown to occur before the dehydration process as shown in Figure 6.

Both xanthylium salts shown in Figure 5 have been detected in wine stored with no care to avoid oxidation, supporting the occurrence of the above described pathways (*33*, *50*, *51*). This constitutes an important support to the occurrence of the above described pathways, since the xanthylium salts is expected to result from anthocyanin-tannin condensation have never been detected.

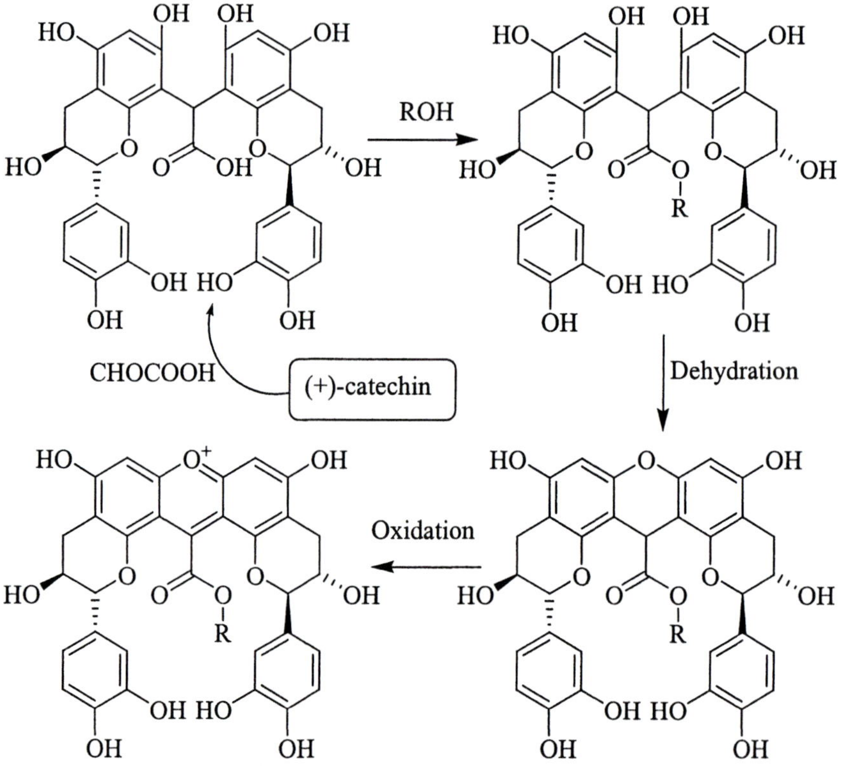

Figure 6: Mechanism of xanthylium salts formation from (+)-catechin, via the colorless and the xanthene derivatives. (Reproduced with permission from reference 39. Copyright 1999 American Chemical Society.)

The mechanism postulated above for xanthylium salts formation was also found to occur more generally. Thus yellow xanthylium salts showing absorption maxima at 440 nm were detected when glyoxylic acid was replaced by furfural and HMF. Using LC/ESI-MS chromatography these compounds were detected at *m/z:* 637 and 667 amu for furfural and HMF respectively. The corresponding xanthenes were also detected at *m/z:* 635 and 669 amu respectively. It must also be noted that no xanthylium was observed in the case of acetaldehyde.

Formation of other pigments from non colored precursors

In addition to xanthylium salts, other new red pigments with strong absorption intensity at 560 nm and presenting maxima at 450, 350, 330 and 280 nm were also detected in the medium containing (+)-catechin and glyoxylic acid (Figure 7). On the basis of their mass spectra showing signals located at *m/z*: 959 and 961 amu (positive ion mode) and *m/z*: 957 and 959 amu (negative ion mode), respectively, for compounds NR2 and NR1, the structures shown in Figure 8 were proposed for these compounds (*41*, *50*). The presence of a quinonoidal skeleton may explain the absorption around 560 nm observed for these compounds and their naturally intense red color. This also supports the implication of bridged polymerized compounds with quinonoid skeletons in the transformations usually occurring during ageing of fruit derived foods as previously postulated (*18*, *26*).

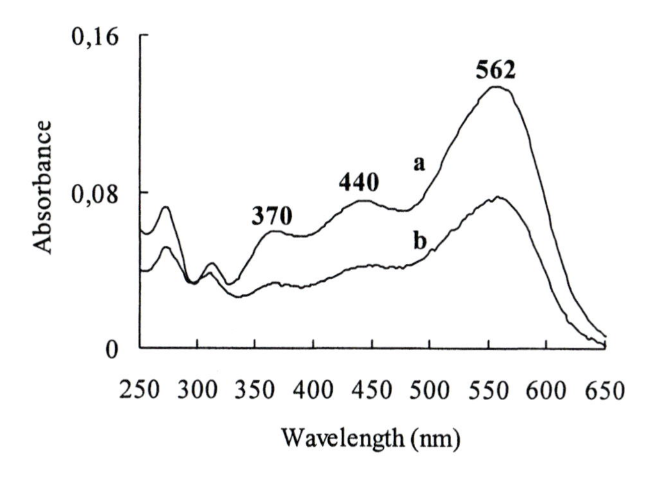

Figure 7: UV-visible spectra of the red detected compounds with strong absorption around 560 nm. (Reproduced with permission from reference 39. Copyright 1999 American Chemical Society.)

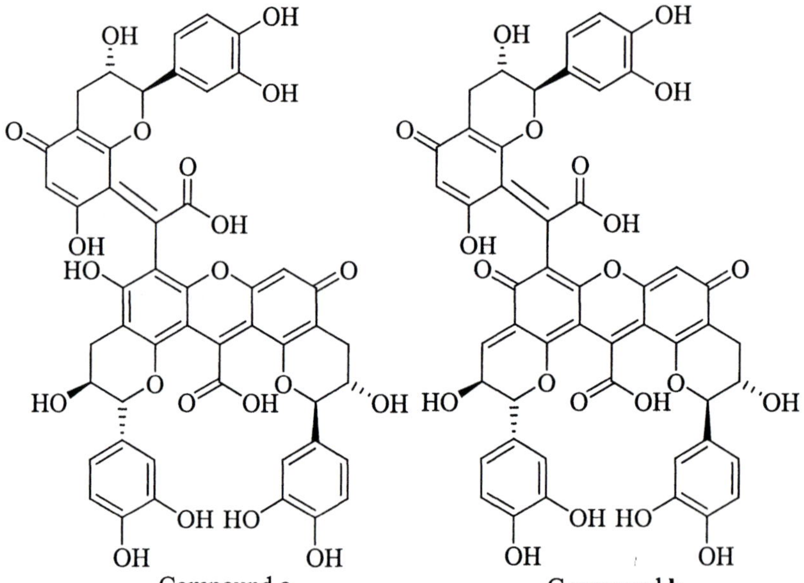

Compound **a**

Compound **b**

Figure 8: Proposed structures of the detected red compounds with strong absorption around 560 nm.

Conclusion

This study illustrates the diversity of purple and yellow pigments formed by reactions of anthocyanins and flavan-3-ols with other non phenolic compounds. The resulting bridged purple pigments are relatively stable with regard to pH or to bleaching by SO_2, probably as they are protected from the hydration reaction by self-association phenomenon, leading to non-covalent dimers associated by sandwich-type stacking (*28*, *52*). They are also much more resistant towards acid-catalyzed cleavage than the equivalent colorless bridged flavanols (*29*).

Practices that increase formation of aldehydes, such as microoxygenation, promote polymerization of anthocyanins and catechins or tannins, thereby improving color intensity and stability. The formation of yellow and red pigments from colorless compounds is a very recent concept in enology since until now it was believed that xanthylium salts could result only from anthocyanins and tannins reactions. The proposed structures of compounds NR1 and NR2 are a new support for the implication of xanthylium salt derivatives in the formation of more polymeric colored compounds and the contribution of quinonoidal derivatives in the color change of grape derived foods as previously postulated (*18*, *26*).

The reactions described in this work were conducted in model solution system and obviously do not match what really occurs during storage and ageing of red wine which has a more complex composition. Nevertheless, it is expected that such experiments may at least enable elucidation of the simpler polymerization products such as those slowly formed during maturation and storage.

Although the reactions described above have been shown to modify the properties of polyphenols, they are not sufficient to explain the variations observed in the organoleptic properties (taste and color) of red wine. As well, the relationships existing between structure and taste (e.g. astringency, bitterness) of proanthocyanidins and proanthocyanidin-derived molecules remain to be established. Answers to these questions will require the use of different complementary approaches, including chemical, sensory and statistical analysis.

The evolution of phenolic compounds remains thus an open research area. The improved understanding of the reaction mechanisms involved in alterations occurring during production, storage and ageing of wine expected from this research should enable to modify and/or better control technological processes so as to better fulfill the consumer demand.

Acknowledgments

Author N. Es-Safi thanks the International Foundation for Science (grant N° E/3261-1) for partial financial support.

References

1. Brouillard, R. In *Anthocyanins as Food Colors*; Markakis, P., Ed.; Academic Press, New York, **1982**, 1-40.
2. Noble, A. C. In *Developments in Food Science*; Rouseff, R.L., Ed.; 25. Bitterness in Foods and Beverages; Elsevier: New York, **1990**, 145-158.
3. Robichaud, J. L.; Noble, A. C. *J. Sci. Food Agric.* **1990**, *53*, 343-353.
4. Mazza, G.; Miniati, E. *Anthocyanins in fruits, vegetables, and grains*; CRC Press: Boca Raton, **1993**.

5. Souquet, J. M.; Cheynier, V.; Brossaud, F.; Moutounet, M. *Phytochemistry* **1996**, *43*, 509-512.
6. Kennedy, J. A.; Troup, G. J.; Pilbrow, J. R.; Hutton, D. R.; Hewitt, D.; Hunter, C. R.; Ristic, R.; Iland, P. G.; Jones, G. P. *Aust. J. Grape Wine Res.* **2000**, *6*, 244-254.
7. Thorngate, J. H.; Singleton., V. L. *Am. J. Enol. Vitic.* **1994**, *45*, 259-262.
8. Prieur, C.; Rigaud, J.; Cheynier, V.; Moutounet, M. *Phytochemistry* **1994**, *36*, 781-784.
9. Ricardo-Da-Silva, J. M.; Rigaud, J.; Cheynier, V.; Cheminat, A.; Moutounet, M. *Phytochemistry* **1991**, *30*, 1259-1264.
10. Czochanska, Z.; Foo, L. Y.; Porter, L. J. *Phytochemistry* **1979**, *18*, 1819-1822.
11. Amrani Joutei, K.; Glories, Y. *J. Int. Sci. Vigne Vin* **1994**, *28*, 303-317.
12. Sun, B. S.; Pinto, T.; Leandro, M. C.; Ricardo-Da-Silva, J. M.; Spranger, M. I. *Am. J. Enol. Vitic.* **1999**, *50*, 179-184.
13. Cheynier, V.; Hidalgo-Arellano, I.; Souquet, J.M.; Moutounet, M. *Am. J. Enol. Vitic.* **1997**, *48*, 225-228.
14. Atanasova, V.; Fulcrand, H.; Cheynier, V.; Moutounet, M. *Anal. Chim. Act.* **2002**, *458*, 15-27.
15. Somers, T. C.; Verette, E. In *Modern Methods of Plant Analysis*; Linskins, H. F.; Jackson, J. F., Ed.; New Series: *Wine Analysis*; Springer-Verlag: Berlin, **1988**; *6*, 219-257.
16. Bakker, J.; Preston, N. W.; Timberlake, C. *Am. J. Enol. Vitic.* **1986**, *37*, 121-126.
17. Nagel, C. W.; Wulf, L. W. *Amer. J. Enol. Vitic.* **1979**, *30*, 111-116.
18. Somers, T. C. *Phytochemistry* **1971**, *10*, 2175-2186.
19. Somers, T. C. *Vitis* **1978**, *25*, 161-167.
20. Somers, T. C.; Evans, M. C. *J. Sci. Food Agric.* **1979**, *30*, 623-633.
21. Somers, T. C.; Evans, M. C. *Vitis* **1986**, *25*, 31-39.
22. Liao, H.; Cai, Y.; Haslam, E. *J. Sci. Food Agric.* **1992**, *59*, 299-305.
23. Timberlake, C. F.; Bridle, P. *Am. J. Enol. Vitic.* **1976**, *27*, 97-105.
24. Jurd, L.; Somers, T. C. *Phytochemistry* **1970**, *9*, 419-427.
25. Ribereau-Gayon, P.; Pontallier, P.; Glories, Y. *J. Sci. Food Agric.* **1983**, *34*, 505-516.
26. Haslam, E. *Phytochemistry* **1980**, *19*, 2577-2582.
27. Bakker, J.; Picinelli, A.; Bridle, P. *Vitis* **1993**, *32*, 111-118.
28. Escribano-Bailon, M. T.; Dangles, O.; Brouillard, R. *Phytochemistry* **1996**, *41*, 1583-1592.
29. Es-Safi, N.; Fulcrand, H.; Cheynier, V.; Moutounet, M. *J. Agric. Food Chem.* **1999**, *47*, 2096-2102.
30. Dallas, C.; Ricardo-Da-Silva, J. M; Laureano, O. *J. Agric. Food Chem.*, **1996**, *44*, 2402-2407.

31. Es-Safi, N.; Cheynier, V.; Moutounet, M. *J. Agric. Food Chem.* **2002**, *50*, 5571-5585.
32. Atanasova, V.; Fulcrand, H.; Cheynier, V.; Moutounet, M. *Tetrahedron Lett.* **2002**, *43*, 6151-6153.
33. Es-Safi, N.; Le Guernevé, C.; Fulcrand, H.; Cheynier, V.; Moutounet, M. *Int. J. Food Sci. Technol.* **2000**, *35*, 63-74.
34. Es-Safi, N.; Cheynier, V.; Moutounet, M. *J. Agric. Food Chem.* **2000**, *48*, 5946-5954.
35. Fulcrand, H. ; Cheynier, V. ; Oszmianski, J.; Moutounet, M. *Phytochemistry* **1997**, *46*, 223-227.
36. Es-Safi, N.; Cheynier, V.; Moutounet, M. *J. Agric. Food Chem.* **2002**, *50*, 5586-5586.
37. Cheynier, V. ; Fulcrand, H.; Sarni, P. ; Moutounet, M. *Analusis* **1997**, *25*, M14-M21.
38. Es-Safi, N.; Le Guernevé, C.; Labarbe, B.; Fulcrand, H.; Cheynier, V.; Moutounet, M. *Tetrahedron Lett.* **1999**, *40*, 5869-5872.
39. Es-Safi, N.; Le Guernevé, C.; Cheynier, V.; Moutounet, M. *J. Agric. Food Chem.* **1999**, *47*, 5211-5217.
40. Es-Safi, N.; Le Guernevé, C.; Cheynier, V.; Moutounet, M. *Tetrahedron Lett.* **2000**, *41*, 1917-1921.
41. Es-Safi, N.; Le Guernevé, C.; Cheynier, V.; Moutounet, M. *J. Agric. Food Chem.* **2000**, *48*, 4233-4240.
42. Es-Safi , N.; Le Guernevé, C.; Cheynier, V.; Moutounet, M. *Magn. Reson. Chem.* **2002**, *40*, 693-704.
43. Liu S. –Q.; Pilone J. *Int. J. Food Sci. Technol.* **2000**, *35*, 49-61.
44. Fulcrand, H.; Docco, T.; Es-Safi, N.; Cheynier, V.; Moutounet, M. *J. Chromatogr. A* **1996**, *752*, 85-91.
45. Atanasova, V.; Fulcrand, H.; Le Guernevé, C.; Dangles, O.; Cheynier, V.; Moutounet, M. In *Polyphenols Communications 2002*; El Hadrami, I., Eds.; Marrakech, **2002**; *2*, 417-418.
46. Oszmianski, J.; Cheynier, V. ; Moutounet, M. *J. Agric. Food Chem* **1996**, *44*, 1712-1715.
47. Es-Safi , N.; Cheynier, V.; Moutounet, M. *Int. J. Food Sci. Technol.* **2003**, *38*, 153-163.
48. Hrazdina, G.; Borzell, A. J. *Phytochemistry* **1971**, *10*, 2211-2213.
49. Jurd, L. *Am. J. Enol. Vitic.* **1969**, *20*, 191-195.
50. Es-Safi , N.; Cheynier, V.; Moutounet, M. *J. Food. Com. Anal.* **2003**, *16*, 535-553;
51. Fulcrand, H.; Es-Safi, N.; Cheynier, V., Moutounet, M. In *Polyphenol communications 1998;* Charbonnier, F.; Delacotte, J. M.; Rolando, C., Eds. ; Lille, **1998**; 259-260.
52. Escribano-Bailon, T.; Alvarez-Garcia, M. ; Rivas-Gonzalo, J. C. ; Heredia, F. J. ; Santos-Buelga C. *J. Agric. Food Chem.* **2001**, *49,* 1213-1217.

Chapter 10

Structural Changes of Anthocyanins during Red Wine Aging: Portisins: A New Class of Blue Anthocyanin-Derived Pigments

V. A. P. de Freitas and N. Mateus

Centro de Investigação em Química, Departamento de Química, Faculdade de Ciências, Universidade do Porto, Rua do Campo Alegre, 687, 4169-007 Porto, Portugal

Different families of anthocyanin-derived pigments including anthocyanin-ethyl-flavanols, pyruvic acid adducts, anthocyanin-vinylphenol pigments and pyranoanthocyanin structures bearing different flavan-3-ol moieties were isolated from a two-year-old Port wine and characterized by UV-vis spectroscopy, mass spectrometry and NMR. Additionally, two compounds belonging to a new class of vinylpyranoanthocyanin blue pigments, named Portisins, were also isolated and structurally characterized. These pigments are proposed to result from the reaction between anthocyanin-pyruvic acid adducts and vinyl-flavanol adducts.

Color is one of the main organoleptic properties of red wine and is of crucial importance for the consumer since it is the first characteristic to be perceived in the glass. The mysteries of red wine pigments have attracted the interest of many researchers over the last years. Nevertheless, the formation mechanisms of anthocyanin-derived pigments remain a matter of interest and constitute a stimulating challenge for the wine chemists. In general, the evolution of red wine color during ageing is a result of different reactions comprising oxidation-reduction reactions and complexation with other compounds such as carbohydrates, proteins, metals or flavanols. Some of these reactions lead to the progressive displacement of anthocyanins by more stable pigments. These red wine pigments were first thought to result mainly from condensation reactions between anthocyanins and flavanols either direct (*1-4*) or mediated by acetaldehyde (*5-7*). Nevertheless, reactions between anthocyanins and/or flavanols with other compounds such as pyruvic acid (8-*13*), vinylphenol (*14,15*) α-ketoglutaric acid (*16*), acetone (*16-18*), 4-vinylguaiacol (*18*) and glyoxylic acid (*19,20*) have recently been demonstrated yielding new families of anthocyanin-derived pigments, namely pyranoanthocyanins and xanthylium derivatives with spectroscopic features that may contribute to a more orange-red color. The advances of mass spectrometry and NMR techniques allowed confirming the occurrence of some newly formed pigments directly in wine. The present work deals with the screening and structural identification of different classes of newly formed pigments recently detected in Port wine.

MATERIALS AND METHODS

Source. The pigments were isolated from a two-year-old Port wine (pH 3.6, 18.5% alcohol (v/v), total acidity 6.5 g/L, total SO_2 20 mg/L), made from grapes of Touriga Nacional (*Vitis vinifera*) grown in the Região Demarcada do Douro (Northern Portugal).

Pigment purification. Port wine samples were directly applied on a 250 x 16 mm i.d. Toyopearl HW-40(s) gel column (Tosoh, Japan) at a flow rate of 0.8 mL/min. A first elution was performed with 20% aqueous ethanol yielding fraction A. When practically no more colored compounds were eluted from the column, the solvent was changed to 99.8% aqueous ethanol yielding fraction B. The pH of all the eluents was set to 2.0 with HCl. The major pigments reported in each fraction were purified by semi-preparative HPLC on a 250 x 4.6 mm i.d. reversed-phase C18 column (Merck, Darmstadt) using an injection volume of 500 μL, as described elsewhere (*21*).

Semi-preparative and analytical HPLC conditions. The fractions yielded from the TSK Toyopearl gel column were analyzed by HPLC (Merck-Hitachi L-7100) using the above indicated reversed-phase C18 column; detection was carried out using a diode array detector (Merck-Hitachi L-7450A). The solvents were A, H_2O/HCOOH (9:1), and B, CH_3CN/H_2O/HCOOH (3:6:1). The gradient consisted of: 20-85% B for 70 min, 85-100% B for 5 min and then isocratic for 10 min at a flow rate of 1 mL/min (*22*).

LC-MS analysis. A Hewlett-Packard 1100 Series liquid chromatograph, equipped with an AQUA™ (Phenomenex, Torrance, CA, USA) reversed-phase column (150 x 4.60 mm, 5 μm, C18) thermostated at 35°C was used. Solvents were (A) aqueous 0.1% trifluoracetic acid and (B) 100% HPLC-grade acetonitrile, establishing the following gradient: isocratic 10%B over 5 min, from 10 to 15%B over 15 min, isocratic 15%B over 5 min, from 15 to 18%B over 5 min, and from 18 to 35%B over 20 min, at a flow rate of 0.5 $ml.min^{-1}$, as described elsewhere (*21*). Double online detection was made in a photodiode spectrophotometer and by mass spectrometry. The mass detector was a Finnigan LCQ equipped with an API source, using an electrospray ionisation (ESI) interface. Both the auxiliary and the sheath gas were a mixture of nitrogen and helium. The capillary voltage was 3V and the capillary temperature 190°C. Spectra were recorded in positive ion mode between m/z 120 and 1500.

NMR analysis. ^1H NMR (500.13 MHz) and ^{13}C NMR (125.77 MHz) spectra were measured in CD_3OD/TFA (98:2) on a Bruker-AMX500 spectrometer at 303 K and with TMS as internal standard. ^1H chemical shifts were assigned using 1D and 2D ^1H NMR (COSY and NOESY), while ^{13}C resonances were assigned using 2D NMR techniques (*g*HMBC and *g*HSQC) (*23,24*). The delay for the long-range C/H coupling constant was optimized to 7 Hz.

Formation of Portisins in model solution. The formation of a Portisin was monitored at 35 °C in 20% aqueous ethanol (pH=2.0) in a 2 mL screw cap vial containing 0.2 mg of malvidin-3-coumaroylglucoside pyruvic acid adduct previously isolated (*11*) and 0.33 mg of (+)-catechin. The total volume of the solution was set to 50% of the vial capacity. After 15 days of reaction, the solution was analyzed by HPLC using the conditions described above.

RESULTS AND DISCUSSION

Fraction A obtained from the elution of a two-year-old Port wine through Toyopearl gel column with 20% aqueous ethanol was found to be mainly comprised of eight pigments which structural identities were ascertained by LC-MS (Table 1). The number attributed to each compound in Table 1 corresponds to the elution order by HPLC in the condition described herein. The three major pigments detected were isolated by semi-preparative HPLC and their structures were fully elucidated by mass spectrometry and NMR. Based on their UV-visible characteristics and structural analysis, these pigments were found to correspond to the pyruvic acid adducts of malvidin 3-glucoside (3A), malvidin 3-acetylglucoside (5A) and malvidin 3-coumaroylglucoside (7A) (Figure 1).

Table 1. Anthocyanin-derived pigments detected in Port wine fractions eluted from Toyopearl gel column with 20% aqueous ethanol (Fraction A).

Peak	Pigment identity	*m/z* (M^+)	λ_{max} (nm)	Structural elucidation
1A	Mv 3-gluc	493	529	*NMR, MS, UV-Vis*
2A	Mv 3-gluc-vinyl (vitisin B)	517		*MS*
3A	Mv 3-gluc-py	561	511	*NMR, MS, UV-Vis*
4A	Pt 3-(acetyl)gluc-py	589		*MS*
5A	Mv 3-(acetyl)gluc-py	603	511	*NMR, MS, UV-Vis*
6A	Pt 3-(coumaroyl)gluc-py	693		*MS*
7A	Mv 3-(coumaroyl)gluc-py	707	511	*NMR, MS, UV-Vis*
8A	Pn 3-(coumaroyl)gluc-py	677		*MS*

Mv = malvidin; Pt = petunidin; Pn = paeonidin; py = pyruvic acid; gluc = glucoside

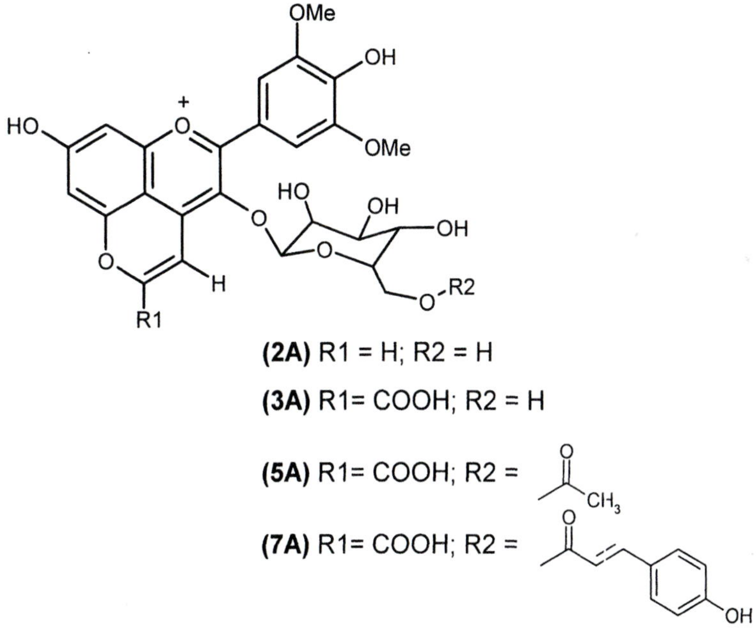

(2A) R1 = H; R2 = H

(3A) R1= COOH; R2 = H

(5A) R1= COOH; R2 = acetyl ($COCH_3$)

(7A) R1= COOH; R2 = coumaroyl

Figure 1. Structures of the major anthocyanin-pyruvic acid adducts detected in fraction A.

These compounds showed a λ_{max} around 511 nm whereas anthocyanins have their maximal absorption at 529 nm. These pyranoanthocyanin pigments were already reported in red wines, grape pomace and model solutions and result from the cyclic-addition of pyruvic acid-mediated onto the C-4 position and the OH group at position 5 of the anthocyanin (*7-11*). The further elution with 99.8% ethanol yielded a second colored fraction (B) that contains a large number of anthocyanin-derived pigments (Table 2). After a careful sub-fractionation of fraction B by Toyopearl gel column chromatography with increasing percentage of ethanol, the major pigments were isolated by semi-preparative HPLC and their structures were characterized by NMR and UV-vis spectroscopy.

Many of the pigments shown in Table 2 were found to arise from the association of malvidin 3-glucoside and 3-flavanols through vinyl or ethyl linkages. Only two pigments with ethyl linkage were detected by mass spectrometry, malvidin 3-gluc-ethyl-cat (1B) and malvidin 3-(coumaroyl)gluc-ethyl-cat (11B). This kind of pigments has been evidenced in model solutions (*5-7,25,26*) and wine (*27*) resulting from the acetaldehyde-mediated condensation of flavanols and anthocyanins. In wine, acetaldehyde is supposed to result essentially from the oxidation of ethanol during aging (*28*). Another group of four pigments detected in fraction B (17B, 18B, 19B, 20B, 21B) correspond to anthocyanins linked to a 4-vinylphenol group, as previously reported in red wines (*14,15*). The *p*-vinylphenol directly involved in the formation of these pigments can be formed in wines from the degradation of *p*-coumaric acid (*29*).

The major anthocyanin-derived pigments found in fraction B correspond to pyranoanthocyanins linked to catechin (cat), epicatechin (epi) or a procyanidin dimer (PC) (Figure 2). These pigments are similar to those previously described by Francia-Aricha *et al.* (*30*) in studies carried out in model solutions. The λ_{max} of these pigments are hypsochromically shifted with regard to those of anthocyanins: pigments 2B and 7B have a λ_{max} at 520 nm, whilst the other four pyranoanthocyanin-flavanol pigments (4B, 9B, 15B and 16B) showed a λ_{max} at 511 nm similar to the one of pyruvic acid adducts. It is interesting to notice that the pigment structure (acylated or non-acylated) that contains a procyanidin dimer unit (2B and 7B) revealed an important bathochromic shift (9 nm) comparatively to the structures that contain a flavanol monomeric unit (4B, 9B, 15B and 16B), despite having the same flavylium moiety. This outcome highlights the importance of the type of flavanol moiety on the color characteristics of these pigments, suggesting that an intramolecular co-pigmentation between the flavanol moiety and the pyranoanthocyanin chromophore may somehow occur. On the other hand, the existence of a coumaroyl group in the glucose moiety did not influence the λ_{max} of the pigments. The spatial conformation of malvidin 3-coumaroylglucoside-vinyl-(+)-catechin (15B) estimated in solution by molecular mechanics (MM3) conjugated with data from nOe experiments, points to a relatively closed structure with the

Table 2. Anthocyanin-derived pigments detected in Port wine fractions eluted from Toyopearl gel column with 99.8 % ethanol (Fraction B).

Peak	Pigment identity	*m/z* (M^+)	λ_{max} (nm)	Structural elucidation
1B	Mv 3-gluc-ethyl-cat	809		*MS*
2B	Mv 3-gluc-vinyl-(+)-cat-(+)-cat	1093	520	*NMR, MS, UV-Vis*
3B	Mv 3-(acetyl)gluc-vinyl-PC dimer	1135		*MS*
4B	Mv 3-gluc-vinyl-(+)-cat	805	511	*NMR, MS, UV-Vis*
5B	Dp 3-gluc-py	533		*MS*
6B	Mv 3-(acetyl)gluc-vinyl-cat	847		*MS*
7B	Mv 3-(coumaroyl)gluc-vinyl-(-)-epi-(+)-cat	1239	520	*NMR, MS, UV-Vis*
8B	Mv 3-gluc-bivinyl-(-)-epi-cat	1119	583	*NMR, MS, UV-Vis*
9B	Mv 3-gluc-vinyl-(-)-epi	805	511	*NMR, MS, UV-Vis*
10B	Mv 3-(coumaroyl)gluc-bivinyl-(-)-epi-cat	1265	583	*NMR, MS, UV-Vis*
11B	Mv 3-(coumaroyl)gluc-ethyl-cat	955		*MS*
12B	Mv 3-(acetyl)gluc-vinyl-cat	847		*MS*
13B	Mv 3-(coumaroyl)gluc	639	529	*NMR, MS, UV-Vis*
14B	Dp 3-acetylgluc-py	575		*MS*
15B	Mv 3-(coumaroyl)gluc-vinyl-(+)-cat	951	511	*NMR, MS, UV-Vis*
16B	Mv 3-(coumaroyl)gluc-vinyl-(-)-epi	951	511	*NMR, MS, UV-Vis*
17B	Mv 3-gluc-vinylphenol	609		*MS*
18B	Mv 3-(caffeoyl)gluc-vinylphenol	771		*MS*
19B	Pn 3-(coumaroyl)gluc-vinylphenol	725		*MS*
20B	Mv 3-(coumaroyl)gluc-vinylphenol	755		*MS*
21B	Mv 3-(acetyl)gluc-vinylphenol	651		*MS*

Mv = malvidin; Dp = delphinidin; Pt = petunidin; Pn = paeonidin; py = pyruvic acid; gluc = glucoside; cat = (+)-catechin or (-)-epicatechin; PC = procyanidins.

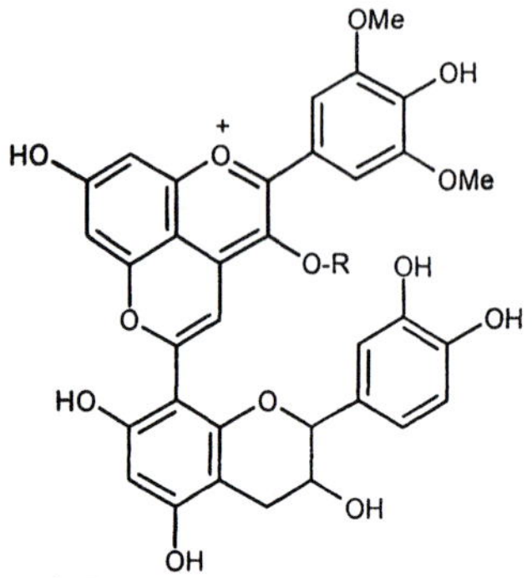

(4B,9B) R=glucose
(6B,12B) R=acetylglucose
(15B,16B) R=coumaroylglucose

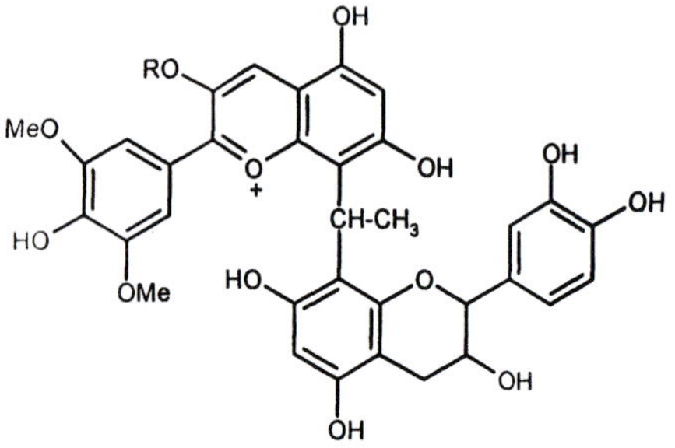

(1B) R=glucose
(11B) R=coumaroylglucose

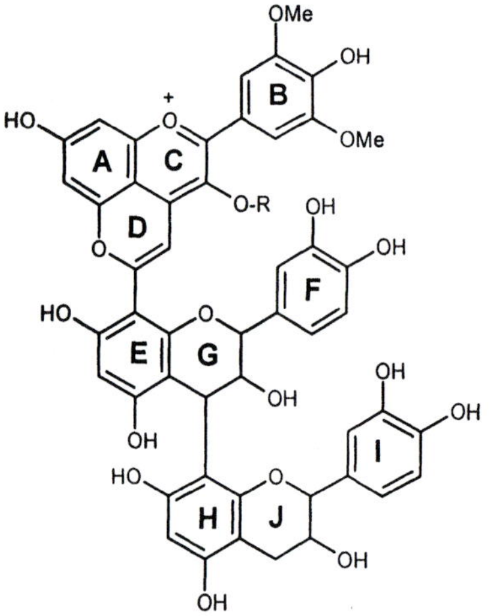

(2B) R=glucose
(3B) R=acetylglucose
(7B) R=coumaroylglucose

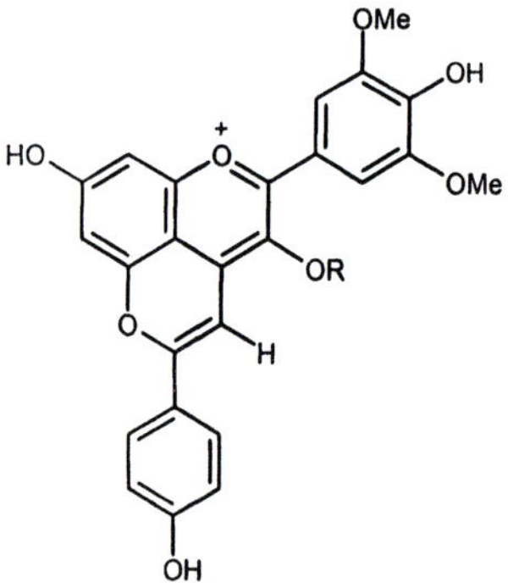

(17B) R=glucose
(21B) R=acetylglucose
(20B) R=coumaroylglucose
(18B) R=caffeoylglucose

Figure 2. Structures of the major anthocyanin-derived pigments detected in fraction B.

coumaroyl group being oriented towards the outside of the bulk structure (Figure 3). The absence of interactions with the pyranoanthocyanin chromophore could explain why the acylation of these pigments did not induce any change on their visible spectra.

The formation of these pyranoanthocyanin-flavanol derivatives (Figure 4) are thought to arise from the reaction between anthocyanins and an 8-vinylflavanol adduct, through a mechanism similar to that proposed by Fulcrand *et al.* for the formation of 4-vinylphenol anthocyanin-derived pigments (*15*). The last step of the formation involves an oxidative process whereby the vinylflavanol adduct binds to the flavylium moiety, giving rise to the aromatization of ring D. The resulting extended conjugation of the π electrons in this newly formed structure is likely to confer a higher stability of the molecule. The vinyl-flavanol adducts may derive either from the cleavage of ethyl-linked flavanol oligomers resulting from the acetaldehyde-induced condensation of flavanols (*31*), from the dehydration of the flavanol-ethanol adduct formed after reaction with acetaldehyde, or from the cleavage of anthocyanin-ethyl-flavanol pigments.

A new class of blue anthocyanin-derived pigments (Portisins) isolated from Port red wines

Additionally, a new class of wine pigments, named here for the first time as Portisins, was found to occur in fraction B. Indeed, two additional pigments (8B and 10B) with maximum absorption in the visible region at 583 nm were detected by HPLC. These two Portisins were isolated by semi-preparative HPLC and characterized by NMR, MS and UV-vis (*32*). During the final purification of each pigment by TSK Toyopearl gel column chromatography in acidic conditions (pH=2.0), a blue band was observed, which is in agreement with the UV-Vis spectra of these pigments.

Analysis of Portisin 8B and 10B by mass spectrometry revealed $[M]^+$ ions at *m/z* 1119 and 1265, respectively. These molecular ion masses and their NMR data fit with the structures of vinylpyranoanthocyanin derivatives shown in Figure 5. The protons of the vinyl group linkage (H_α and H_β) are correlated in the NMR spectra with a large coupling constant (J=15.8 Hz) suggesting a *trans* stereochemistry. Nevertheless, the NMR data did not allow determining the full identity of the procyanidin dimer that constitutes the flavanol moiety. The NMR data suggests that the upper flavanol unit for both portisins 8B and 10B consisted of one (-)-epicatechin molecule, but the identity of the lower flavanol unit could not be deduced. In addition, the interflavanoid linkage of dimer moiety can not be fully ascertained from the data yielded from the NMR experiments, likewise for pyranoanthocyanin-PC dimer pigments previously

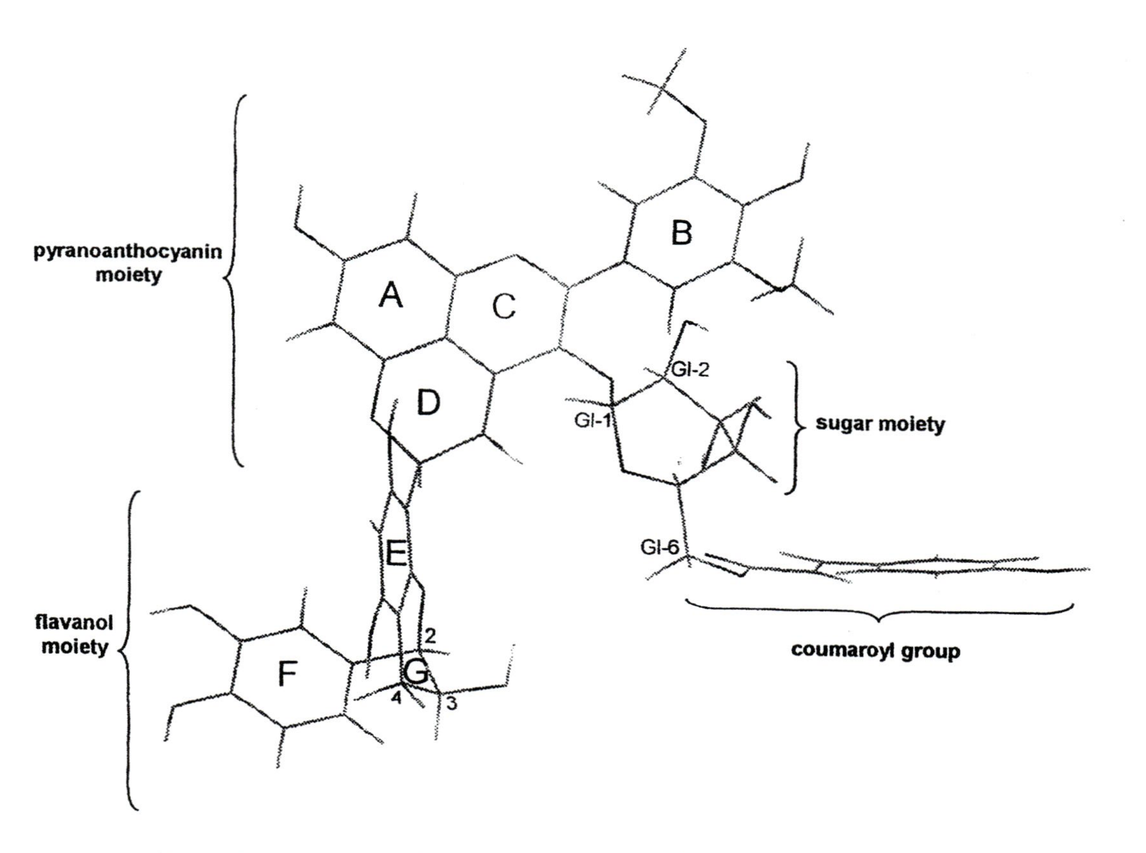

Figure 3. Preferred conformation of pigment 15B determined by computer-assisted model building (MacroModel) and molecular mechanics (MM3) (31).

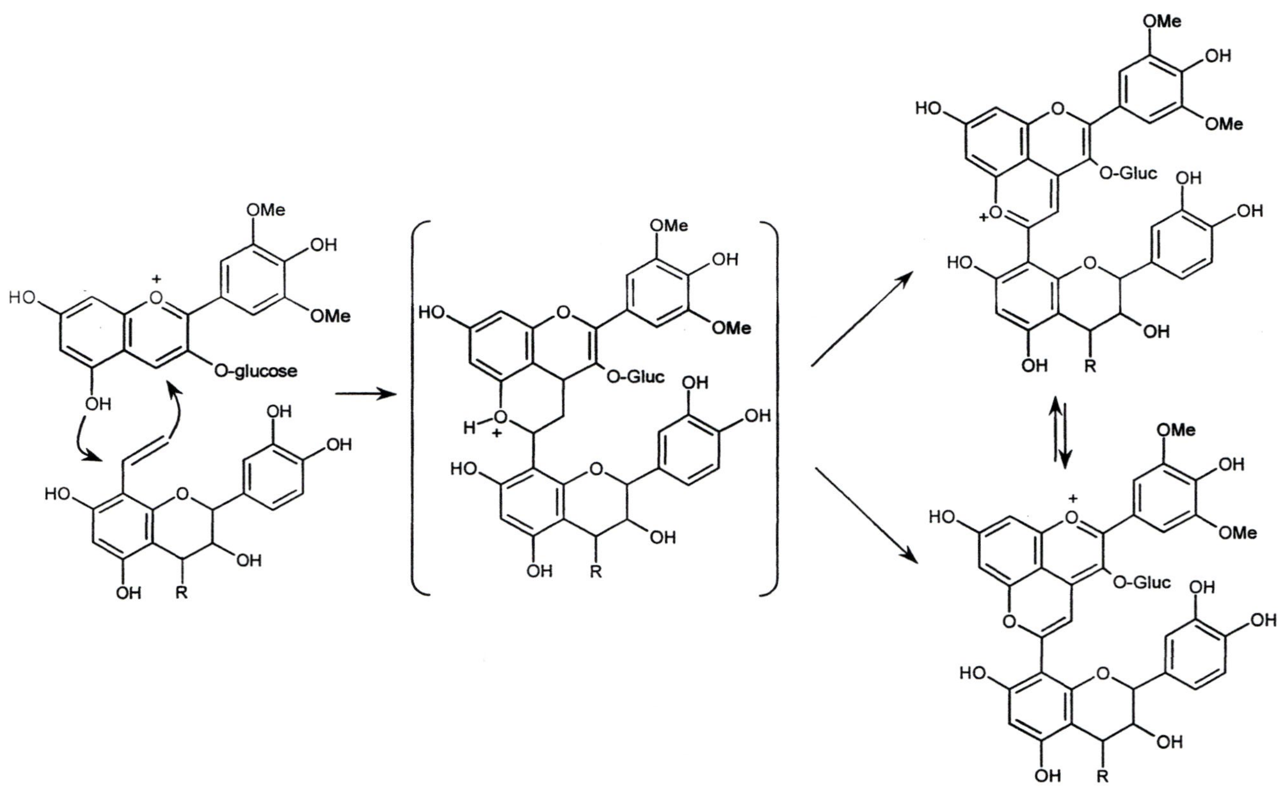

Figure 4. Mechanism proposed for the formation of vinyl-linked pigments.

referred (2B and 7B). Nevertheless, a C4-C8 interflavanoid linkage is expected since the C4-C8 procyanidin dimers (B1 to B4) are more abundant in grapes and in the resulting Port wines than their respective C4-C6 counterparts (B5 to B8) (*33,34*).

The formation mechanism of Portisins is proposed in Figure 5. Anthocyanin-pyruvic acid derivatives, which are formed in a previous step, are thought to react through their C-10 position with the vinyl group of an 8-vinylflavanol adduct. The last step of the formation involves the loss of a formic acid group and oxidation, yielding the new vinylpyranoanthocyanin pigment. The extended conjugation of the π electrons in this newly formed structure is likely to confer a higher stability of the molecule and is probably at the origin of its blue color. In order to obtain further evidences to support this mechanism, model solutions of malvidin-3-coumaroylglucoside pyruvic acid adduct and (+)-catechin were prepared in 20% aqueous ethanol (v/v) (pH=2.0) and maintained at 35°C in oxidative conditions (*32*). After 15 days of reaction, a new peak was observed in the HPLC chromatogram. Its UV-Vis spectrum was similar to those of portisins 8B and 10B with a λ_{max} at 583 nm. The LC/MS analysis of this newly formed pigment showed a $[M]^+$ ion at *m/z* 977, which fits with the structure of pigment 8B but with a catechin unit in the flavanol moiety instead of a procyanidin dimer.

In order to provide an unambiguous chemical identification of these structures, the anthocyanin-derived pigments detected in Port wines were named according to the chemical nomenclature rules recently proposed in literature (*31*):

Pyranoanthocyanins:

- 3A: 1-carboxyvinyl-[1V,2V:5O,4]-malvidin-3-glucoside;
- 5A: 1-carboxyvinyl-[1V,2V:5O,4]-malvidin-3-acetylglucoside;
- 7A: 1-carboxyvinyl-[1V,2V:5O,4]-malvidin-3-coumaroylglucoside;
- 2B: 8-vinyl-cat-(4-8)-cat-[1V,2V:5O,4]-malvidin-3-glucoside;
- 4B: 8-vinyl-cat-[1V,2V:5O,4]-malvidin-3-glucoside;
- 7B: 8-vinyl-epi-(4-8)-cat-[1V,2V:5O,4]-malvidin-3-coumaroylglucoside;
- 9B: 8-vinyl-epi-[1V,2V:5O,4]-malvidin-3-glucoside;
- 15B: 8-vinyl-cat-[1V,2V:5O,4]-malvidin-3-coumaroylglucoside;
- 16B: 8-vinyl-epi-[1V,2V:5O,4]-malvidin-3-coumaroylglucoside.

Vinylpyranoanthocyanins (Portisins):

- 8B: 8-bivinyl-epi-(4-8)-cat-[3V,4V:5O,4]-malvidin-3-glucoside;
- 10B: 8-bivinyl-epi-(4-8)-cat-[3V,4V:5O,4]-malvidin-3-coumaroylglucoside;

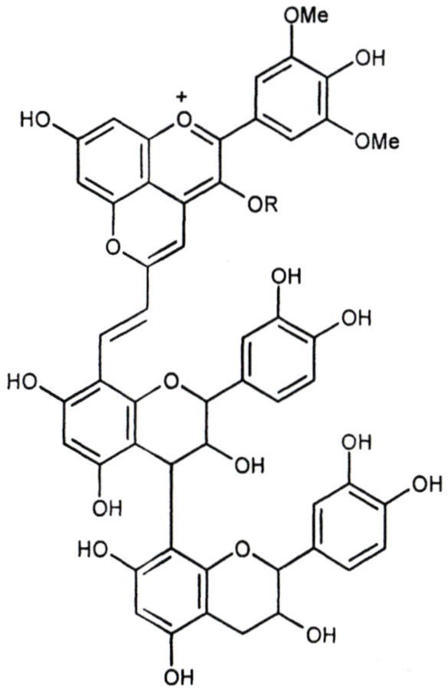

(8B) R= glucose
(10B) R= coumaroylglucose

Figure 5. Structures of Portisins 8B and 10B.

Evolution of anthocyanins and respective pyruvic acid adducts during wine ageing

In the Port winemaking process, when about half of the original sugar content has been converted to alcohol, the must is usually taken off-skins and the fermentation is stopped by addition of wine spirit (ratio wine/wine spirit ≈ 5) in order to maintain its natural sweetness and approximately 20% of alcohol.

In order to study the initial evolution of anthocyanins and their respective pyruvic acid adducts, their concentration was followed immediately after red wine fortification (approximately after two days of fermentation) (*13*). The initial concentration of anthocyanin-pyruvic acid adducts probably results from their formation during the period between grape crushing and fortification (day zero in the graphs) (Figure 7). The anthocyanin-pyruvic acid adducts, mv 3-gluc-py (3A), mv 3-(acetyl)gluc-py (5A) and mv 3-(coumaroyl)gluc-py (7A) started to increase up to a maximum concentration near 60 days and started to decrease after 100 days. Despite the great difference between their concentrations, the initial formation of anthocyanin-pyruvic acid adducts occurred simultaneously with the degradation of anthocyanidin monoglucosides. The levels of anthocyanins were not significant after 200 days for mv 3-gluc and after 120 days for its acylated forms. The pyruvic acid derivatives levels seem to be less affected during aging. This feature was also observed in a monovarietal Port wine stored in oak barrels during 38 months (*13*). Indeed, after one year of oak ageing, the major grape anthocyanins underwent a decrease between 80 and 90%, whilst mv 3-gluc-py decreased only between 15 and 25% during the same period. After 30 months of ageing practically no anthocyanins were detected, whereas 20 to 35% of the original content of anthocyanin-pyruvic adducts were still present in the wine.

Anthocyanin-pyruvic acid adducts are known to be more abundant in Port wines than in red table wines, as seen from previous analysis in our laboratories (data not shown) and as referred by other authors (*11*). This feature may be related to the higher levels of pyruvic acid expected in fortified wines as a result of a shortened fermentation. In fact, when wine spirit is added in order to stop fermentation, the pyruvic acid concentration is expected to be higher than when the fermentation is allowed to go to dryness. Effectively, the pyruvic acid excreted by the yeast at the beginning of the fermentation is further used in the yeast metabolism (*35*). Therefore, could favor the formation of anthocyanin-pyruvic acid adducts.

CONCLUSION

The structural diversity of anthocyanin-derived pigments found in two-year-old Port wine suggests different reactions whereby original grape anthocyanins may associate to other colourless compounds. The detection and structural characterization of these pigments provides important information regarding the

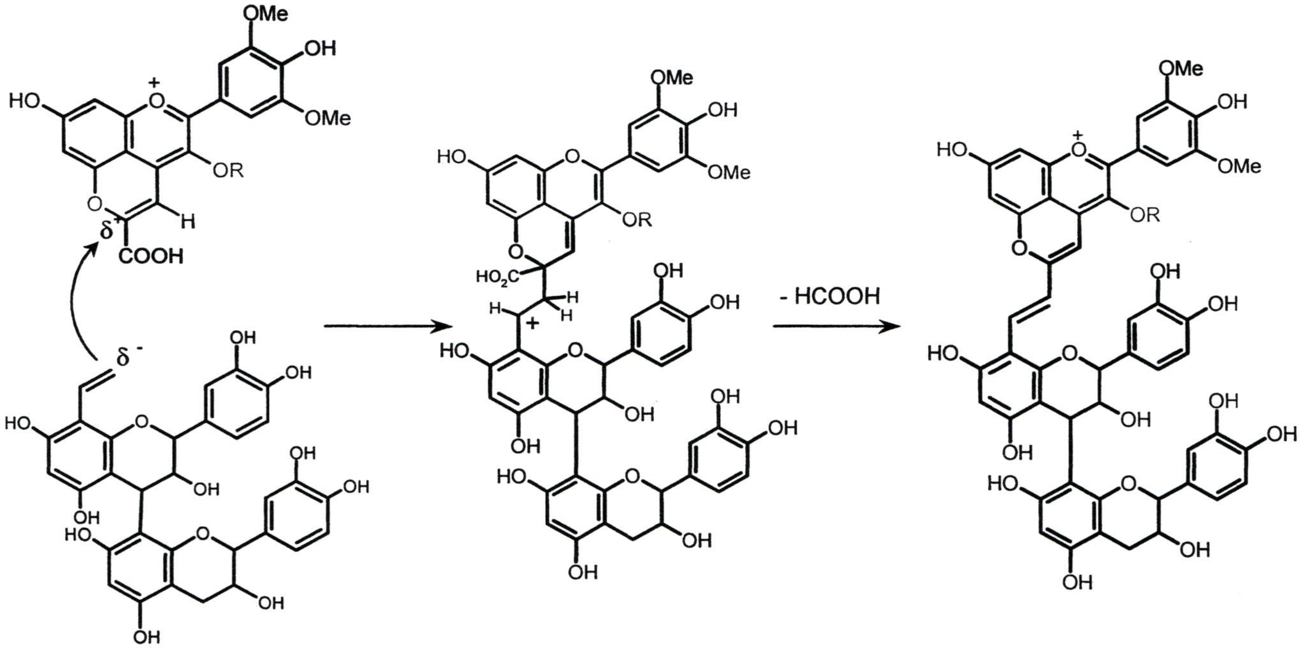

Figure 6. Mechanism proposed for the formation of Portisins (32).

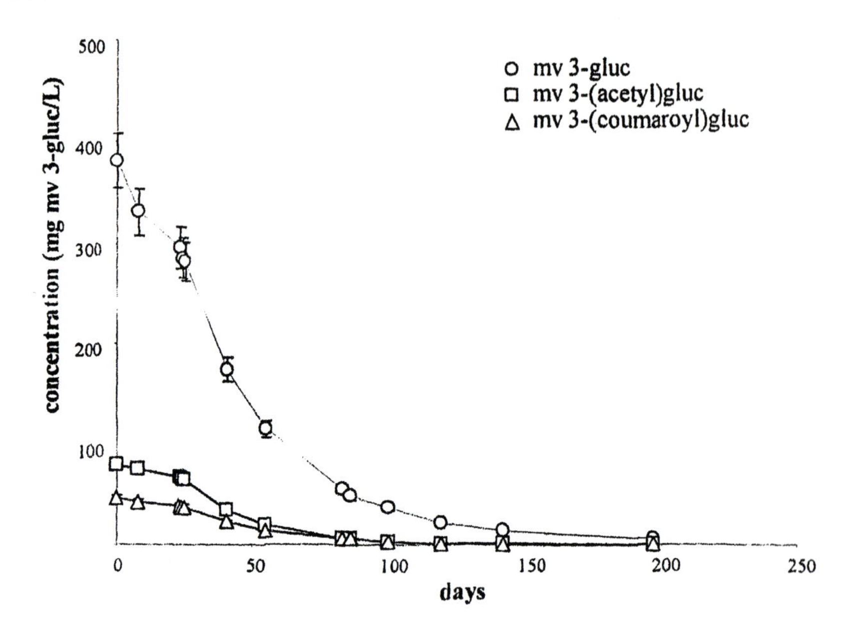

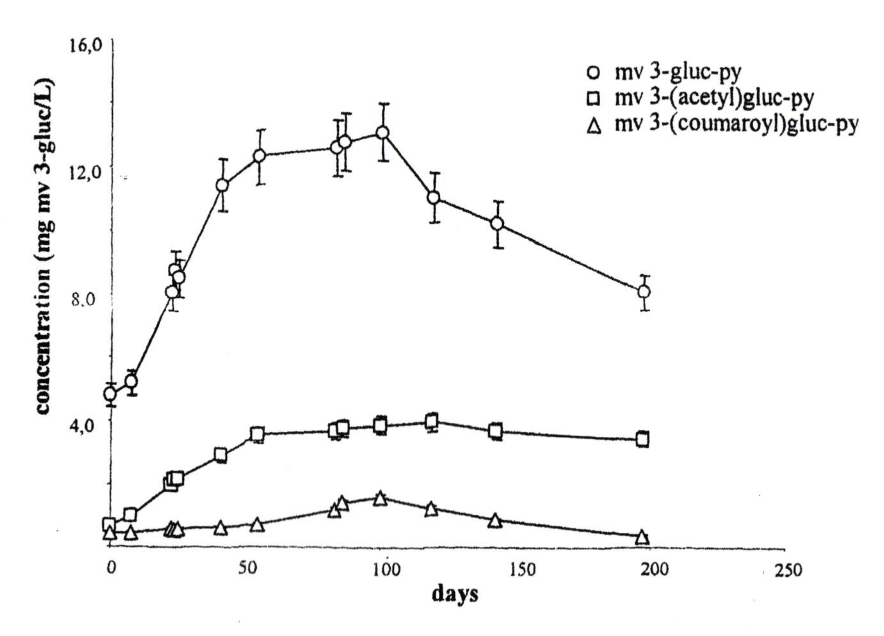

Figure 7. Evolution of the level of malvidin 3-glucoside, malvidin 3-(acetyl)glucoside and malvidin 3-(coumaroyl)glucoside, and their pyruvic acid derivatives in fermented must during the first 200 days after fortification with wine spirit (13). ***(Reproduced with permission from reference 13.*** ***Copyright 2001American Chemical Society.)***

chemical transformations involved in the complex evolution of the color of red wines. Most of these newly formed pigments have different UV-visible spectral characteristics, showing a hypsochromic shift in the visible spectrum with regard to that of original anthocyanins. Overall, the formation of all these anthocyanin-derived pigments may play a crucial role in the changing color of Port wine from purple-red (in young wines) to a more orange-red hue. Indeed, the λ_{max} of the pyruvic acid derivatives is situated near 511 nm whereas the λ_{max} of the anthocyanin-flavanol vinyl-linked pigments is near 511 or 520 nm.

The detection of newly formed Portisins represents the first evidence of blue pigments occurring in red wines. This outcome provides important information regarding the further steps of chemical transformations involving anthocyanin secondary products (pyruvic acid derivatives). Despite the fact that these blue pigments were only detected in very small quantities in fortified wines, they present unique spectroscopic features that may somehow contribute to the changing color of aged red wines. Nevertheless, further studies are needed in order to assess the real contribution of these anthocyanin-derived pigments to the resulting wine color.

ACKNOWLEDGMENTS

The authors thank the Portugal-Spain Programme of 'Acciones integradas' and the Portuguese-American Foundation for Development (FLAD) for the financial support to the mobility of researchers (ref. HP01-24 and E-23/02). This work was also funded by a research project grant POCTI/40124/QUI/2001. The authors also wish to thank Prof. A.M.S. Silva, Prof. C. Santos-Buelga, Prof. J. Rivas-Gonzalo, Prof. A. Gonzalez-Paramás and Prof. J. Vercauteren for their participation as co-authors in the papers related to the results presented herein.

LITERATURE CITED

1. Somers, T. C. The polymeric nature of wine pigments. *Phytochemistry* 1971, *10*, 2175-216.
2. Jurd, L. Review of polyphenol condensation reactions and their possible occurrence in the aging of wines. *Am. J. Enol. Vitic.* 1969, *20*, 191-195.
3. Liao, H.; Cai, Y.; Haslam, E. Polyphenols interactions. Anthocyanins: Copigmentation and colour changes in red wines. *J. Sci. Food Agric.* 1992, *59*, 299-305.
4. Remy, S.; Fulcrand, H.; Labarbe, B.; Cheynier, V.; Moutounet, M. First confirmation in red wine of products resulting from direct anthocyanin-tannin reactions. *J. Sci. Food Agric.* 2000, *80*, 745-751.

5. Bakker, J.; Picinelli, A.; Bridle, P. Model wine solutions: Colour and composition changes during ageing. *Vitis* 1993, 32, 111-118.
6. Rivas-Gonzalo, J. C.; Bravo-Haro, S.; Santos-Buelga, C. Detection of compounds formed through the reaction of malvidin-3-monoglucoside and catechin in the presence of acetaldehyde. *J. Agric. Food Chem.* 1995, *43*, 1444-1449.
7. Timberlake, C. F.; Bridle, P. Interactions between anthocyanins, phenolic compounds and acetaldehyde and their significance in red wines. *Am. J. Enol. Vitic.* 1976, *27*, 97-105.
8. Bakker, J.; Bridle, P.; Honda, T.; Kuwano, H.; Saito, N.; Terahara, N.; Timberlake, C. F. Isolation and identification of a new anthocyanin occurring in some red wines. *Phytochemistry* 1997, *44*, 1375-1382.
9. Bakker, J.; Timberlake, C. F. Isolation, identification, and characterization of new color-stable anthocyanins occurring in some red wines. *J. Agric. Food Chem.* 1997, *45*, 35-43.
10. Fulcrand, H.; Benabdeljalil, C.; Rigaud, J.; Cheynier, V.; Moutounet, M. A new class of wine pigments generated by reaction between pyruvic acid and grape anthocyanins. *Phytochemistry* 1998, *47*, 1401-1407.
11. Romero, C.; Bakker, J. Interactions between grape anthocyanins and pyruvic acid, with effect of pH and acid concentration on anthocyanin composition and color in model solutions. *J. Agric. Food Chem.* 1999, *47*, 3130-3139.
12 Mateus, N.; Silva, A. M. S.; Vercauteren, J.; De Freitas, V. A. P. Occurrence of anthocyanin-derived pigments in red wines. *J. Agric. Food Chem.* 2001, *49*, 4836-4840.
13 Mateus, N.; De Freitas, V. A. P. Evolution and stability of anthocyanin-derived pigments in Port wine during ageing. *J. Agric. Food Chem.* 2001, *49*, 5217-5222.
14. Cameira dos Santos, P. J.; Brillouet, J. M.; Cheynier, V.; Moutounet, M. Detection and partial characterisation of new anthocyanin-derived pigments in wine. *J. Sci. Food Agric.* 1996, *70*, 204-208.
15. Fulcrand, H.; Cameira dos Santos, P. J.; Sarni-Manchado, P.; Cheynier, V.; Bonvin, J. F. Structure of a new anthocyanin-derived wine pigment. *J. Chem. Soc. Perkin Trans. 1* 1996, 735-739.
16. Benabdeljalil, C.; Cheynier, V.; Fulcrand, H.; Hakiki, A.; Mosaddak, M.; Moutounet, M. Mise en évidence de nouveaux pigments formés par réaction des anthocyanes avec des métabolites de levure. *Sci. Aliment.* 2000, *20*, 203-220.
17. Lu, Y.; Foo, L. Y. Unusual anthocyanin reaction with acetone leading to pyranoanthocyanin formation. *Tetrahedron Lett.* 2001, *42*, 1371-1373.
18. Hayasaka, Y.; Asenstorfer, R. E. Screening for potential pigments derived from anthocyanins in red wine using nanoelectrospray tandem mass spectrometry. *J. Agric. Food Chem.* 2002, *50*, 756-761.
19. Es-Safi, N. E.; Le Guernevé, C.; Labarbe, B.; Fulcrand, H.; Cheynier, V.; Moutounet M. Structure of a new xanthylium salt derivative. *Tetrahedron Lett.* 1999, *40*, 5869-5872.

20. Fulcrand, H.; Cheynier, V.; Oszmiansky, J.; Moutounet, M. An oxidised tartaric acid residue as a new bridge potentially competing with acetaldehyde in flavan-3-ol condensation. *Phytochemistry* 1997, *46*, 223-227.
21. Mateus, N.; Silva, A. M. S.; Rivas-Gonzalo, J. C.; Santos-Buelga, C.; De Freitas, V. A. P. Identification of anthocyanin-flavanol pigments in red wines by NMR and mass spectrometry. *J. Agric. Food Chem.* 2002, *50*, 2110-2116.
22. Roggero, J. P.; Coen, S; Archier, P.; Rocheville-Divorne, C. Étude par C.L.H.P. de la reaction glucoside de malvidine-acétaldéhyde-composé phénolique. *Conn. Vigne Vin* 1987, *21*, 163-168.
23. Bax, A.; Subramanian, S. Sensitivity enhanced two-dimensional heteronuclear shift correlation NMR spectroscopy. *J. Magn. Reson.* 1986, *67*, 565-569.
24. Bax, A.; Summers, M. F. ^{1}H and ^{13}C assignments from sensitivity-enhanced detection of heteronuclear multiple-bond connectivity by 2D multiple quantum NMR. *J. Am. Chem. Soc.* 1986, *108*, 2093-2094.
25. Es-Safi, N.E.; Fulcrand H.; Cheynier, V.; Moutounet, M. Studies on the acetaldehyde induced condensation of (-)-epicatechin and malvidin 3-*O*-glucoside. *J. Agric. Food Chem.* 1999, *47*, 2096-2102.
26. Escribano-Bailon, T.; Dangles, O.; Brouillard, R. Coupling reactions between flavylium ions and catechin. *J. Agric. Food Chem.* 1996, *41*, 1583-1592.
27. Vivar-Quintana, A. M.; Santos-Buelga, C.; Francia-Aricha, E.; Rivas-Gonzalo, J.C. Formation of anthocyanin-derived pigments in experimental red wines. *Food Sci. Tech. Int.* 1999, *5(4)*, 347-352.
28. Wildenradt, H. L.; Singleton, V. L. The production of aldehydes as a result of oxidation of polyphenolic compounds and its relation to wine aging. *Am. J. Enol. Vitic.* 1974, *26*, 25-29.
29. Chatonnet, P., Dubourdieu, D., Boidron, J. & Lavigne, V. Synthesis of volatile phenols by Saccharomyces cerevisiaein wines. *J. Sci. Food Agric.*1993, *62*, 191-202.
30. Francia-Aricha, E. M.; Guerra, M. T.; Rivas-Gonzalo, J. C.; Santos-Buelga, C. New anthocyanin pigments formed after condensation with flavanols. *J. Agric. Food Chem.* 1997, *45*, 2262-2265.
31. Mateus, N.; Carvalho, E.; Carvalho, A. R. F.; Melo, A.; González-Paramás, A. M.; Santos-Buelga, C.; Silva, A. M. S.; Rivas-Gonzalo, J. C.; De Freitas, V. A. P. Isolation and structural characterization of new acylated anthocyanin-vinyl-flavanol pigments occurring in aging red wines *J. Agric. Food Chem.* 2003, *51*, 277-282.
32. Mateus, N.; Silva, A. M. S.; Rivas-Gonzalo, J. C.; Santos-Buelga, C.; De Freitas, V. A. P. A new class of blue anthocyanins-derived pigments isolated from red wines. *J. Agric. Food Chem.* 2003, *51*, 1919-1923.

33. Mateus, N.; Proença, S.; Ribeiro, P.; Machado, J. M.; De Freitas, V. A. P. Grape and wine polyphenolic composition of red *Vitis vinifera* varieties concerning vineyard altitude. *Cienc. Tecnol. Aliment.* 2001, *3*, 102-110.
34. Mateus, N.; Marques, S.; Gonçalves, A. C.; Machado, J. M.; De Freitas, V. A. P. Proanthocyanidin composition of red *Vitis vinifera* varieties from the Douro Valley during ripening: Influence of cultivation altitude. *Am. J. Enol. Vitic.* 2001, *52(2)*, 115-121.
35. Whiting, G. C.; Coggins, P. A. Organic acid metabolism in cider and perry fermentations. III.-Keto-acids in cider-apple juices and ciders. *J. Sci Food Agric.* 1960, *11*, 705-709.

Chapter 11

Novel Aged Anthocyanins from Pinotage Wines: Isolation, Characterization, and Pathway of Formation

Michael Schwarz and Peter Winterhalter*

Institute of Food Chemistry, Technical University of Braunschweig, Schleinitzstrasse 20, 38106 Braunschweig, Germany

Wines from the South African grape variety Pinotage (Pinot Noir × Cinsault) show an unusual anthocyanin profile with several additional peaks eluting in the same region as the acetylated and coumaroylated anthocyanins. By means of mass spectrometric analyses these pigments were tentatively identified as the vinylcatechol derivatives of malvidin 3-glucoside (Pinotin A), malvidin 3-(6''-acetylglucoside) and malvidin 3-(6''-coumaroyl-glucoside). After isolation by adsorption on Amberlite XAD-7 resin and purification using high-speed countercurrent chromatography (HSCCC) in combination with semi-preparative HPLC, the proposed structures were confirmed by one- and two-dimensional NMR measurements. As pathway of formation, a direct reaction of anthocyanins with cinnamic acid derivatives was elucidated. Importance of this reaction for the stabilization of red wine color is shown and the potential use of Pinotin A for a rough estimation of the age of Pinotage wines will be discussed. The newly discovered reaction scheme also provides a better understanding of the chemistry involved in color stabilization/enhancement by copigmentation.

Typical red wine pigments are the 3-O-glucosides of malvidin and peonidin as well as their 6''-acetylated and coumaroylated derivatives. The 3-O-glucoconjugates of petunidin, delphinidin, and cyanidin are also widespread. During wine aging, these pigments polymerize to a large extent and form a heterogeneous not well characterized group of compounds, which is considered to be of major importance for the color of aged wines (*1-4*). Multiple other reactions take place during the aging of red wines and several low molecular weight reaction products have been isolated and characterized in the last decade. Some of the aged pigments bear an additional pyran ring between C-4 and the hydroxyl group in position 5 of the aglycon moiety. Among these, the vitisin-type pigments originate from the reaction of malvidin 3-glucoside and acylated derivatives with pyruvic acid (*5,6*). Another pyranoanthocyanin, the malvidin 3-glucoside 4-vinylphenol adduct was first detected on membranes used for cross-flow microfiltration of Carignane red wines (*7*) and later isolated and structurally identified through comparison with a synthetic sample (*8*). In the course of a study on the anthocyanin composition of Pinotage wines (Pinot Noir × Cinsault; a grape variety autochthonous to South Africa), mass spectrometric analyses (HPLC-ESI-MS^n) indicated the presence of anthocyanin derivatives of so far unknown structure and origin. In the following, isolation and complete structural characterization of these novel pigments is presented together with a new pathway of formation of the pyranoflavylium skeleton.

Experimental

Isolation and Purification of Pinotage Pigments. An anthocyanin enriched extract from a 1997 Pinotage (Zonnebloem, Stellenbosch, South Africa) was prepared by solid phase extraction on Amberlite XAD-7 resin according to the method previously published (*9,10*). A pre-separation of the XAD-7 extract was achieved by high-speed countercurrent chromatography (HSCCC 1000-apparatus, Pharma-Tech Research Corp., USA) equipped with three coils (total volume 850 mL). Solvent system I consisted of n-butanol/TBME/acetonitrile/water (2/2/1/5, v/v/v/v, acidified with 0.1% TFA). Solvent system II was n-butanol/TBME/acetonitrile/water (0.65/3.35/1/5, v/v/v/v, acidified with 0.1% TFA). Mobile phases (lower layers) were delivered by a Biotronik BT 3020 HPLC pump (Jasco, Germany) at a flow rate of 3.0 mL/min. Separation was monitored with a UV/Vis-detector (Knauer, Germany) at λ = 510 nm. Fractions were collected every 4 min with a Super Frac fraction collector (Pharmacia LKB, Sweden), combined according to the chromatogram and lyophilized prior to HPLC analysis. Final purification was achieved by semi-preparative HPLC with water/acetonitrile/formic acid (70/25/5, v/v/v, 6.0 mL/min) on a Luna

RP-18 (Phenomenex, Germany) column (250 × 10 mm) equipped with a guard column (50 × 10 mm) of the same material.

Synthesis of Pinotin A and Related Compounds. Malvidin 3-glucoside of approx. 90% purity was isolated from various red wines by countercurrent chromatography according to methods previously described (*9-11*) and dissolved in model wine medium (0.02 M solution of potassium hydrogen tartrate in deionized water, adjusted to pH 3.2 with hydrochloric acid). To 27 mL of the malvidin 3-glucoside solution (concentration approx. 400 mg/L) 3 mL of an ethanolic solution of each of the following *E*-configured cinnamic acid derivatives, i.e. cinnamic acid, *p*-coumaric acid, ferulic acid, caffeic acid, sinapic acid (approx. 400 mg/L) as well as 4-dimethylaminocinnamic acid, 4-nitrocinnamic acid, caffeic acid methyl ester (approx. 1200 mg/L) were added. The solutions were filled into 30 mL amber glass bottles, air in the headspace was replaced with argon, and the solutions were stored in the dark at 15°C in a climatized room. Samples were analyzed in regular intervals by HPLC-DAD and HPLC-ESI-MS^n for 4 months. Relative reaction rates were determined by HPLC-DAD analyses comparing the concentrations of the newly formed pigments (expressed as Pinotin A; authentic reference standard used for calibration between 0 and 15 mg/L, detection wavelength 510 nm).

Determination of Visual Detection Limits. Visual detection limits were determined for Pinotin A, Vitisin A, malvidin 3-glucoside, malvidin 3-(6''-acetylglucoside) and the polymeric fraction isolated from a 1997 red wine. The compounds were dissolved in a saturated aqueous solution of potassium hydrogen tartrate containing 10% EtOH (pH 3.60) and diluted to various concentrations above and below the visual detection limit. Four mL of the solutions were pipetted into plastic cuvettes (1 × 1 cm; liquid layer thickness: 4 cm) and placed into a white polystyrene holder with spaces for ten cuvettes. Blank model wine and sample solutions were arranged in decreasing concentration from left to right in such a way that the diluted sample was always placed between two blanks. Five judges were asked to specify the last solution which displayed any color in comparison to the surrounding blanks under ambient light conditions.

HPLC-ESI-MS^n: Analyses were done on an Esquire Ion Trap LC-MS system (Bruker, Germany). Equipment and chromatographic conditions were the same as previously published (*10,12*). Alternatively, the sample solution was delivered directly by a syringe pump 74900 (Cole-Parmer, USA) into the ESI source at a flow rate of 300 µL/h. ESI-spectra were measured in positive mode. MS parameters: dry gas: N_2, 4.0 L/min, dry temperature: 300°C, nebulizer: 10 psi, capillary: –3500 V, end plate offset:

–500 V, skimmer 1: 30 V, skimmer 2: 10 V, capillary exit offset: 60 V, trap drive: 60, accumulation time: 200 ms, scan range: 50-1000 m/z.

Nuclear Magnetic Resonance (NMR). ^{1}H-, ^{13}C-NMR and DEPT-spectra were measured on a Bruker AMX 300 spectrometer at 300.1 MHz and 75.4 MHz, respectively. HMQC and HMBC experiments (*13*) were performed on a Bruker AM 360 instrument at 360.1 MHz. Solvent was a mixture of methanol-d_4 and TFA-d_1 (19:1, v/v).

Results and Discussion

Isolation of Novel Pyranoanthocyanins from Pinotage Wines

The HPLC chromatogram of the Pinotage wine under investigation showed an intense peak with a retention time of 44.7 min, eluting just prior to malvidin 3-(6''-*p*-coumaroylglucoside). Molecular mass $[M]^+$ of the unknown compound was determined to be 625 mass units (in negative mode 623 $[M-2H]^-$). Upon fragmentation, a loss of *m/z* 162 indicated the cleavage of a hexose moiety. The remaining aglycon had a mass of *m/z* 463. Additional peaks were observed at retention times of 46.4 min and 48.2 min, exhibiting molecular masses of *m/z* 667 and *m/z* 771, respectively. With the fragment *m/z* 463 they had the same aglycon mass which was explained by cleavage of an acetylated or coumaroylated hexose. ESI-MS^n spectra recorded for the unknown pigments were in line with data published by Hayasaka and Asenstorfer (*14*). These authors tentatively identified a series of aged anthocyanins in red wines and grape skin extracts solely by mass spectrometric analyses. Importantly, the tiny amounts of the pigments did not allow an isolation. Hence, their complete structural characterization/confirmation still necessitates thorough NMR spectroscopic studies.

In order to confirm the structures of the novel Pinotage pigments, the anthocyanin fraction was enriched by adsorption on XAD-7 resin. After elution with methanol/acetic acid (19:1), the XAD-7 extract was lyophilized and subjected to a first clean-up step using countercurrent chromatography. Polymeric pigments eluted first, followed by malvidin 3-O-β-D-glucoside, after which the CCC-separation was stopped. The remaining liquid phases on the coil were pushed out using N_2 gas; upper and lower layer were collected separately. After concentration and lyophilization, the deeply red-colored upper layer was re-chromatographed by CCC using the slightly less polar solvent system II. For final purification of the unknown pigments semi-preparative HPLC was used. In this way, 70 mg, 10 mg, and 12 mg of the target compounds **1-3** were obtained from eight bottles (i.e. total of 6 L) of Pinotage wine. The ^{1}H-NMR spectrum of the major compound **1** showed signals for H-6, H-8, H-2', H-6' and two methoxy groups, thus suggesting the structure to be derived from malvidin 3-O-β-glucoside. Importantly, the signal for H-4 was missing and the anomeric glucose proton shifted upfield to 4.81 ppm with a coupling constant of 7.5 Hz, thus proving β-configuration. The remaining glucose resonances appeared between 3.19 and 3.68 ppm. Further signals observed were a singlet at 7.84 ppm, assigned to H-11, and three additional aromatic proton signals of the catechol ring system. The ^{13}C-NMR spectrum displayed a total of 31 resonances, six of them being related to β-D-glucose and another eight to the unchanged B ring of the malvidin-based structure. A long range CH-correlation starting from the anomeric proton H-1' (4.81 ppm) to C-3 (134.9 ppm) is clearly indicating the linkage of the sugar moiety to the malvidin aglycon. By means of 2D-NMR techniques (HMQC, HMBC), we were able to assign all ^{1}H- and ^{13}C-NMR connectivities in structures **1-3**. Some of the long-range CH-correlations important for structure elucidation of the major compound are illustrated in Figure 1 verifying the proposed structure of the novel Pinotage pigments which were named Pinotin A **1**, Acetyl-Pinotin A **2**, and Coumaroyl-Pinotin A **3**. Details for isolation and structure elucidation of compounds **1-3** have been published separately (*15*).

It is also noteworthy that two further pigments could be obtained from the anthocyanin fraction of Pinotage wine, namely the 4-vinylphenol adduct of malvidin 3-glucoside **4** (10 mg) as well as the 4-vinylguaiacol derivative of malvidin 3-glucoside **5**. In the latter case, identification has to be considered as tentative because the trace amounts obtained of compound **5** (< 1 mg) did not enable measurement of a complete set of NMR spectroscopic data.

Pathway of Formation

Up to now, the formation of pyranoanthocyanin derivatives, such as e.g. the malvidin 3-glucoside 4-vinylphenol adduct **4,** is considered to be due to a direct reaction of malvidin 3-glucoside with 4-vinylphenol (Figure 2). Obviously through an enzymatic side activity of *Saccharomyces cerevisiae*, formation of 4-vinylphenol takes place during fermentation *via* enzymatic decarboxylation of *p*-coumaric acid (*8*). It has been suggested that the formation of the 4-vinylcatechol, 4-vinylguaiacol and 4-

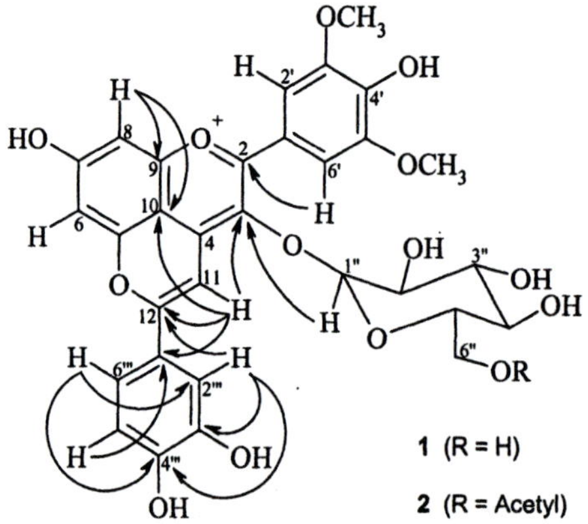

1 (R = H) Pinotin A

2 (R = Acetyl) Acetyl-Pinotin A

3 (R = Coumaroyl) Coumaroyl-Pinotin A

Figure 1. Structures and long-range correlations of novel Pinotage pigments 1-3.

vinylsyringol derivatives follows the same mechanism involving the respective vinylphenols emerging from the enzymatic decarboxylation of caffeic, ferulic and sinapic acid (*14,16*).

First doubts concerning the suggested formation pathway arose when we observed that the concentration of Pinotin A was approximately ten times higher in aged wines (5-6 years old) compared to young (< 1 year) wine samples. This finding suggested that the reaction between 4-vinylcatechol and malvidin 3-glucoside proceeds rather slowly and requires years of storage to complete. In order to check the reactivity of vinylphenols, model experiments with 4-vinylcatechol and malvidin 3-glucoside were carried out. As 4-vinylcatechol is not commercially available, a synthesis was performed by the method of Bücking (*17*). Synthesized 4-vinylcatechol and malvidin 3-glucoside were reacted in model wine medium. Subsequent HPLC-DAD and HPLC-ESI-MS^n analyses revealed that malvidin 3-glucoside had been almost quantitatively converted to Pinotin A after stirring overnight (*15*). It should be mentioned that our findings are in line with earlier observations of Fulcrand *et al.* (*8*) and Sarni-Manchado *et al.* (*18*) who also reported that the reaction of malvidin 3-glucoside and 4-vinylphenol went to completion within hours.

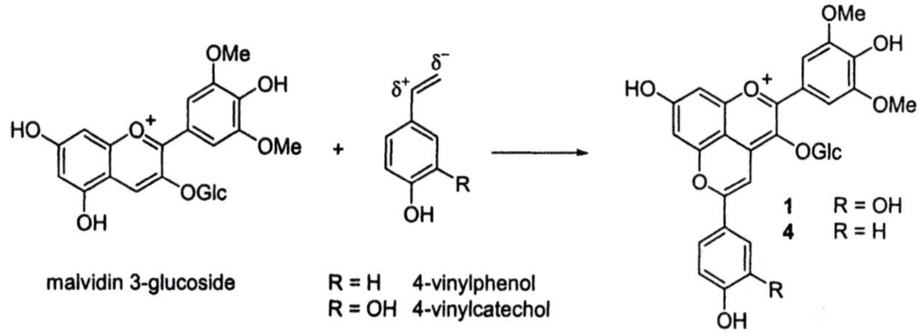

*Figure 2. Formation of the malvidin 3-glucoside 4-vinylphenol adduct **4** as proposed by Fulcrand et al. (8) and – in analogy – the likely formation of compound **1** (Pinotin A) through reaction with vinylcatechol.*

Consequently, the increase of Pinotin A during storage of Pinotage wines requires another explanation. It seems to be clear that 4-vinylcatechol, if present at all, would be consumed rapidly and the gradual formation of **1** over prolonged periods could not be rationalized in this way. Our suspicion was strengthened by the fact that despite a comprehensive literature search we were unable to find evidence that either 4-vinylcatechol or 4-vinylsyringol had ever been detected in red or white wines, whereas the presence of 4-vinylphenol and 4-vinylguaiacol is well documented (*19-23*). Although red wines contain much more of the precursors coumaric and ferulic acid, the amount of the corresponding volatile vinylphenols in white wines was determined to be orders of magnitude higher and beyond a certain limit responsible for unpleasant 'phenolic' off-flavors. A possible explanation for this phenomenon could be seen in the reaction of anthocyanins with vinylphenols (*8*). In this context, studies by Chatonnet *et al.* (*21*) concerning the ability of different strains of *Saccharomyces cerevisiae* to decarboxylate cinnamic acids are of importance. The authors found that certain wine constituents, especially catechin, epicatechin and oligomeric procyanidins, but not anthocyanins and polymeric tannins, strongly inhibited the decarboxylation of *p*-coumaric acid. The concentration of these inhibitors is much higher in red wines. Hence, it can been concluded that the cinnamate carboxy-lyase is largely inactive during red wine fermentation and the enzyme-mediated synthesis of 4-vinylphenols in red wine is not significant with regard to anthocyanin-vinylphenol adduct formation. The authors also discovered that all of the yeast strains investigated were able to decarboxylate *p*-coumaric and ferulic acid, but none of them was capable of transforming either caffeic or sinapic or unsubstituted cinnamic acid into 4-vinylcatechol, 4-vinylsyringol or styrene, respectively.

The decisive hint for a novel pathway of anthocyanin-vinylphenol adduct formation was finally provided by the unusually high concentrations of caffeic acid in Pinotage wines. The caffeic acid content in most of the common red wine varieties does not exceed 10 mg/L (*24-26*). Only in Pinotage wine, we were able to detect as much as 77 mg/L of caffeic acid with a mean value around 35 mg/L. Hereupon we started to consider the possibility of a direct reaction between malvidin 3-glucoside and caffeic acid - or anthocyanins and cinnamic acids in general. Hence, another experiment was performed with malvidin 3-glucoside and caffeic acid in model wine solution. This solution was stored in the dark at 15°C and analyzed at regular intervals. Just after five days we were able to detect the formation of a new pigment with the same retention time and mass spectral properties as Pinotin A. The concentration of this peak constantly increased during the following months as shown by HPLC-DAD analyses.

These results showed for the first time that pyranoanthocyanins can be formed directly in a reaction between intact cinnamic acid derivatives and anthocyanin. In order to obtain a mechanistic rationale for this unexpected conversion further experiments were conducted, including investigations on the reactivity of other cinnamic acid derivatives and the possible influence of oxygen on the reaction rate. Based on our findings the following reaction sequence (cf. Figure 3) can be suggested. A more detailed discussion can be found in Ref. (*12*).

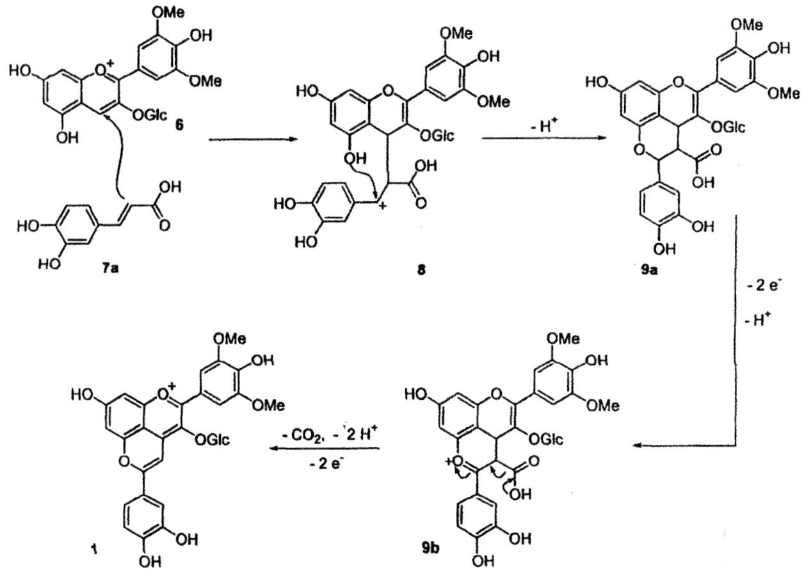

*Figure 3. Postulated pathway of Pinotin A (**1**) formation involving caffeic acid (**7a**) and malvidin 3- glucoside **6**.*

The initial bond formation between the C-4 position of malvidin 3-glucoside **6** and the C-2 position of caffeic acid **7a** is in line with the strongly electrophilic nature of the benzopyrylium unit and the nucleophilicity of the α-carbon atom of acid **7a**. Given the electron-deficient character of the resulting intermediate **8**, it can be expected that electron-donating substituents on the aromatic ring of the cinnamic acid moiety facilitate this reaction due to stabilisation of the generated carbenium ion **8**. This has indeed been observed in a series of conversion experiments. Whereas *p*-coumaric acid **7e**, ferulic acid **7c**, sinapic acid **7d** and 4-dimethylaminocinnamic acid **7b** could be successfully reacted, no adduct formation at all was observed with 4-nitrocinnamic acid **7g** or the parent compound cinnamic acid **7f**. Also in the case of the ester **7h** no reaction was observed (Figure 4 and 5).

Mass spectral properties of the generated anthocyanin derivatives were in line with data published for the isolated 4-vinylphenol and 4-vinylcatechol derivatives of malvidin 3-glucoside (*8,15*) and the tentatively identified 4-vinylguaiacol and 4-vinylsyringol adducts (*14*). The *p*-dimethylamino cinnamic acid product has not been reported before (Table I).

Table I. Mass Spectral Properties of the Generated Malvidin 3-glucoside Cinnamic Acid Derivatives

Malvidin 3-glucoside Adduct	*Molecular Ion* [M^+] (m/z)	*Aglycon* (m/z)
4-Vinylphenol	609	447
4-Vinylcatechol (Pinotin A)	625	463
4-Vinylguaiacol	639	477
4-Vinylsyringol	669	507
4-Dimethylaminostyrene	636	474

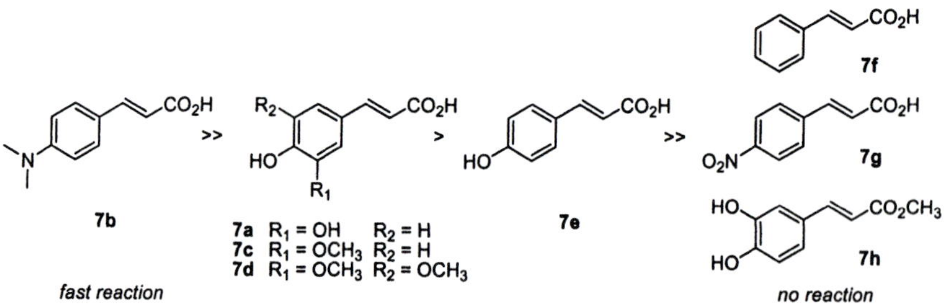

Figure 4. Reactivity of cinnamic acid derivatives with regard to anthocyanin adduct formation with malvidin 3- glucoside.

Kinetics of reaction of the di- and trisubstituted acids **7a**, **7c** and **7d** with anthocyanins were moderately enhanced compared to **7e** (Figure 5). With the amino moiety in **7b**, however, rate of formation was accelerated by a factor larger than 100 (data not shown). These observed reactivities coincide with the expected order of reaction rates for *para*-substituent effects when electron-deficient transition states are involved, as given by Brown's σ_p^+ constants (*27,28*). It is known that electron-donating substituents can accelerate reactions involving the olefinic double bond of cinnamates. Negative Hammett reaction parameters ρ have been obtained, e.g. in the bromate oxidation of cinnamic acids (*29-32*). These data confirm that electrophilic attack on cinnamic acids as shown here can be favored over nucleophilic pathways such as conjugate addition to the double bond, which can be accelerated by electron-withdrawing substituents (*33*).

Subsequently, the final product **1** is formed *via* oxidation and decarboxylation of the intermediate **9**. Few details are known about the precise mechanism of pyran oxidation involving reactive benzylic carbon-

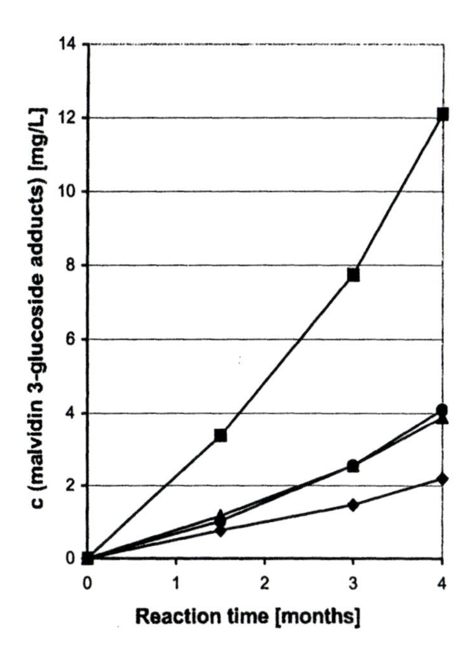

Figure 5. Course of the reaction of malvidin 3-glucoside with coumaric acid (◆), ferulic acid (●), sinapic acid (▲) and caffeic acid (■).

hydrogen bonds under the conditions used here, but these conversions can generally be achieved with remarkable ease (*34,35*). Hydride abstractions (e.g. by unreacted **6**) (*36-38*), enzymatic/coenzymatic oxidations (*39,40*) or involvement of other known oxidants such as nitrite (*41*) are generally conceivable in authentic wine samples. Radical pathways (autoxidation) in the presence of oxygen and a catalytic species (*42*) are less likely due to the antioxidative properties of phenolic compounds.

The direct reaction between anthocyanins and hydroxycinnamic acids readily explains the formation of anthocyanin-vinylphenol-type adducts in red wines. At the time of writing, this is the only experimentally verified mechanism leading to the development of 4-vinylcatechol and 4-vinylsyringol pigments, as the free vinylphenols have neither been detected in wines nor was it possible to generate these compounds *via* enzymatic decarboxylation using yeasts commonly applied to red wine fermentation (*21*). Small amounts of 4-vinylphenol and 4-vinylguaiacol were detected in experimental Shiraz red wines only if the grape juice was treated with pectic enzymes possessing cinnamoyl esterase activity prior to fermentation, thus increasing the amount of free coumaric and ferulic acids (*23*). However, even in this case the concentrations of the respective vinylphenols were in the low μg-range and they would have to react quantitatively with malvidin 3-glucoside to generate detectable amounts of the derived pigments. Given the high reactivity of vinylphenols towards other constituents of young red wine, including different anthocyanins, it is extremely unlikely that high levels of anthocyanin-vinylphenol adducts can arise in this way.

Importance of the Novel Pigments for Color Stability, Estimation of the Age of Pinotage Wine as well as Copigmentation

Stability of Pyranoanthocyanins

Stability of anthocyanins is strongly dependent on the pH value of the solution. In highly acidic media anthocyanins occur mainly as their red to bluish-red flavylium cations. An increase in pH leads to the formation of colorless carbinol bases, which in turn are in an equilibrium with the open chalcone form. In neutral and alkaline medium finally the blue quinonoidal base is responsible for the perceived color. The importance of substitution at C-4 on the stability of anthocyanins in solution has been reported (*1,43,44*). The newly introduced pyran ring between C-4 and the hydroxyl group attached to C-5 of the malvidin based structure shields pyranoanthocyanins from hydration, thus delaying the formation of the colorless carbinol base, which at wine pH is already the predominant form of malvidin 3-glucoside. Pinotin A stability was checked in model solution at different pH-values (cf. Figure 6). The color expression of Pinotin A remains almost unaffected upon an increase in pH from 1.0 to 4.5. The maximum absorbance value at 505 nm decreases by only 7%. A further increase in pH to 5.1 results in a loss of approximately 23% of the initial color of the Pinotin A solution, however, the hue remains unchanged. At pH 5.9 minor amounts of the blue quinonoidal base of Pinotin A have formed as can be concluded from the slight increase in absorbance between 550 and 750 nm. At pH 7.5 flavylium cation and quinonoidal base are in an equilibrium, the latter one finally dominating at pH 11.9 with an absorbance maximum at 570 nm (spectrum not shown). In contrast, the absorbance of a solution of malvidin 3-glucoside decreases by 44% after adjusting the pH from 0.9 to 2.9 (maximum absorbance at 521 nm) and 96% of the color is lost at pH 4.8. Instead of the blue quinonoidal base only the colorless carbinol base and chalcone are formed until pH 5.5 (*45*).

The color activity concept allows identification of key colorants in mixtures and their percentage contribution to total color after the concentration and visual detection limits of the single compounds under investigation have been determined (*46*). Results for the visual detection limits of anthocyanin and derivatives are summarized in Table II. The data shows that at a typical wine pH of 3.60, the visual detection limit for Pinotin A is well below its mean concentration in Pinotage wine, approximately 2 times lower than for malvidin 3-glucoside and about 3 times lower in regard to malvidin 3-(6''-acetylglucoside). In comparison, the detection limit for the polymeric fraction from a 1997 red wine with 0.94 mg/L is more than 13fold higher (*47*). These findings are not directly comparable to the results published by Degenhardt et al. (*9*) as the liquid layer thickness used for determination of the visual detection limits was not specified in this study. However, the ratios of their detection limits for polymers: malvidin 3-(6''-acetylglucoside) : malvidin 3-glucoside were 6.8 : 2.2 : 1 and thus very similar to ours with 6.8 : 1.3 : 1. Comparative values for the visual detection limit of Pinotin A and Vitisin A were not available.

Table II. Visual Detection Limits for Selected Pigments

Pigment	*Visual Detection Limit [mg/L]*
Malvidin 3-glucoside	0.138
Malvidin 3-(6''-acetylglucoside)	0.182
Pinotin A	0.068
Vitisin A	0.034
Polymeric pigments	0.939

Estimation of the Age of Pinotage Wines

Due to the pure chemical nature of the generation of anthocyanin-vinylphenol adducts (without the need for any enzymatic support) this reaction can take place during years of storage, thus making the products potentially attractive as possible aging indicators for Pinotage wines. Preliminary results of a screening of 50 Pinotage wines from 1996 to 2002 are shown in Figure 7. It is obvious that a constant increase in Pinotin A concentration within the first five years of storage can be observed. We are presently working on a formula for a rough estimation of the age of Pinotage wine which will include the initial concentrations of caffeic and caftaric acid as well as of malvidin 3-glucoside (*48*). The most likely reason for the observed decrease in Pinotin A content in wines 6 years and older is that the concentration of malvidin 3-glucoside and consequently the rate of Pinotin A synthesis in these wines is too low to compensate for the constant incorporation of Pinotin A into polymeric pigments.

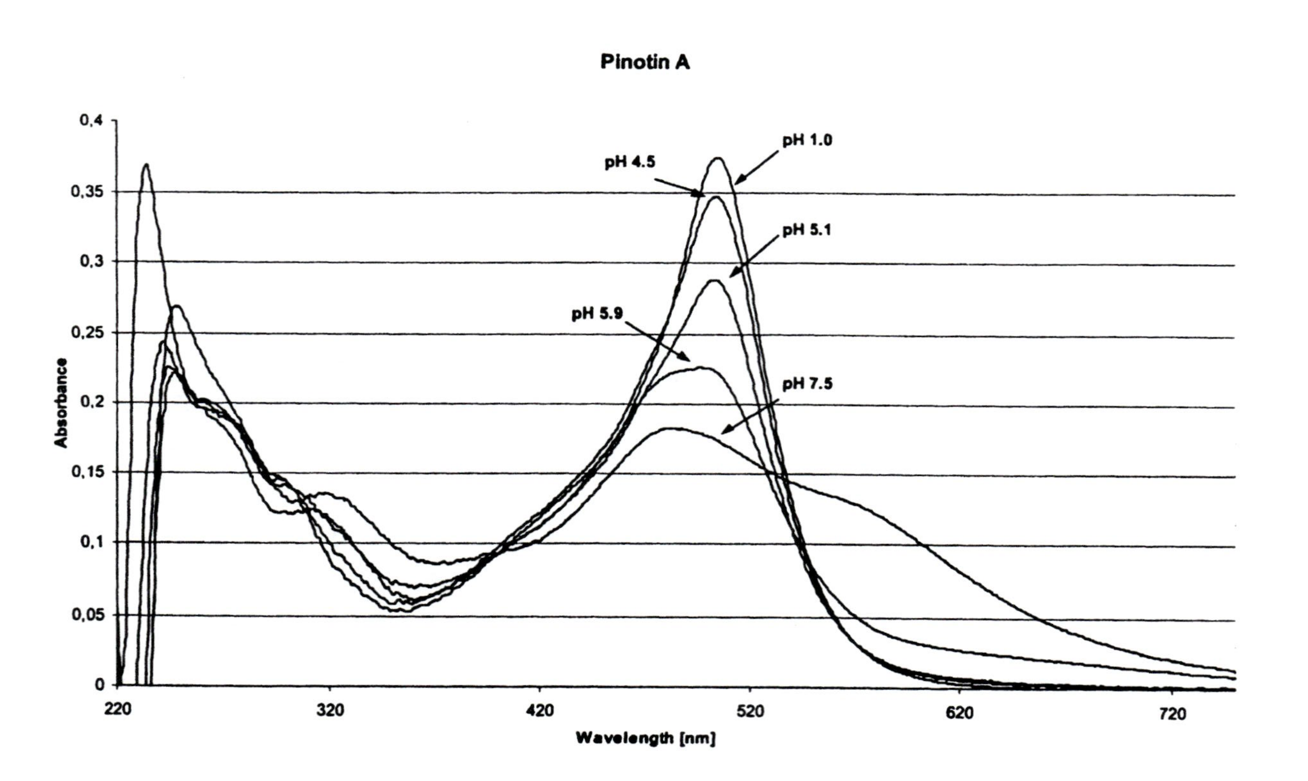

Pinotin A
pH 1.0
pH 4.5
pH 5.1
pH 5.9
pH 7.5
Absorbance
0,4
0,35
0,3
0,25
0,2
0,15
0,1
0,05
0
220
320
420
520
620
720
Wavelength [nm]

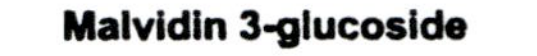

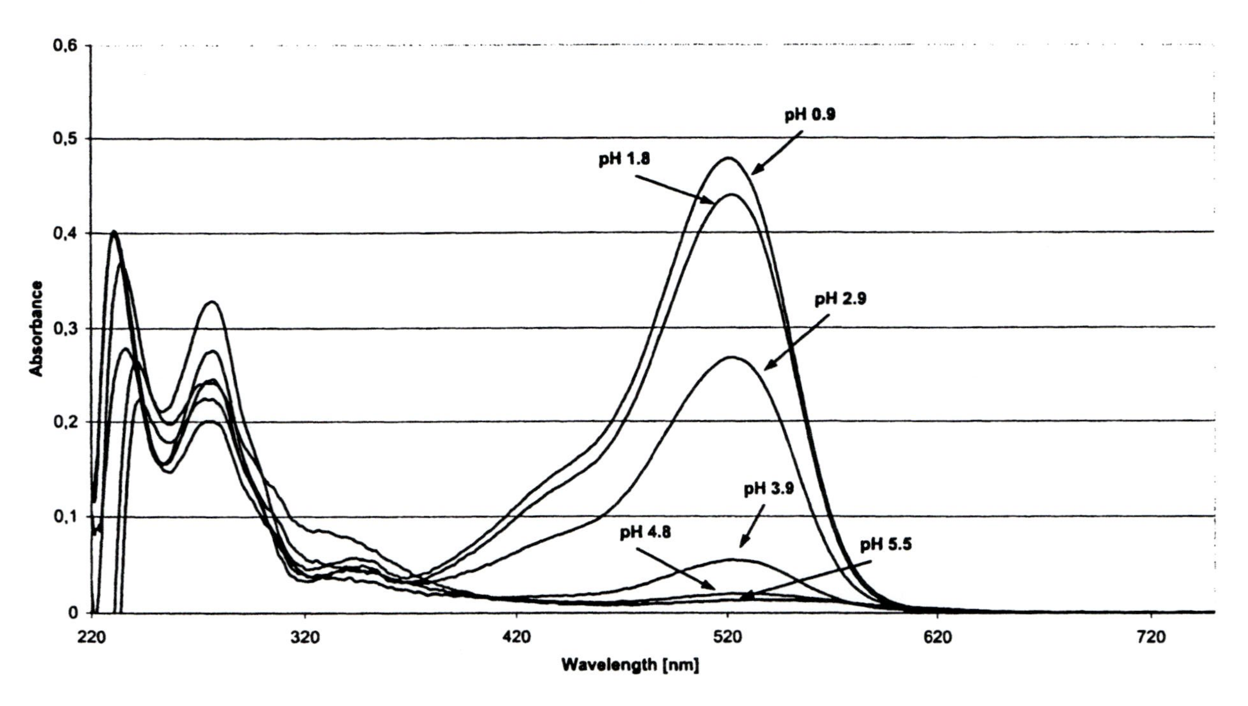

Figure 6. Stability of Pinotin A (top) at different pH values compared to malvidin 3-glucoside (bottom).

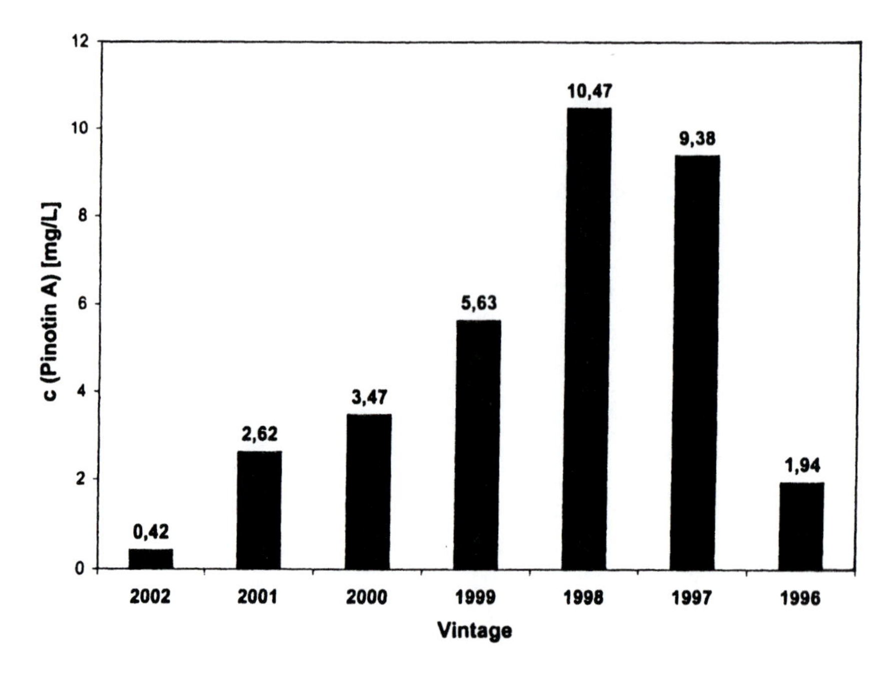

Figure 7. Concentration of Pinotin A in Pinotage wines (50 samples) from the years 2002-1996

Relevance for Copigmentation

In a recent study, the effect of pre-fermentation additions of different doses of caffeic acid on the color of red wine has been reported. An increase of up to 75 % in color was observed at the end of fermentation, a further increase was measurable during storage (*49*). Besides the hyperchromic effect a bathochromic effect was observed for the red grape cultivar Listán negro. The stabilization of color was solely interpreted by assuming the formation of non-covalent anthocyanin-copigment complexes in a ratio 1:1. With our present knowledge about the reaction between hydroxy cinnamic acids and anthocyanins, another explanation for the observed copigmentation phenomenon can be given - namely the formation of color stable pyranoanthocyanins – such as Pinotin A. Further work is underway in order to identify additional covalent structures involved in color stabilization during wine aging.

Summary

The presented pathway of stable pigment formation starting from free hydroxycinnamic acids is with utmost certainty responsible for the formation of the vast majority of anthocyanin-vinylphenol adducts in red wine and also explains part of the effects observed in copigmentation experiments. The reaction rate in wines is expected to be slower compared to our model solutions because of the lower concentration of the reactants. This makes the vinylphenol derived pigments potentially attractive to be used as aging indicators for red wines as their concentration will constantly increase for approximatly the first five years of storage or as long as sufficient amounts of free anthocyanins and cinnamic acids are available. During wine storage, the latter can be constantly replenished through slow hydrolysis of the corresponding tartaric esters, which are usually present in higher concentrations than the free acids.

Acknowledgements

We gratefully acknowledge the support of G. Jerz, M. Messerer and R. Waibel in performing the NMR experiments. We also thank H. Otteneder and M. Zimmer of LUA Trier for helpful discussions and for supplying some of the Pinotage samples. This work was supported by the FEI (Forschungskreis der Ernährungsindustrie e.V., Bonn), the AiF and the Ministry of Economics and Labour. Project No.: AiF-FV 12896 N.

References

1. Somers, T. C. *Phytochemistry* **1971,** *10,* 2175-2186.
2. Francia-Aricha, E. M.; Guerra, M. T.; Rivas-Gonzalo, J. C.; Santos-Buelga, C. *J. Agric. Food Chem.* **1997,** *45,* 2262-2266.
3. Rivas-Gonzalo, J. C.; Bravo-Haro, S.; Santos-Buelga, C. *J. Agric. Food Chem.* **1995,** *43,* 1444-1449.
4. Mateus, N.; Silva, A. M. S.; Santos-Buelga, C.; Rivas-Gonzalo, J. C.; de Freitas, V. *J. Agric. Food Chem.* **2002,** *50,* 2110-2116.
5. Bakker, J.; Bridle, P.; Honda, T.; Kuwano, H.; Saito, N.; Terahara, N.; Timberlake, C. F. *Phytochemistry* **1997,** *44,* 1375-1382.
6. Fulcrand, H.; Benabdeljalil, C.; Rigaud, J.; Cheynier, V.; Moutounet, M. *Phytochemistry* **1998,** *47,* 1401-1407.
7. Cameira dos Santos, P.-J.; Brillouet, J.-M.; Cheynier, V.; Moutounet, M. *J. Sci. Food. Agric.* **1996,** *70,* 204-208.
8. Fulcrand, H.; Cameira dos Santos, P.-J.; Sarni-Manchado, P.; Cheynier, V.; Favre-Bonvin, J. *J. Chem. Soc., Perkin Trans. 1* **1996,** 735-739.

9. Degenhardt, A.; Hofmann, S.; Knapp, H.; Winterhalter, P. *J. Agric. Food Chem.* **2000,** *48,* 5812-5818.
10. Schwarz, M.; Hillebrand, S.; Habben, S.; Degenhardt, A.; Winterhalter, P. *Biochem. Eng. J.* **2003,** *14,* 179-189.
11. Degenhardt, A.; Knapp, H.; Winterhalter, P. *Vitis* **2000,** *39,* 43-44.
12. Schwarz, M.; Wabnitz, T. C.; Winterhalter, P. *J. Agric. Food Chem.* **2003,** *51,* 3682-3687.
13. Braun, S.; Kalinowski, H.-O.; Berger, S. *150 and more basic NMR experiments. A practical course.*, 2nd ed.; Wiley-VCH: Weinheim, Germany, 1998.
14. Hayasaka, Y.; Asenstorfer, R. E. *J. Agric. Food Chem.* **2002,** *50,* 756-761.
15. Schwarz, M.; Jerz, G.; Winterhalter, P. *Vitis* **2003,** *42,* 105-106.
16. Håkansson, A. E.; Pardon, K.; Hayasaka, Y.; de Sa, M.; Herderich, M. *Tetrahedron Lett.* **2003,** *44,* 4887-4891.
17. Bücking, W. Ph.D. thesis, University of Leipzig, Germany, 1928.
18. Sarni-Manchado, P.; Fulcrand, H.; Souquet, J.-M.; Cheynier, V.; Moutounet, M. *J. Food Sci.* **1996,** *61,* 938-941.
19. Etievant, P. X. *J. Agric. Food Chem.* **1981,** *29,* 65-67.
20. Baumes, R.; Cordonnier, R.; Nitz, S.; Drawert, F. *J. Sci. Food. Agric.* **1986,** *37,* 927-943.
21. Chatonnet, P.; Dubourdieu, D.; Boidron, J.-N.; Lavigne, V. *J. Sci. Food. Agric.* **1993,** *62,* 191-202.
22. Lao, C. L.; López-Tamames, E.; Lamuela-Raventós, R. M.; Buxaderas, S.; de la Torre-Boronat, M. d. C. *J. Food Sci.* **1997,** *62,* 1142-1144, 1149.
23. Dugelay, I.; Gunata, Z.; Sapis, J.-C.; Baumes, R.; Bayonove, C. *J. Agric. Food Chem.* **1993,** *41,* 2092-2096.
24. Rodríguez-Delgado, M. A.; González-Hernández, G.; Conde-González, J. E.; Pérez-Trujillo, J. P. *Food Chem.* **2002,** *78,* 523-532.
25. Soleas, G. J.; Dam, J.; Carey, M.; Goldberg, D. M. *J. Agric. Food Chem.* **1997,** *45,* 3871-3880.
26. Landrault, N.; Poucheret, P.; Ravel, P.; Gasc, F.; Cros, G.; Teissedre, P. L. *J. Agric. Food Chem.* **2001,** *49,* 3341-3348.
27. Okamoto, Y.; Inukai, T.; Brown, H. C. *J. Am. Chem. Soc.* **1958,** *80,* 4969-4972.
28. Brown, H. C.; Okamoto, Y. *J. Am. Chem. Soc.* **1958,** *80,* 4979-4987.
29. Reddy, C. S.; Sundaram, E. V. *Tetrahedron* **1989,** *45,* 2109-2126.
30. Sabapathy Mohan, R. T.; Gopalakrishnan, M.; Sekar, M. *Tetrahedron* **1994,** *50,* 10945-10954.
31. Aruna, K.; Manikyamba, P. *Indian J. Chem., Sect. A* **1995,** *34,* 822-825.

32. Lee, D. G.; Brown, K. C. *J. Am. Chem. Soc.* **1982,** *104,* 5076-5081.
33. Ogino, T.; Watanabe, T.; Matsuura, M.; Watanabe, C.; Ozaki, H. *J. Org. Chem.* **1998,** *63,* 2627-2633.
34. Roehri-Stoeckel, C.; Gonzalez, E.; Fougerousse, A.; Brouillard, R. *Can. J. Chem.* **2001,** *79,* 1173-1178.
35. Ishii, Y.; Nakayama, K.; Takeno, M.; Sakaguchi, S.; Iwahama, T.; Nishiyama, Y. *J. Org. Chem.* **1995,** *60,* 3934-3935.
36. Fernandez, I.; Pedro, J. R.; Rosello, A. L.; Ruiz, R.; Castro, I.; Ottenwaelder, X.; Journaux, Y. *Eur. J. Org. Chem.* **2001,** 1235-1247.
37. Sakai, A.; Hendrickson, D. G.; Hendrickson, W. H. *Tetrahedron Lett.* **2000,** *41,* 2759-2763.
38. Nikalje, M. D.; Sudalai, A. *Tetrahedron* **1999,** *55,* 5903-5908.
39. Ohshiro, H.; Mitsui, K.; Ando, N.; Ohsawa, Y.; Koinuma, W.; Takahashi, H.; Kondo, S.; Nabeshima, T.; Yano, Y. *J. Am. Chem. Soc.* **2001,** *123,* 2478-2486.
40. Fukuzumi, S.; Itoh, S.; Komori, T.; Suenobu, T.; Ishida, A.; Fujitsuka, M.; Ito, O. *J. Am. Chem. Soc.* **2000,** *122,* 8435-8443.
41. Napolitano, A.; d'Ischia, M. *J. Org. Chem.* **2002,** *67,* 803-810.
42. Ishii, Y.; Sakaguchi, S.; Iwahama, T. *Adv. Synth. Catal.* **2001,** *343,* 393-427.
43. Mazza, G.; Brouillard, R. *J. Agric. Food Chem.* **1987,** *35,* 422-426.
44. Brouillard, R. In *Anthocyanins as food colors;* Markakis, P., Ed.; Academic Press: New York, 1982; pp 1-40.
45. Schwarz, M.; Winterhalter, P. *Tetrahedron Lett.* **2003,** *44,* 7583-7587.
46. Hofmann, T. *J. Agric. Food Chem.* **1998,** *46,* 3912-3917.
47. Schwarz, M.; Quast, P.; von Baer, D.; Winterhalter, P. *J. Agric. Food Chem.* **2003,** *51,* 6261-6267.
48. Schwarz, M.; Hofmann, G.; Winterhalter, P. *J. Agric. Food Chem.* **2004,** in press.
49. Darias-Martín, J.; Martín-Luis, B.; Carrillo-López, M.; Lamuela-Raventós, R.; Díaz-Romero, C.; Boulton, R. *J. Agric. Food Chem.* **2002,** *50,* 2062-2067.

Chapter 12

Anthocyanin Transformation in Cabernet Sauvignon Wine during Aging

Haibo Wang, Edward J. Race, and Anil J. Shrikhande

Research and Development, Canandaigua Wine Company, 12667 Road 24, Madera, CA 93639

Anthocyanins in Cabernet Sauvignon grapes and wines were elucidated by HPLC–MS/MS. Major anthocyanins in Cabernet Sauvignon grape extract are malvidin 3-*O*-glucoside and malvidin 3-*O*-acetylglucoside. In matured wine, anthocyanins are transformed to pyranoanthocyanins, ethyl bridged anthocyanin-flavanol adducts, and anthocyanin-flavanol adducts. The major anthocyanin pigments identified are 5-carboxyl-pyranomalvidin 3-*O*-glucoside (vitisin A), 5-carboxyl-pyranomalvidin 3-*O*-acetylglucoside (acetyl-vitisin A), 5-carboxyl-pyranomalvidin 3-*O*-coumaroylglucoside (coumaroyl-vitisin A), 5-(4′-hydroxyphenyl)-pyranomalvidin 3-*O*-glucoside, 5- (4′-hydroxyphenyl)-pyranomalvidin 3-*O*-acetylglucoside, and 5-(4′-hydroxyphenyl)- pyranomalvidin 3-*O*-coumaroylglucoside. The presence of syringetin 3-*O*-glucoside and syringetin 3-*O*-acetylglucoside has been established for the first time in grape and wine.

INTRODUCTION

During red wine aging, anthocyanins undergo reactions to form polymeric pigments. This transformation was thought to be responsible for the color change during wine maturation. Three mechanisms have been postulated to be involved in the reaction (*1*). The first two mechanisms involve direct reaction between proanthocyanidin and anthocyanin, generating colorless compounds or orange xanthylium salts (*2–8*). The third mechanism involves acetaldehyde-mediated condensation of anthocyanin and proanthocyanidin, yielding purple pigments with an ethyl bridge (*9–14*). Glyoxylic acid, furfural and HMF were found to react in the same way as acetaldehyde-mediated condensation with anthocyanins or flavanols in the model systems (*15–17*). Recent studies further revealed the presence of new pigments other than previously suggested. These pigments were formed from a new type of reaction between anthocyanin and compounds with a polarizable double bond (*18–21*). The mechanism was postulated to be a cycloaddition reaction at carbon 4 and the hydroxyl group of carbon 5 of an anthocyanin, with various components possessing a polarizable double bond (*22*). In this reaction, a fourth ring was formed, which was considered to be responsible for their higher stability than the original anthocyanins (*20*). The components which reacted with anthocyanins include 4-vinylphenol, hydroxycinnamic acids, acetone, and several yeast metabolites such as acetaldehyde and pyruvic acid (*20-28*).

Direct evidence for the reaction between anthocyanin and flavanol was recently identified in matured Cabernet Sauvignon wine (*6*, *11*, *29*). Evidence of the existence of cycloaddition products was reported in matured port wine (*19*), grape pomace (*21*) and aged Shiraz and Cabernet Sauvignon wines (*23*, *25*, *30–31*). Formation of anthocyanin-proanthocyanidin adducts was commonly proposed to explain the loss of astringency during wine maturation, and wine quality was speculated to have something to do with anthocyanin to procyanidin ratio due to interaction between anthocyanins and proanthocyanidins during wine maturation (*22*). However, complete anthocyanin profiles in matured wine was not reported. Little is known about mechanisms of anthocyanin transformation, and the roles of anthocyanins in wine's reduced astringency during wine maturation.

The objective of this work is to establish anthocyanin profiles and individual anthocyanins in Cabernet Sauvignon wines in selected vintages by LC-MS/MS, and to elucidate the mechanisms of anthocyanin transformation. Based on anthocyanin profiles established, the involvement of anthocyanins in wine's reduced astringency during maturation is discussed.

MATERIALS AND METHODS

Wines and Grapes. *Vitis vinifera* var. Cabernet Sauvignon grape samples were harvested from different viticultural areas in Central Valley (Fresno-Madera), North Valley (Lodi-Sacramento) and Napa Valley, CA. Young wines were prepared in Research and Development, Canandaigua Wine Company (Madera, CA) following standard wine making procedures from those grapes. Thirty commercial Cabernet Sauvignon wines from different viticultural areas in California at vintage years ranging from 1997 to 2002 were purchased from local supermarkets.

Standards. Malvidin 3, 5-*O*-diglucoside was purchased from Aldrich (Saint Louis, MO), Cyanidin-3-*O*-glucoside, malvidin 3-*O*-galactoside and syringetin 3-*O*-glucoside were purchased from Indofine (Hillsborough, NJ).

Extraction of anthocyanins in Cabernet Sauvignon wines. Five milliliters of Cabernet sauvignon wine was passed separately through a C-18 Sep-Pak cartridge preconditioned with acidified methanol and water (Waters Corporation). The adsorbed pigments were washed with 5 mL of water and eluted with 2 mL of 0.01% HCl methanol. The eluant was stored at -20 °C prior to HPLC analysis. Wine samples were passed through 0.45 μM glass microfiber filter (GMF) syringe filter (Whatman Inc, Clifton, NJ) before injected onto HPLC.

Extraction of anthocyanins in Cabernet Sauvignon grape skins. Cabernet Sauvignon grape skins (50g) were extracted twice with 100 mL of methanol. Extract was combined and evaporated under reduced pressure to remove methanol. Residue was reconstituted to 20 mL with water. Five milliliters of water solution was passed through a preconditioned C-18 Sep-Pak cartridge (Waters Corporation). The adsorbed pigments were washed with 5 mL of water and eluted with 2 mL of 0.01% HCl methanol. The eluant was stored at -20 °C prior to HPLC analysis. Cabernet Sauvignon grape skins were also extracted with 75% acetone with 0.2% acetic acid to compare anthocyanin profiles in different extraction procedures.

Preparation of model solution. Malvidin 3, 5-diglucoside (10,00 mg/L) and catechin (4,400 mg/L) in the presence of acetaldehyde (0.1%, v/v) were prepared in a medium that contained 5 % tartaric acid (w/v) and 10% ethanol (v/v) and were adjusted to pH 3.2 with NaOH. The mixtures were stored at 20°C in sealed dark glass vials. Samples were taken periodically and diluted 1→5 with model wine solution before injected onto HPLC.

HPLC/DAD/ESI-MS/MS analyses: LC/ESI-MS/MS experiments were performed on Agilent 1100 LC/MSD Trap-SL mass spectrometer (Palo Alto, CA) equipped with an electrospray ionization (ESI) interface, 1100 HPLC, a DAD detector and Chemstation software (Rev.A.09.01). Anthocyanins were monitored at 520 nm. The column used was a 150 cm × 2.0 mm i.d., 4 μm

Synergi hydro-RP 80 Å (Phenomenex, Torrance, CA). Solvents were: (A) 10% acetic acid/0.1% TFA/5% acetonitrile/84.8%water (v/v/v/v), and (B) acetonitrile. Solvent gradient was 0-30 minute, 0-10% B; 30-40 min, 10-30% B; 40-50 min, 30-40% B. Flow rate was 0.2 mL/min and injection volume was 3 μL, and column temperature was 25 °C. The ESI parameters were: nebulizer: 30 psi; dry gas (N_2): 12 L/min; dry temp: 350 °C; trap drive: 50; skim 1: 40 v; skim 2: –5.0 v; octopole RF amplitude: 150 vpp; capillary exit: 103.4 v. The ion trap mass spectrometer was operated in positive ion mode scanning from *m/z* 150 to *m/z* 2000 at a scan resolution of 13,000 amu/s. Trap ICC was 20,000 units and maximal accumulation time was 200 ms. MS-MS was operated at a fragmentation amplitude of 1.2V and threshold ABS was 3,000,000 units.

RESULTS AND DISCUSSION

Anthocyanin profiles in Cabernet Sauvignon wine and grape skin extract were established and elucidated by LC-MS/MS analysis. Molecular ions and product ions of anthocyanins were included in **Table 1.** After comparing anthocyanin profiles of grape skin extracts using different extraction solvent system with that of grape juice, several artificial peaks were found to be generated using 75% acetone with 0.2% acetic acid as extracting solvent. These artificial peaks with trace quantity might be due to interaction between acetone and anthocyanins at acidic media as previously reported (*24*). Therefore, 100% methanol was selected as extracting solvent. Anthocyanin profiles of grape skin extract with 100% methanol from Central Valley, North Valley and Napa Valley were identical, even though the ratio of individual anthocyanins in these three grape skin extracts was different. Grape skin extract with 100% methanol from Napa Valley was selected as the reference to illustrate anthocyanin transformation during wine making and wine maturation. Anthocyanin profile of Cabernet Sauvignon skin extract from Napa Valley was included in **Figure 1A**. Anthocyanins in Cabernet Sauvignon grape skin extract were in agreement with previous reports (*32–33*). However, peaks 4, 8 and 15 with trace quantities were not included in the previous reports. These three peaks had the same molecular and product ions as malvidin 3-*O*-glucoside (peak 6), malvidin 3-*O*-acetylglucoside (peak 11) and malvidin 3-*O*-coumarylglucoside (peak 17), respectively. Also, peaks 4, 8, and 15 were eluted earlier than their glucosidic counterparts. UV-vis spectra also could not distinguish differences among peaks 4, 8, 15 from peaks 6, 11, 17. These results suggested that the differences between peaks 4, 8, 15 and peaks 6, 11, 17 were glycosidic pattern. Malvidin 3-*O*-galactoside along with malvidin-3-*O*-glucoside was identified in black bilberry and blue berry extracts, and anthocyanin 3-*O*-galactosides were eluted earlier on reverse-phase HPLC than their glucosidic counterparts (*34-35*). Peaks

Table 1. Molecular ions and product ions of anthocyanins in Cabernet Sauvignon wine and grape skin extract

Peak	Retention time	Compounds	Molecular and product ions
1	5.8	delphinidin 3-*O*-glucoside	465 (M^+), 303
2	8.9	cyanidin 3-*O*-glucoside	449 (M^+), 287
3	12.3	petunidin 3-*O*-glucoside	479 (M^+), 317
4	15.0	malvidin 3-O-galactoside	493 (M^+), 331
5	18.4	peonidin 3-*O*-glucoside	463 (M^+), 301
6	20.7	malvidin 3-*O*-glucoside	493 (M^+), 331
7	25.5	delphinidin 3-*O*-acetylglucoside	507 (M^+), 303
8	28.8	malvidin 3-*O*-acetylgalactoside	535 (M^+), 331
9	34.4	petunidin 3-*O*-acetylglucoside	521 (M^+), 317
10	39.6	peonidin 3-*O*-acetylglucoside	505 (M^+), 301
11	40.2	malvidin 3-*O*-acetylglucoside	535 (M^+), 331
12	41.3	peonidin 3-*O*-caffeoylglucoside	625 (M^+), 301
13	41.5	malvidin 3-*O*-caffeoylglucoside	655 (M^+), 331
14	42.1	petunidin 3-*O*-coumaroylglucoside	625 (M^+), 317
15	42.3	malvidin 3-*O*-coumaroylgalactoside	639 (M^+), 331
16	43.6	peonidin 3-*O*-coumaroylglucoside	609 (M^+), 301
17	43.8	malvidin 3-*O*-coumaroylglucoside	639 (M^+), 331
18	4.4	malvidin-3-*O*-glucoside-(epi)-catechin	781(M^+), 619, 467
19	7.1	5-carboxyl-pyranodelphinidin 3-*O*-glucoside	533 (M^+), 371
20	15.1	5-carboxyl-pyranopetunidin 3-*O*-glucoside	547 (M^+), 385
21	24.9	5-carboxyl-pyranomalvidin 3-*O*-glucoside	561 (M^+), 399

22	28.4	5-carboxyl-pyranomalvidin 3-*O*-acetylglucoside	603 (M^+), 399
23	29.4	pyranomalvidin 3-*O*-glucoside	517 (M^+), 355
24	35.6	syringetin 3-O-glucoside	509 (M^+), 347
25	36.7	5-methyl-pyranomalvidin 3-*O*-glucoside	531 (M^+), 369
26	37.0	malvidin 3-*O*-glucoside-ethyl-(epi)catechin	809 (M^+), 519, 357
27	39.1	malvidin 3-*O*-coumaroylglucoside-pyruvate	707 (M^+), 399
28	41.5	malvidin 3-*O*-acetylglucoside-ethylflavanol	851 (M^+), 561, 357
29	42.2	syringetin 3-*O*-acetylglucoside	551 (M^+), 347
30	42.4	malvidin 3-*O*-glucoside-4-vinyl-(epi)catechin	805 (M^+), 643, 491
31	43.6	5-(4′-hydroxyphenol)-pyranopetunidin 3-*O*-glucoside	595 (M^+), 433
32	45.0	5-(4′-hydroxyphenol)-pyranomalvidin 3-*O*-glucoside	609 (M^+), 447
33	45.3	5-(3′,4′-hydroxyphenol)-pyranomalvidin 3-*O*-glucoside	625 (M^+), 463
34	46.2	5-(4′-hydroxyphenol)-pyranomalvidin 3-*O*-acetylglucoside	651 (M^+), 447
35	47.2	5-(4′-hydroxyphenol)-pyranomalvidin 3-*O*-coumaroylglucoside	755 (M^+), 447

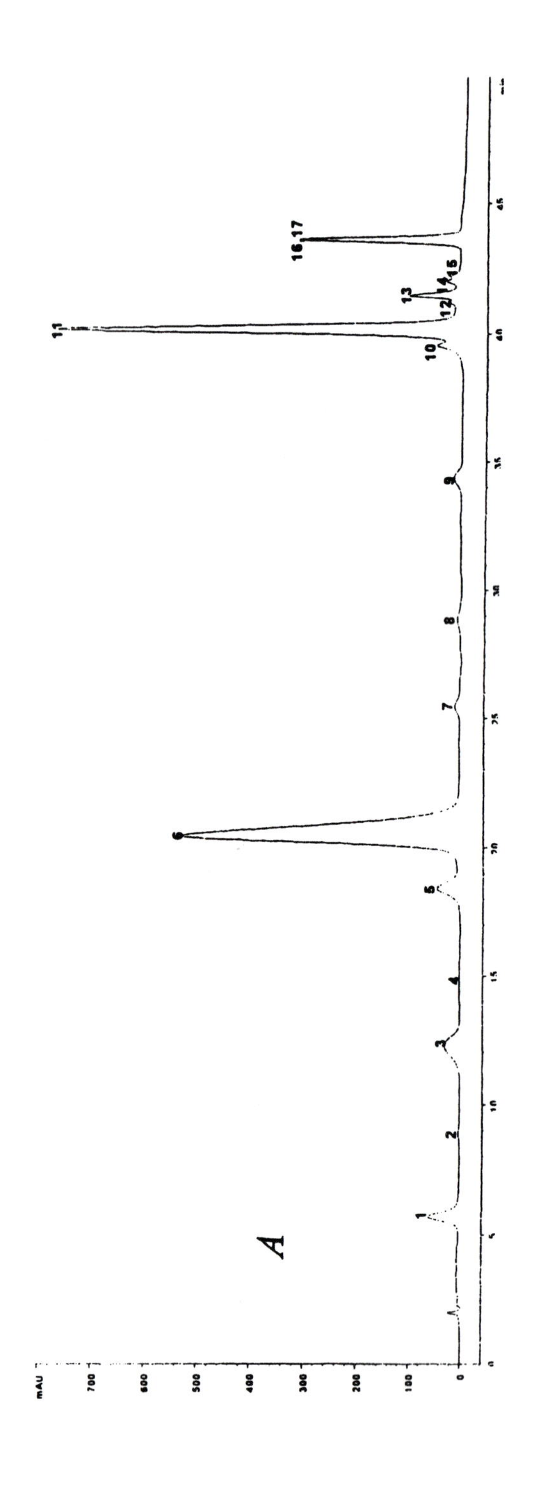
A
mAU
700
600
500
400
300
200
100
0
1
2
3
4
5
6
7
8
9
10
11
12
13
14
15
16,17

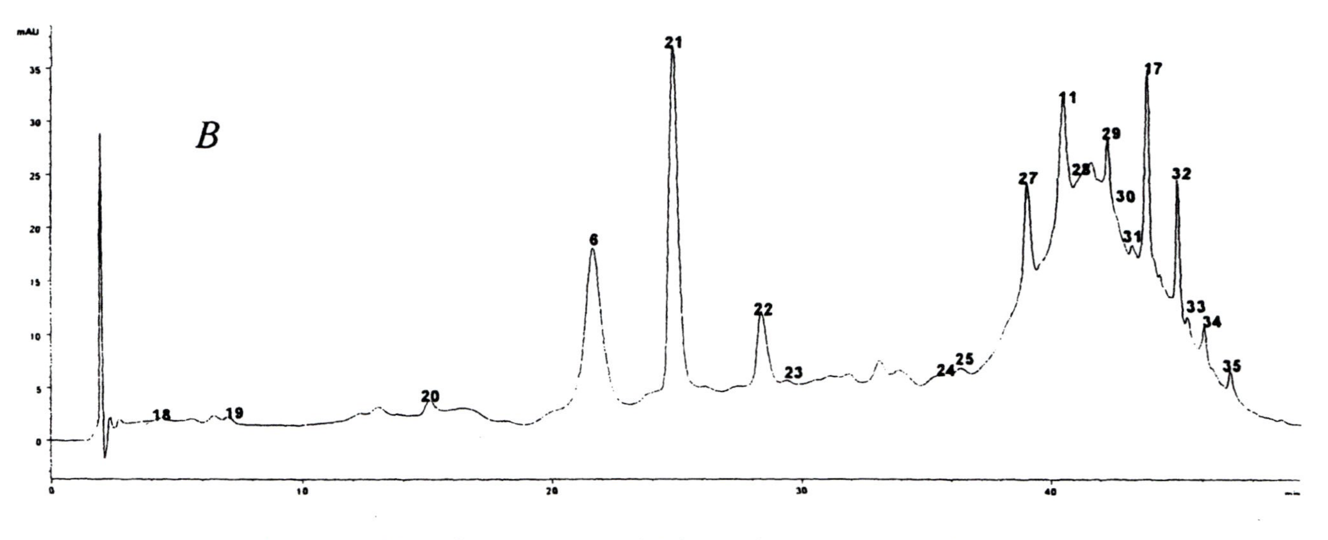

Figure 1. HPLC chromatograms of Cabernet Sauvignon grape skin extract (A); young wine Cabernet Sauvignon wine make in R&D (B); 1997 Cabernet Sauvignon wine from Napa valley (C).

Continued on next page.

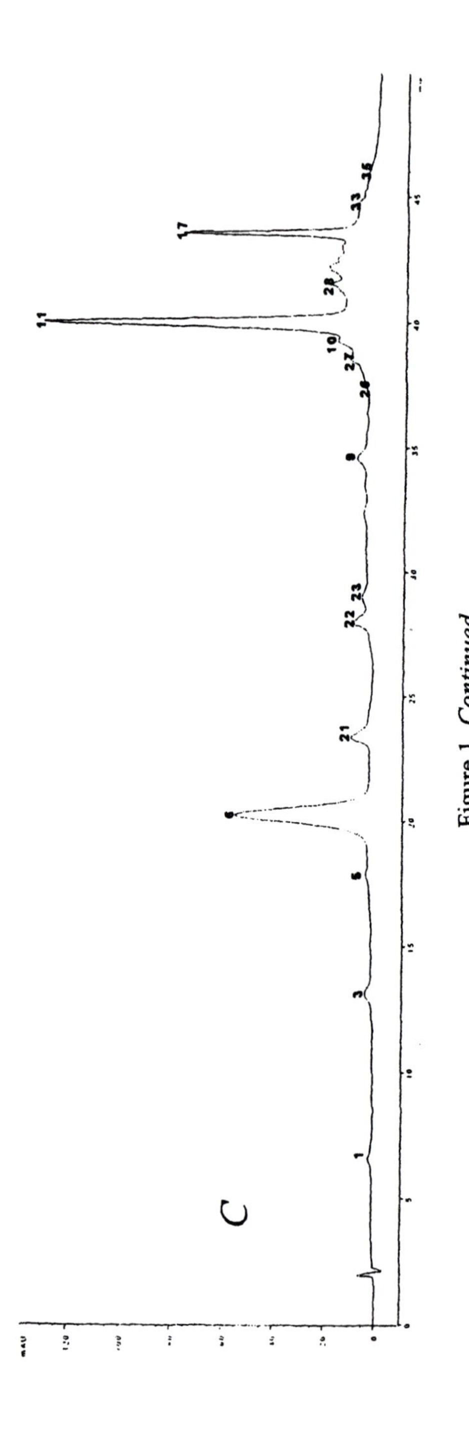

Figure 1. *Continued.*

4, 8 and 15 could well be malvidin 3-*O*-galactoside, malvidin 3-*O*-acetylgalactoside and malvidin 3-*O*-coumaroylgalactoside, respectively. Peak 4 was further confirmed by comparison of malvidin 3-*O*-galactoside chloride standard based on their retention times and mass spectral.

During winemaking and wine maturation, new anthocyanin pigments were developed and original anthocyanins were decreased. In 30 commercial Cabernet Sauvignon wines analyzed, anthocyanin profiles were very similar. However, matured wines had a higher percentage of newly formed anthocyanin-derived pigments than young wines even though absolute amount of these anthocyanin-derived pigments are lower (**Table 2**). Therefore, Cabernet Sauvignon wine from Napa Valley in 1997 and two-month old Cabernet Sauvignon wine made from Napa Valley grapes in Research and Development, Canandaigua Wine Company, Inc., were selected as representatives of matured and young wine, respectively. During winemaking, new anthocyanin pigments (peaks 18 –35) were formed (**Figure 1B**). With the progress of wine maturation, these anthocyanin-derived pigments became dominant (Peaks 21, 22, 27, 32 and 34) (**Figure 1C**). Peak 21, with molecular ion at *m/z* 561 and product ion at *m/z* 399, was assigned to be 5-carboxyl-pyranomalvidin 3-*O*-glucoside (Vitisin A) (*20–21*). Similarly, peaks 19 and 20 were 5-carboxyl-pyranodelphinidin 3-*O*-glucoside and 5-carboxyl-pyranopetunidin 3-*O*-glucoside. Peaks 22 and 27 were 5-carboxyl-pyranomalvidin 3-*O*-acetylglucoside (acetyl-vitisin A) and 5-carboxyl-pyranomalvidin 3-*O*-coumaroylglucoside (coumaroyl-vitisin A). Peak 23 was identified as pyranomalvidin 3-*O*-glucoside (vitisin B) after compared with reference data (*20*). Peaks 32, 34 and 35, had product ions at *m/z*=447, were elucidated as 5-(4′-hydroxyphenyl)-pyranomalvidin 3-*O*-glucoside, 5-(4′-hydroxyphenyl)-pyranomalvidin 3-*O*-acetylglucoside and 5-(4′-hydroxyphenyl)-pyranomalvidin 3-*O*-coumaroylglucoside, respectively (25, 36). Peak 33, had molecular and product ions at *m/z* 625 and 463, was elucidated as 5-(3′,4′-dihydroxyphenyl)-pyranomalvidin 3-*O*-glucoside, which agreed with previous reports (*25, 28*). However, it is important to point out that the concentration of 5-(3′,4′-dihydroxyphenyl)-pyranomalvidin 3-*O*-glucoside in wine was much lower than 5-(4′-hydroxyphenyl)-pyranomalvidin 3-*O*-glucoside, even though significant amount of caffeic acid was monitored in the wine and grape extract (data not shown). The explanation can be that *p*-coumaric acid, but not caffeic acid, was decarboxylated during fermentation (37), and reaction between malvidin 3-glucoside and decarboxylated product of *p*-coumaric acid, 4-vinylphenol, went to completion rather fast (18, 28). The reaction of caffeic acid and malvidin 3-O-glucoside took place directly during wine storage, and reaction proceeded very slowly (28).

Peak 18 at RT 4.4 min had molecular ion at *m/z* 781, and product ions at *m/z* 619 and *m/z* 467. Product ion at *m/z* 619 was a fragment of *m/z* 781 by loss of a glucose and *m/z* 467 was a fragment of *m/z* 619 resulting from retro-Diels-Alder

Table 2: Major anthocyanins in selected Cabernet Sauvignon wines[a]

Vintage	M-3-G	PM-3-G	PM-3-AG	M-3-AG	M-3-CG	VM-3-G	polymers	total
1997	3	3	1	0	2	3	111	175
1997	23	12	3	4	3	2	91	138
1997	17	6	1	2	4	3	97	165
1997	24	5	1	5	4	1	86	269
1998	24	3	1	6	4	4	76	130
1998	63	12	4	20	6	2	86	208
1998	72	6	2	22	7	2	59	224
1998	29	3	0	7	3	3	57	161
1999	44	4	2	13	6	3	74	170
1999	28	4	1	7	4	2	76	146
1999	54	3	1	16	6	3	81	121
1999	53	3	1	14	7	3	70	126
1999	88	7	3	25	9	2	108	151
1999	58	4	2	17	5	2	87	241
1999	51	5	1	14	5	3	83	124
1999	93	7	3	27	10	2	82	101
1999	78	3	1	20	8	2	95	193
1999	70	3	1	19	8	1	62	119
2000	128	4	1	32	12	4	87	164
2002	115	20	14	64	5	ND	198	416
2002	211	20	14	102	13	ND	168	528
2002	79	18	12	44	3	ND	137	293
2002	274	14	9	128	20	ND	108	533

[a] Quantity of individual anthocyanins was expresses as cyanidin 3-glucoside equivalent. Malvidin 3-glucoside (M-3-G); Malvidin 3-acetylglucoside (M-3-AG); Malvidn 3-coumaroylglucoside (M-3-CG); 5-carboxyl-pyranomalvidin 3-glucoside (PM-3-G); 5-carboxyl-pyranomalvidin 3-acetylglucoside (PM-3-AG), 5-(4′-hydroxyphenol)-pyranomalvidin 3-glucoside (VM-3-G). ND, not determined.

cleavage (−152 amu) in (+)-catechin or (−)-epicatechin (38) (**Figure 2**). Since MS data itself could not distinguish the isomers of a compound, peak 18 was assigned to be 5-malvidin 3-*O*-glucoside-(+)-catechin or malvidin 3-*O*-glucoside-(−)-epicatechin. Peak 30 at RT 42.4 with molecular ion at *m/z* 805 and product ions at *m/z* 643 and 491 was assigned as either 5-(+)-catechin-yl-pyranomalvidin 3-*O*-glucoside or 5-(−)-epicatechin-yl-pyranomalvidin 3-*O*-glucoside based on report recently in the literature (26). Peak 26 had a molecular ion at *m/z* 809, and product ions at *m/z* 519 and *m/z* 357. Product ion at *m/z* 519 was a fragment from *m/z* 809 by loss of either (+)-catechin or (−)-epicatechin. Product ion at *m/z* 519 could be further fragmented to yield a product ion at *m/z* 357 by loss of a glucose moiety (**Figure 3**). Peak 26 was suggested to be malvidin 3-*O*-glucoside-ethyl-(+)-catechin or malvidin 3-*O*-glucoside-(−)-epicatechin (29, 39). In the model solution containing catechin, malvidin 3, 5-diglucoside and acetaldehyde, two major compounds were produced. Two enantiomers had identical molecular and products ion at *m/z* 971, 681, 519 and 357, which corresponded to a structure in which malvidin 3, 5-diglucoside and catechin were linked by an ethyl bridge. Enantiomers might be due to presence of an asymmetric carbon in the ethyl- bridge (29). The fragmentation pattern of compounds produced in the model system was the same as that of peak 26. In both cases, the C-C bond connected ethyl group to catechin moiety was cleaved before sugar moiety was cleaved. This fragmentation pattern was quite unique because in most cases, only sugar moiety was cleaved under this ESI condition (40-41). Similarly, peak 28 had molecular ion at *m/z* 851 and product ions at *m/z* 561 and 357, which was assigned to be malvidin 3-*O*-acetylglucoside-ethyl-(+)-catechin or malvidin 3-acetylglucoside-ethyl-(−)-epicatechin. To date, this compound has not been reported in the literature.

Peak 24 at RT 35.6 min had molecular ion at *m/z* 509 and product ion at *m/z* 347. Product ion at *m/z* 347 was a fragment from molecular ion at *m/z* 509 by loss of a glucose moiety. This compound did not have maximal absorbance around 520 nm, but it showed strong signal on MS detector. Molecular ion at *m/z* 509 was 16 amu higher than malvidin 3-*O*-glucoside, which was suggested to be a hydroxyl group attached to C-4 of malvidin 3-*O*-glucoside, or an equilibrium form of malvidin 3-*O*-glucoside (25). However, UV spectrum showed this compound had maximal absorbance at 360 nm instead of 520 nm, suggesting a flavonol component instead of an anthocyanin. Peak 24 was then elucidated to be syringetin 3-*O*-glucoside after further comparison with standard. Similarly, peak 29 had molecular and product ions at *m/z* 551 and *m/z* 347. The molecular and product ions of peak 29 were 42 amu higher than those of peak 24, which suggested an acetylglucoside moiety in the compound. Peak 29 was assigned to be syringetin 3-*O*-acetylglucoside. This is the first time confirming the existence of syringetin 3-*O*-glucoside and syringetin 3-*O*-acetylglucoside in grape and wine.

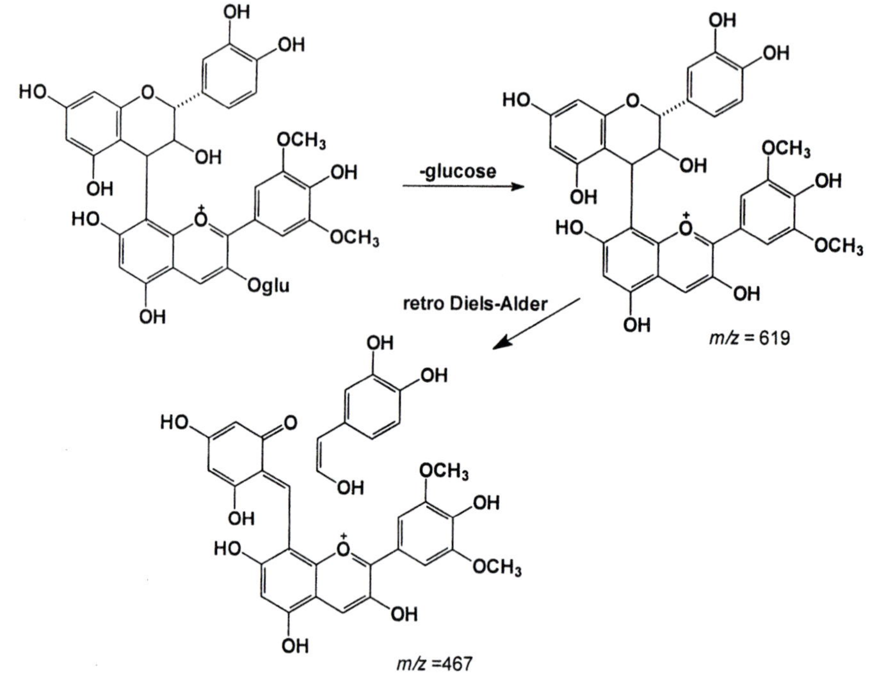

Figure 2. Fragmentation pathway of peak 18

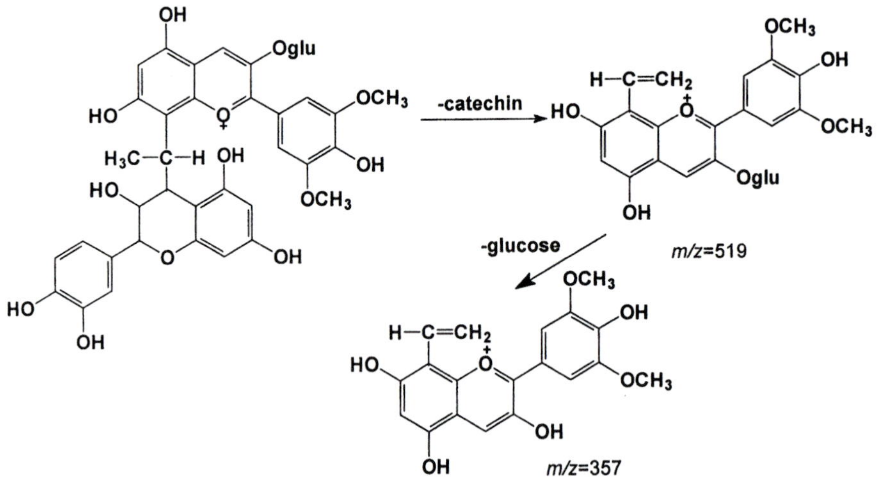

Figure 3. Fragmentation pathway of peak 26

Based on the anthocyanin profiles of Cabernet Sauvignon wines obtained by LC-MS, our data clearly confirmed that the different mechanisms of anthocyanin conversions compete in the course of wine aging, including cycloaddition (peaks 19–23, 27, 30–35), direct reaction (peak 18), and condensation (peaks 26 and 28). In aged Cabernet Sauvignon wines we selected under our conditions, 5-carboxyl-pranomalvidin 3-*O*-glucoside, 5-carboxyl-pyranomalvidin 3-*O*-acetylglucoside, 5-(4′-hydroxyphenol)-pyranomalvidin 3-*O*-glucoside, and 5-(4′-hydroxyphenol)-pyranomalvidin 3-*O*-acetylglucoside were dominant pigments. Recently, researchers further found that the formation of 5-(3′,4′-dihydroxyphenol)-pyranomalvidin 3-*O*-glucoside was resulted from direct interaction between anthocyanins and caffeic acid (28). Since hydroxycinnamic acid, such as caffeic acid, might contribute to bitterness and harshness in wine from sensory evaluation (42), the reaction between anthocyanins and hydroxycinnamic acids might contribute to the reduction of harshness of wine during wine maturation.

The amount of flavanol-yl-pyranoanthocyanin, anthocyanin–flavanol and anthocyanin-ethyl-flavanol adducts identified in wine was not significant. The formation of flavanol-yl-pyranoanthocyanin, anthocyanin-vinylflavanol and anthocyanin–ethyl-flavanol adducts was speculated to follow the acetaldehyde-induced polymerization reaction first proposed by Timberlake et al to give an unstable ethanol adduct (10). The ethanol adduct could then be protonated to form ethyl-flavanol cation or produce vinyl-flavanol by loss of a water molecular. Vinyl-flavanol reacted with malvidin 3-*O*-glucoside to form 5-flavanol-yl-pyranomalvidin-3-*O*-glucoside following the same mechanism as the formation of 5-(4′-hydroxyphenol)-pyranoanthocyanin. Ethyl-flavanol could further react with malvidin 3-*O*-glucoside to form malvidin 3-*O*-glucoside-ethyl-(+)-catechin. Since acetaldehyde was suggested to be produced by the oxidation of ethanol with air in the presence of phenolic compounds during wine maturation (43), a condensation reaction would have occurred only when oxygen was present. Es-Safi et al (44) further discovered that in the mixture containing malvidin 3-*O*-glucoside, epicatechin and acetaldehyde, two reactions competed for the formation between epicatechin-ethyl-epicatechin and malvidin 3-*O*-glucoside-ethyl-epicatechin. Epicatechin-ethyl-epicatechin derivatives were formed first and subsequently transformed to colored derivatives, which seem to be more stable. The formation of the trimeric and tetrameric colored derivatives was also monitored. From a sensory point of view, the formation of anthocyanin-ethyl-flavanol adducts was commonly proposed to explain loss of astringency during wine aging. If ethyl-linked flavanol could be converted to ethyl-flavanol or vinyl-flavanol cation, the reduced astringency of procyanidin could be achieved without oxygen present. In Cabernet wines analyzed, only a trivial amount of individual flavanol-yl-pyranoanthocyanin, anthocyanin–flavanol and anthocyanin-ethyl-flavanol adducts were found in young and matured wines. However, besides peaks identified, a broad band at retention time 37–48 min underneath peaks 27-35 remained to be identified. This broad band had absorbance at around 520 nm, indicating that it contained the flavylium units. This band could be polymeric anthocyanin and flavanol derivatives. This polymeric peak counted for more than 50% of total absorbance at 520 nm (**Table 1**). Interestingly, the young wines made in 2002 from our research facility have consistently higher amount of polymeric peak. However, the nature of the polymeric peak in young wine may be different to that in aged wine since polymer peak in young wine has maximal absorbance at 530-540 nm and polymer peak in aged wine has maximal absorbance at 510-520 nm. This could be that anthocyanin-ethyl-flavanol adducts was first formed during wine making, and then transformed to flavanol-yl-pyranoanthocyanins during wine aging. In a total ion chromatogram, no signals were monitored in that region. which might be due to complexity of these compounds and low concentration of each compound. In addition to monomer flavanols, the acetaldehyde-induced condensation occurred between anthocyanin and proanthocyanidin oligomers or polymers (*44*); this further increased the complexity of the polymer peak. Wines with higher amount of polymers did not tend to be smoother than others with significant amount of 5-carboxyl-pranoanthocyanin and 5-(4′-hydroxyphenol)-pyranoanthocyanin derivatives left from the evaluation of our taste panel (data not shown). This can be explained that after certain stage, free anthocyanin may not be further available to reduce the bitterness and astringency of proanthocyanidins. The only reaction is degradation of anthocyanin derivatives. Other factors, such as direct interaction between anthocyanins and hydroxycinnamic acids, may also play a role in reduced astringency in matured wine. Acetaldehyde-induced polymerization between proanthocyanidins may be another mechanism participating in loss of astringency during wine maturation (45). However, the more logical explanation I believe is that the total polymers may not correlate to sensory character of wine. Only those pyranonthocyanins converted from anthocyanin-ethyl-flavanol adducts may contribute to smoothness of wine. It will be interesting to further separate flavanol-yl-pyranoanthocyanins from anthocyanin-ethyl-flavanol adducts and examine the sensory properties of these components to further address how anthocyanins and their derivatives are involved in reducing astringency during wine maturation.

LITERATURE CITED

1. Ribéreau-Gayon, P. The anthocyanins of grapes and wines, in *Anthocyanins as food colors*, Markakis, P., Ed., Academic Press: New York, NY, 1982; pp. 209-244.
2. Somers, T. C. Wine tannins isolation of condensed flavanoid pigments by gel-filtration. *Nature* **1966**, *209*, 368-370.
3. Jurd, L. Anthocyanins and related compounds XI. Catechin-flavylium salt condensation reaction. *Tetrahedron* **1967**, *23*, 1057-1064
4. Somers, T.C. The polymeric nature of wine pigments. *Phytochemistry* **1971**, *10*, 2175-2186
5. Bishop, P. D.; Nagel, C. W. Characterization of the condensation product of malvidin 3,5-diglucoside and catechin. *J. Agric. Food Chem.* **1984**, *32*, 1022-1026.
6. Remy, S.; Fulcrand, H., Labarbe, B.; Cheynier, V.; Moutounet, M. First confirmation in red wine of products resulting from direct anthocyanin-tannin reactions. *J. Sci. Food Agric.* **2000**, *80*, 745-751.
7. Jurd, L.; Somers, T. C. The formation of xanthylium salts from proanthocyanidins. *Phytochemistry* **1970**, *9*, 419-427
8. Hrazdina, G.; Borzell, A. J. Xanthylium derivatives in grape extracts. *Phytochemistry* **1971**, *10*, 2211-2213
9. Berg, H. W.; Akiyoshi, M. A. On the nature of reactions responsible for color behavior in red wine: a hypothesis. *Am. J. Enol. Vitrc.* **1975**, *26*,134-143.
10. Timberlake, C. F.; Bridle, P. Interactions between anthocyanins, phenolic compounds and acetaldehyde and their significance in red wine. *Am. J. Enol. Vitic.* **1976**, *27*, 97-105.
11. Saucier, C.; Little, D.; Glories, Y. First evidence of acetaldehyde-flavanol condensation products in red wine. *Am. J. Enol. Vitic.* **1997**, *48*, 370-373.
12. García-Viguera, C., Bridle P., Bakker, J. The effect of pH on the formation of colored compounds in model solution containing anthocyanins, catechin and acetaldehyde. *Vitis* **1994**, *33*, 37-40.
13. Dallas, C.; Ricardo-da-Silva, J. M; Laureano, O. Products formed in model wine solutions involving anthocyanins, procyanidin B2, and acetaldehyde. *J. Agric. Food Chem.* **1996**, *44*, 2402-2407.
14. Fulcrand, H.; Cheynier, V.; Oszmianski, J; Moutounet, M. An oxidized tartaric acid residue as a new bridge potentially competing with acetaldehyde in flavan-3-ol condensation. *Phytochemistry* **1997**, *46*, 223-227.
15. Es-Safi, N.; Le Guernevé, C.; Cheynier, V.; Moutounet, M. New polyphenolic compounds with xanthylium skeletons formed through

reaction between (+)-catechin and glyoxylic acid. *J. Agric. Food Chem.* **1999**, *47*, 5211-5217.

16. Es-Safi, N.; Cheynier, V.; Moutounet, M. Study of the reaction between (+)-catechin and furfural derivatives in the presence of anthocyanins and their implication in food color change. *J. Agric. Food Chem.* **2000**, *48*, 5946-5954.
17. Es-Safi N.; Cheynier V.; Moutounet, M. Interactions between cyanidin 3-*O*-glucoside and furfural derivatives and their impact on food color changes. *J. Agric. Food Chem.* **2002**, *50*, 5586-5595.
18. Fulcrand, H; Cameira dos Santos, P. J.; Sarni-Manchado, P.; Cheynier, V; Favre-Bonvin, J. Structure of new anthocyanin-derived wine pigments. *J. Chem. Soc. Perkin trans. 1* **1996**, 735-739
19. Bakker, J.; Bridle, P.; Honda, T.; Kuwano, H.; Saito, N.; Terahara, N.; Timberlake, C. F. Isolation and identification of a new anthocyanin occurring in some red wines. *Phytochemistry* **1997**, *44*, 1375-1382
20. Bakker, J.; Timberlake, C. F. Isolation, identification and characterization of new color-stable anthocyanins occurring in some red wines. *J. Agric. Food Chem.* **1997**, *45*, 35-43
21. Fulcrand, H.; Benabdeljalil, C.; Rigaud, J.; Cheynier, V.; Moutounet, M. A new class of wine pigments generated by reaction between pyruvic acid and grape anthocyanins. *Phytochemistry* **1998**, *47*, 1401-1407
22. Cheynier, V; Fulcrand, H.; Brossaud, F.; Asselin C.; Moutounet, M. Phenolic composition as related to red wine flavor in *Chemistry of wine flavor*. ACS symposium series 714, Eds: Waterhouse, A. L. and Ebeler, S. E. Oxford University Press, pp 124-141.
23. Mateus, N.; Silva, A. M. S; Vercauteren, J.; Freitas-de, V.; Occurrence of anthocyanin-derived pigments in red wines. *J. Agric. Food Chem.* **2001**, *49*, 4836-4840.
24. LU, Y.; Foo, L. Y. Unusual anthocyanin reaction with acetone leading to pyranoanthocyanin formation. *Tetrahedron Lett.* **2001**, *42*, 1371-1373
25. Hayasaka, Y.; Asenstorfer, R. E. Screening for potential pigments derived from anthocyanins in red wine using nanoelectrospray tandem mass spectrometry. *J. Agric. Food Chem.* **2002**, *50*, 756-761.
26. Mateus, N.; Silva, A. M. S.; Santos-Buelga, C.; Rivas-Gonzalo, J. C.; De Freitas, V. Identification of anthocyanin-flavanol pigments in red wines by NMR and mass spectrometry. *J. Agric. Food Chem.* **2002**, *50*, 2110-2116
27. Mateus, N.; Carvalho, E.; Carvalho, A. R. F.; Melo, A.; González-Paramás, A. M. Santos-Buelga, C.; Silva, A. M. S.; De Freitas, V. Isolation and structural characterization of new acylated anthocyanin-

vinyl-flavanol pigments occurring in aging red wines. *J. Agric. Food Chem.* **2003**, 51, 277-282.

28. Schwarz, M; Wabnitz, T. C. Winterhalter, P. Pathway leading to the formation of anthocyanin-vinylphenol adducts and related pigments in red wine. *J. Agric. Food Chem.* **2003**, 51, 3682-3687.
29. Francia-Aricha, E. M.; Guerra, M. T.; Rivas-Gonzalo, J. C; Santos-Buelga, C. New anthocyanin pigments formed after condensation with flavanols. *J. Agric. Food Chem.* **1997**, *45*, 2262-2266
30. Peng, Z.; Iland, P. G.; Oberholster, A.; Sefton, M. A.; Waters, E. J. Analysis of pigmented polymers in red wine by reverse phase HPLC. *Austral. J. Grape and Wine Res.* **2002,** *8*, 70-75.
31. Revilla I. Pérez-Magariño, S.; González-SanJosé, M. L.; Beltrán, S. Identification of anthocyanin derivatives in grape skin extracts and red wines by liquid chromatography with diode array and mass spectrometric detection. *J. Chromatog. A* **1999**, *847*, 83-90.
32. Wulf, L. W.; Nagel, C. W. High-pressure liquid chromatography of anthocyanins in *Vitis vinifera*. *Am. J. Enol. Vitic.* **1978**, *29*, 42-49
33. Burns, J.; Mullen, W.; Landrault, N.; Teissedre, P.; Lean, M. E. Crozier, A. Variation in the profile and content of anthocyanins in wines made from Cabernet Sauvignon and hybrid grapes. *J. Agric. Food Chem.* **2002,** *50*, 4094-4102.
34. Dugo P.; Mondello, L.; Errante, G.; Zappia, G.; Dugo, G. Identification of anthocyanins in berries by narrow-bore high-performance liquid chromatography with electrospray ionization detection. *J. Agric Food Chem.* **2001**, *49*, 3987-3992
35. Cabrita, L; Anderson, Ø. M. Anthocyanins in blue berries of *Vaccinium pasifolium*. *Phytochemsitry* **1999**, 52, 1693-1696.
36. BenAbdeljalil, C.; Cheynier, V.; Fulcrand, H.; Hakiki, A.; Mosaddak, M.; Moutounet M. Mise en évidence de nouveaux pigments formés par réaction des anthocyanes avec des métabolites de levures. Sci. des Aliments, **2000**, 20, 203-220.
37. Chatonnet, P.; Dubourdieu, D.; Boidron, J.; Lavigne, V. Synthesis of volatile phenols by *Saccharomyces cerevisiae* in wines. *J. Sci. Food Agric.* **1993**, *62*, 191-202
38. Karchesy, J. J; Hemingway, R. W; Foo, Y. L. Barofsky, E.; Barofsky, D. F. Sequencing procyanidin oligomers by fast atom bombardment mass spectrometry Anal. Chem. **1986**, 58, 2563-2567.
39. Escribano-Bailón, T.; Álvarez-García, M.; Rivas-Gonzalo, J. C.; Heredia, F. J.; Santos-Buelga, C. Color and stability of pigments derived from the acetaldehyde-mediated condensation between malvidin 3-*O*-glucoside and (+)-catechin. *J. Agric. Food Chem.* **2001**, 49, 1213-1217.

40. Giusti, M; Rodriguez-Saona, L. E.; Griffin, D.; Wrolstad, R. Electrospray and tandem mass spectroscopy as tools for anthocyanin characterization. *J. Agric. Food Chem.* **1999**, *47*, 4657-4664.
41. Wang, H; Race, E. J.; Shrikhande, A. J. Characterization of anthocyanins in grape juices by ion trap liquid chromatography-mass spectrometry. *J. Agric. Food Chem.* **2003,** *51,* 1839-1844.
42. Auw, J. M.; Blanco, V.; O'keefe, S.F.; Sims, C.A. Effects of processing on the phenolics and color of Cabernet Sauvignon, Chambourcin and Noble wines and juices. *Am. J. Enol. Vitic.* **1996**, 47, 279-286.
43. Wildenradt, H. L.: Singleton, V. L. The production of aldehydes as a result of oxidation of polyphenolic compounds and its relation to wine aging. *Am. J. Enol. Vitic.* **1974,** *25*, 119-126
44. ES-Safi N.; Cheynier V.; Moutounet, M. Role of aldehydic derivatives in the condensation of phenolic compounds with emphasis on the sensorial properties of fruit-derived foods. *J. Agric. Food Chem.* **2002,** *50*, 5571-5585.
45. Tanaka, T.; Takahashi, R.; Kouno, I; Nonaka, G. I. Chemical evidence for the de-astringency (insolubilization of tannins) of persimmon fruit. *J. Chem Soc. Pekin Trans. 1* **1994**, 3013-3022.

Chapter 13

The Fate of Malvidin-3-glucoside in New Wine

Andrew L. Waterhouse[1] and Alejandro Zimman[1,2]

[1]Department of Viticulture and Enology, University of California, Davis, CA 95616
[2]Current address: Pathology and Laboratory Medicine, University of California at Los Angeles, Los Angeles, CA 90095–1732

Numerous chemical reactions have been described in the formation of polymeric pigment from malvidin-3-glucoside and condensed tannin. To date it has not been possible to monitor the fate of the majority of the anthocyanin products, but here it was possible to monitor the products in toto, because the anthocyanin was radiolabeled. In a fermentation experiment there was a rapid exchange of the anthocyanin into the solids, followed by a period of no loss of radioactivity with decreasing concentration of anthocyanins. In an aging experiment the water soluble pigmented tannins showed multiple forms with differential reactions to pH changes. Temperature had a dramatic effect on the observed reactions, but most surprisingly, pH had negligible effect. This latter result shows that many of the proposed reaction pathways to polymeric pigment may not be significant routes. The data does show that anthocyanin is efficiently and almost completely converted to polymeric pigment, though there were some losses attributed to precipitation which must be accounted for in the future.

Introduction

The reactions between anthocyanins and other wine components that lead to stable red color in aged red wine are not fully understood. There are many possible reactions that could lead to covalent bonds between an anthocyanin and condensed tannin, the presumed co-reactant. Many model reactions have been elucidated and the presence of these linkage types has been shown in wine in some cases. This will not be reviewed further as the first chapter of this monograph covers prior art and the other chapters discuss other current results. In large part, the actual linkage or linkages responsible for the majority of pigmented tannin are not known. This is due in large part to the high molecular weight of these products, and the difficulty in purifying individual compounds from a complex mixture of very similar substances.

To approach an understanding of these products a recent study analyzed the incorporation of radiolabeled malvidin-3-glucoside (M-3-G) into wine in both a fermentation and an aging experiments. The methods and results were described in an original report (Zimman and Waterhouse, submitted); the purpose here is to expand on the possible significance of those results.

Discussion

Fermentation Experiment

In the first experiment, tritium-labeled M-3-G was added to an ongoing fermentation of Cabernet Sauvignon grapes after the third day of the fermentation. This coincided with the maximum level of M-3-G in the must, Figure 1. After 24 hours, a new analysis of both malvidin-3-glucose by chromatography and the concentration of radioactivity by liquid scintillation counting showed decreases. However, the radioactivity had dropped almost by half, while the M-3-G level had decreased only by about 15%.

The loss of radioactivity observed here suggests that there is a very rapid reaction of anthocyanin at this point, but the small reduction of total M-3-G disproves this possibility. Another possibility is that the observed M-3-G in solution is equilibrating with a another pool that is not quantified in the analysis. For instance, if a substantial amount of M-3-G is bound to the grape solids, but equilibrating with the solution, then the hot M-3-G will be distributed into these two pools in the same proportion to their chemical concentrations. To test this theory we first have to know the total amount of anthocyanins present in the original fruit. This was determined by extraction and HPLC analysis of a replicate sample of grapes to be 681 mg/kg. At the 100 hour mark, the first time that the must was analyzed after the addition, the concentration of M-3-G was

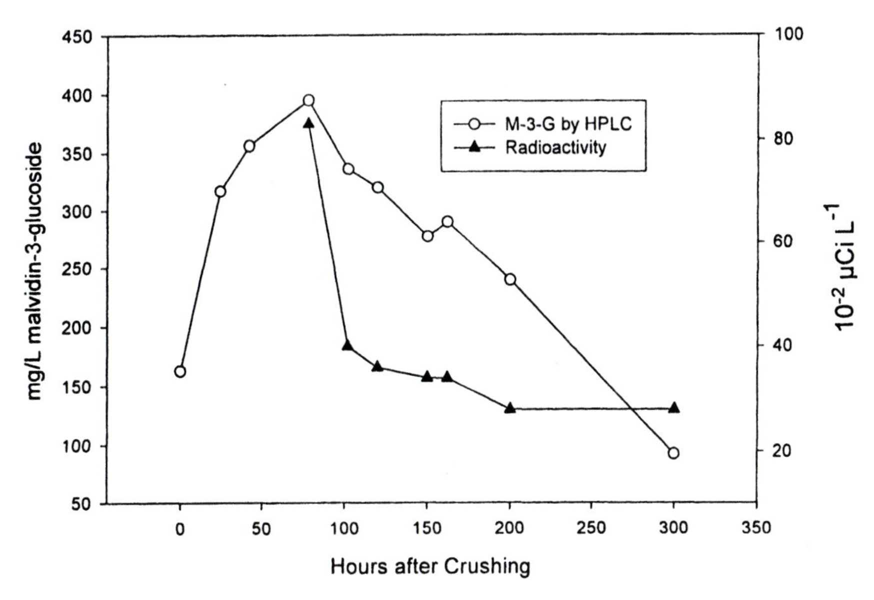

Figure 1. Concentration during and after fermentation

336 mg/L, but the radioactivity level was 48% of the initial addition. So if there was a complete exchange at that point between the two pools of M-3-G, the total amount present in the system would be 697 mg/L, an amount very close to that observed by direct extraction of the berries. Therefore, at that point in time, it appeared that virtually all the anthocyanin from the skin was available, either in solution or in rapid equilibrium with an association to the solids.

The fact that grape solids as well as yeast lees adsorb anthocyanins is no secret to winemakers, as these materials are always darkened by pigment when removed from a fermentation. Yeast lees also have been shown to adsorb anthocyanins (1), and grape solids have been shown to re-adsorb pigment (D. Block and D. Bone, personal communication). However, an estimate of the proportion of these losses as well as an estimate of the proportion of anthocyanin extracted has not previously been possible.

In the next stage of the fermentation, from about 100 hours to 200, it is not possible to demonstrate whether the slopes of the continued slow decrease of both M-3-G and radioactivity levels are significantly different from each other, but it is also not possible to establish that they are in fact parallel to each other. If, in fact, M-3-G levels are dropping more quickly than the radioactivity, this would show that chemical transformation of anthocyanins is already happening over this short time span. On the other hand, if the two lines follow parallel declines, it shows that there is no significant chemical change, and the decrease in M-3-G levels at this point is due to increasing loss due to solids binding caused by higher affinity. By the end of this period, the decrease with respect to the total M-3-G in grapes has increased from 48% to ~60%. One explanation for this decrease could be precipitation of M-3-G or, more likely, co-precipitation with tannins or other lees. Recent reports have established that compounds like caffeic acid can increase the concentration of anthocyanins in solution (2), thus copigmentation could compete with the adsorption interaction described here and increase the amount of anthocyanins that remain in solution.

The last stage of this experiment actually measured the changes in the first stage of aging, from pressing to 10 months later. Over this period of time there was virtually no change in radioactivity levels. This in itself was a bit surprising, as it might be expected that some lees formation during this time would have resulted in an adsorptive loss of M-3-G. As there was virtually no change in radioactivity, all the M-3-G present at press, as well as any transformed products present at that time, remained in solution. The actual chemical levels of M-3-G however, dropped dramatically. This clearly shows that there were significant chemical transformations during this period. However, since it was not possible to monitor the chemical forms of the radiolabeled substances, due to the high dilution of the labeled M-3-G in this experiment, the specific fate of the M-3-G cannot be determined.

In summary, an adsorptive equilibrium between solids and liquid appears to be a major factor in limiting the concentration of pigments in young wine, and in the first few months of aging there are few losses of the original anthocyanin fragment, but significant transformation of them to different forms.

Aging Experiment

In this experiment, the 3H-M-3-G was added to new Cabernet Sauvignon wine that had just completed malolactic fermentation. The wine was stored in small glass vials of approximately 2 mL volume, using duplicates for the control and each treatment. The vials were opened at each sampling point, but there was one extra set of single vials of each treatment that were opened at the end of the experiment, after 8 months aging. Samples were taken for radioactivity counts, spectral measurements, as well as chromatographic analyses where molecular size fraction were tested for radioactivity. The different treatments included temperature, with the control at 20°C, and treatments at 5°C and 35°C, seed tannin addition, chlorogenic acid addition, and pH adjustment from the control at 3.6 to both 3.1 and 4.1.

Total tritium levels were quantified after 8 months of aging. The control wine lost 37% of its counts in solution demonstrating the large losses of the original pigment molecules to the wine over this time period. Much more, 54%, was lost in the wine kept warm at 35°C, and much less, 14%, in the wine kept cool at 5°C. It is presumed that these losses were caused by precipitation of polymeric material, but other precipitations may have occurred, such as small amounts of lees settling or perhaps tartrate crystallization where these precipitates may have entrapped some of the original pigment. This result shows that losses over this period are significant, but are slightly different from the fermentation result above which showed no loss over a similar period, from press to 10 months of aging, where the above "fermentation" result was aged at 5°C but never opened during that time. The chlorogenic acid addition, seed addition, and the pH shift to 3.1 had no significant difference from the control experiments, but higher pH had higher losses.

The liquid-liquid extraction procedure partitions the starting material (3H-malvidin-3-glucoside in the iso-amyl alcohol fraction) from some of the products (polymeric pigments in the aqueous fraction). The proportions in the aqueous fraction, which is free of anthocyanin, varied from a low of 29% in the 5°C experiment to 64% in the seed experiment, Figure 2. The 35°C experiment, where no monomeric anthocyanin was present, only had 51% in the aqueous fraction. However, since prior work has shown by thiolysis of an iso-amyl

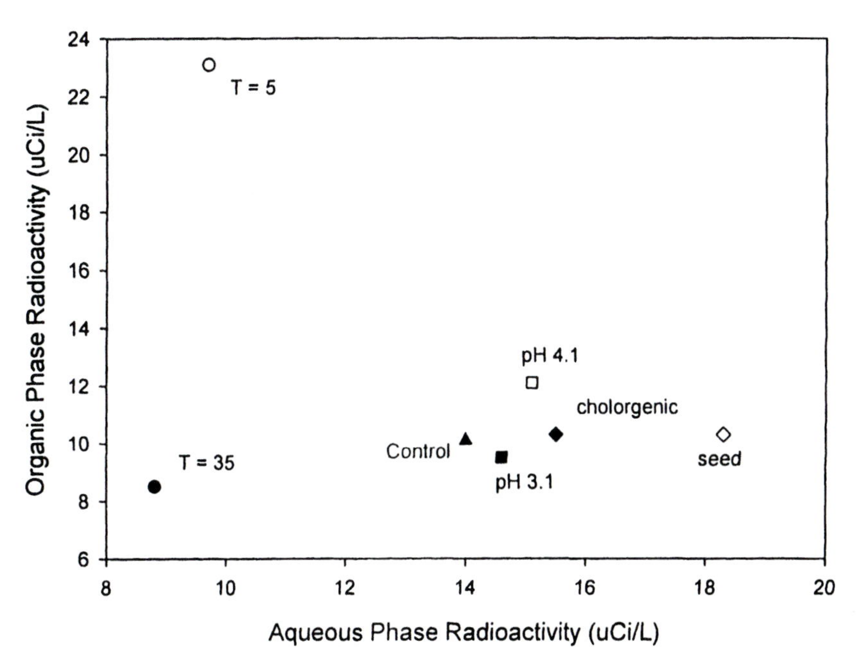

Figure 2. Fractionation of Radioactivity into two phases after 8 months aging

alcohol extract have a degree of polymerization >1 (3), it is certain that some of the tritiated polymeric pigment products are extracted into the iso-amyl alcohol fraction. The fact that the 35°C treatment had such a large amount of counts in the organic fraction confirms that here the organic fraction was contaminated with much polymeric pigment.

Alternatively, is also possible that at 35°C, the polymeric pigment attains a form that is less polar, and thus more organic soluble than in the other treatments. One would expect that, in general, the increase in radioactivity in the polymer fractions would be reflected with comparable increases in absorbance at 520 nm due to increased incorporation of malvidin-3-glucoside as a linked pigment. That trend is observed in general, but there are significant deviations. For example, the pH experiments retained similar amounts of radioactivity but the absorbance of these products at pH 3.6 was different, Figure 3. Because the response factor (absorbance per mol of malvidin-3-glucoside incorporated) is different, then the chemical composition of the pigmented polymers cannot be identical, either because there are different forms of polymeric pigments, or, more likely, different mixtures of these products.

Looking at the 520 nm absorbance of the aqueous fraction under different pH conditions, one would expect that since all the anthocyanin has been

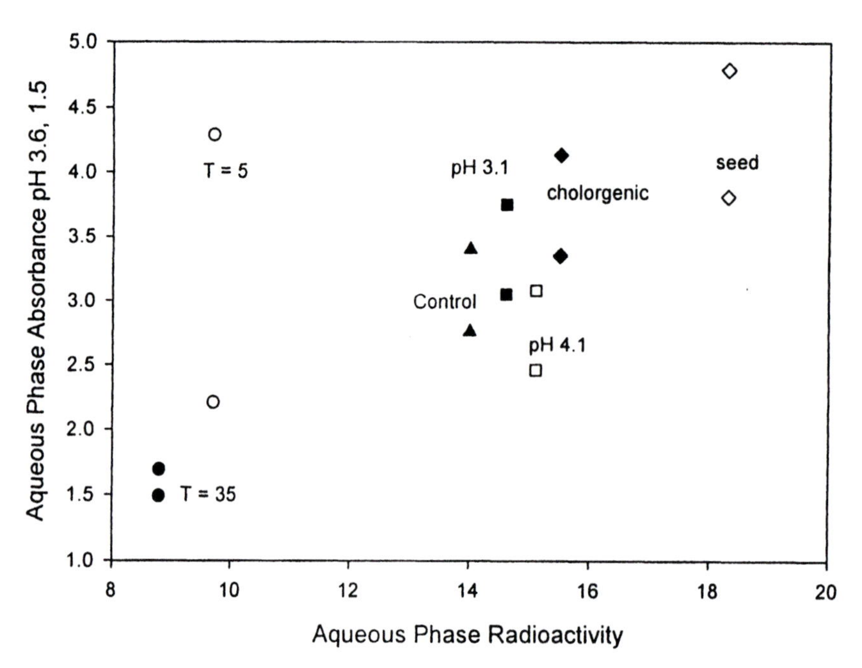

Figure 3. Change in absorbance of aqueous phase fractions from pH 3.6 to 1.5

removed, there would be little pH response, an expected characteristic of polymeric pigment. However, as shown in Figure 3, there is a large variation in the response of different treatments to pH changes.

Therefore these various results show a clear conversion to polymeric pigment, and that there are differences in the chemical structures of the polymeric pigment from the original anthocyanin. In particular, the molar absorptivity of the "anthocyanin" molecule drops dramatically when it is incorporated into a polymeric pigment. In addition, a comparison of different treatments shows that there must be at least two forms of pigment in the condensed anthocyanin, and these also differ in absorptivity. Two different structures have been identified in wine where malvidin-3-glucoside incorporated as a terminal unit and as an electrophile (flavene structure) (3). The absorbance of the first one should be pH dependent whereas the second one is colorless. Therefore, if these were the only possible structures, a young wine would be rich in structures with anthocyanin as terminal unit. As the wine ages, it would require condensation of the pigmented polymers where anthocyanin is a terminal unit into a flavene to produce a loss of absorbance while maintaining the amount of radioactivity. Other scenarios involve initial incorporation as a terminal anthocyanin unit, but followed by conversion to D-ring (vitisin) type linkages where pH sensitivity is lost.

Using a normal-phase HPLC, which nominally separates tannins based on molecular size, we can observe very large differences in the proportions of small and large molecules which incorporate radioactivity, Figure 4. At the low temperature treatment, 5°C, there were very large proportions found in the monomer fraction, with decreasing proportions in progressively larger fractions. This was particularly notable at 4 months. After 7 months at 5°C, there was a shift to the next larger fraction, but still very small amounts at the highest molecular weight. In contrast, the warmest treatment, 35°C, even at 4 months, has very small amounts of counts in the monomer fraction, while the bulk of counts were in the highest molecular weight fraction. It was notable that the 7 month, 20°C sample was similar to the 4 month 35°C sample.

A few assumptions must be made in order to interpret the differences between levels of radioactivity incorporation and absorbance of samples at 520 nm. First, it is assumed that the radioactivity is the true measure of anthocyanin incorporation. Second, when considering the absorbance, one has to take into account: 1) polymer pigments were already present when 3H-malvidin-3-glucoside was added, 2) in the analysis by chromatography, the absorbance values are obtained at a much lower pH than wine pH. We presume this will enhance the absorbance of the monomer fraction so that its apparent contribution to wine color is magnified beyond its real value. These results clearly show that the fate of most anthocyanins during aging is the formation of higher molecular weight structures by two analytical procedures, chromatography and liquid extraction. The maximum amount of material incorporated into different

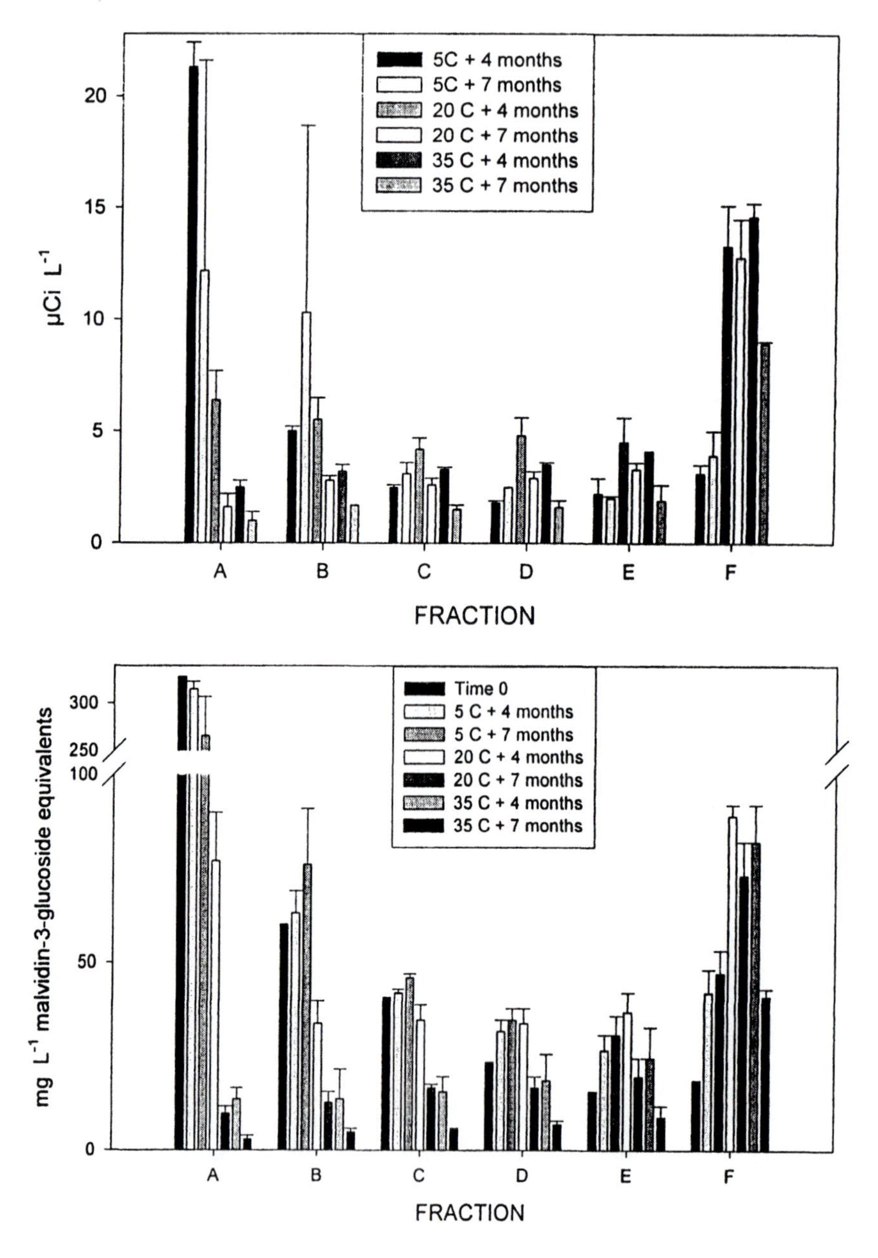

Figure 4. Normal phase fractions of wines aged at different temperatures, comparing radioactivity content and absorbance. (Reproduced with permission from Am. J. Enol. *Vitriculture, in press. Copyright 2004 American Society of Enology and Viticulture.)*

oligomer and polymer structures was 85% at 20°C after 4 months although this value does not include material lost to precipitation. In addition, there are monomers still available for incorporation. It is likely that some of the compounds included in the oligomer fractions are the result of condensation of anthocyanins with small molecules such as pyruvate. However, others have observed that even in fortified wine those amounts found are below 50 mg/L (4). The causes of tannin loss by precipitation has been discussed by Haslam (5) and the presence of polymerized anthocyanin in precipitate has been studied by Waters et al. (6). Therefore it is not surprising to observe a decrease of tritium in solution and particularly in the polymer fraction for wine aged at 35°C for more than 7 months. To compare the effects of other treatments, the results shown in Figure 4 are derived from NP-HPLC and are taken after 4 months of aging, because at 20°C all treatments still have anthocyanins available for reaction, thus it is possible to compare the rate of reaction for the treatments.

The absorbance of some of the fractions dropped even below the values at time 0 (when 3H-malvidin-3-glucoside was added), showing that the substances in these fractions are being either degraded or converted to low absorbing materials. The response factor parameter can also be used to compare monomer, oligomer and polymer fractions by HPLC. [need response factors in graph] Fraction A (containing monomers) had a higher response than the oligomer and polymer fractions even without considering that there was pigmented material already present before the addition of tritiated material. Another example is to compare the 5°C and 35°C samples after 7 months. There are clear differences in the amounts incorporated in the polymer fraction F, but they have similar values of absorbance.

The first result to consider is that the transformation from monomeric anthocyanin to polymeric pigment is not affected substantially by pH, Figure 5. This is a striking result because many of the proposed reactions towards polymeric pigment are dependent on acid-catalyzed reactions. For instance, it is important to note that based on the equilibrium distribution of malvidin-3-glucoside at 25°C calculated by Brouillard and Delaporte (7) the proportions of flavylium and the pseudobase (or carbinol) species depends on the pH. At pH 3.1, 21.8% of malvidin-3-glucoside is in the flavylium form while 68.5% is in the carbinol form. At pH 3.6, the values for flavylium and pseudobase forms are 8.1% and 80.4%, and at pH 4.1 2.7 % and 85.1% respectively. Changes in the chalcone and quinonoidal species are much smaller. With the lower pH having 8 times as much flavylium form present than the higher pH, it would be reasonable to expect a much faster rate of conversion if flavylium was a reactant, but the difference observed here is negligible, so it appears that the key reactions to form polymeric pigment do not involve the flavylium species (or it is not the limiting step).

Another pH sensitive reaction intermediate would be the carbocation at the 4-position of a flavan-3-ol, formed by acid-catalyzed cleavage of a

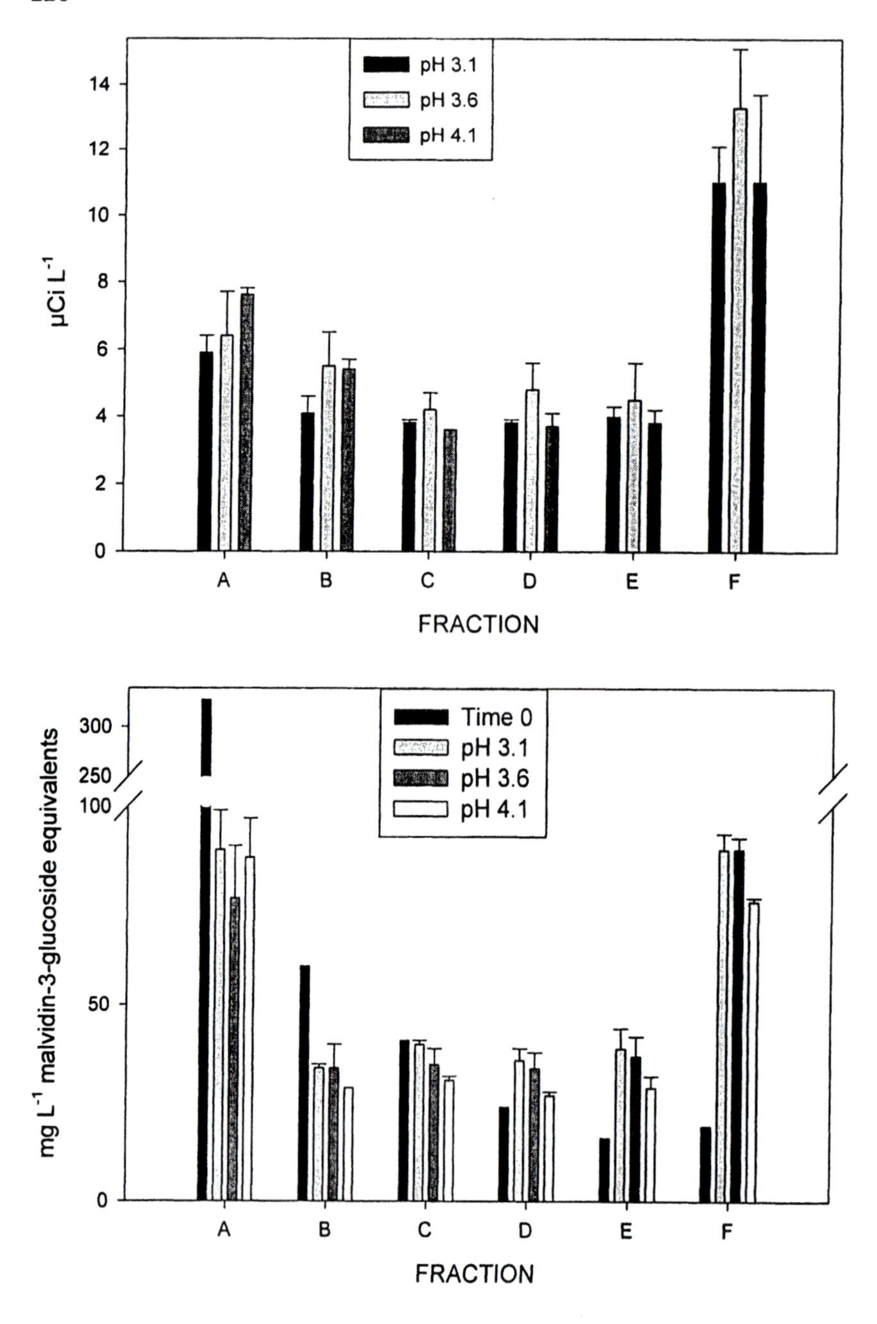

Figure 5. Normal phase fractions of wines aged at pH levels, comparing radioactivity content and absorbance. (Reproduced with permission from Am. J. Enol. *Vitriculture, in press. Copyright 2004 American Society of Enology and Viticulture.)*

proanthocyanindin linkage. If this cleavage was a rate-limiting step, its formation should be accelerated by about an order of magnitude between pH 4.1 and 3.1. Again, because the data clearly shows a very small effect of pH, the formation of such a carbocation cannot be a rate-limiting step. An additional intermediate that would be sensitive to pH is acetaldehyde. Its reaction with flavonoid is proposed to involve acid catalysis of a two-stage hydroxyalkylation reaction. Since a higher hydronium ion concentration does not enhance the reaction rate, the data suggest that the acetaldehyde bridged products are not significant components in pigmented tannin.

The addition of seed extract containing tannins yielded higher levels of total soluble radioactivity after 8 months of aging. The amount of radioactivity was also higher in the aqueous fraction after the extraction procedure, by about the same amount, 4 μCi/L, while there was not increase in the organic fraction, so the increased retention of the original anthocyanin appears to be in the polymeric pigment fraction, Figure 2. This result concurs with previous reports which have shown that higher amount of tannins yielded more polymeric pigments (??Singleton and Trousdale 1992) in experiments with tannin-anthocyanin additions to white wine. It should be remembered that the control wine was able to incorporate 85% of the material after 4 months, probably because it had "adequate" tannin content (measured concentration of tannins at time 0 was 1.6 g/L). On the other hand, the amount measured chromatographically at the earlier time of 4 months in Figure 6 reveals lower levels of polymer, and higher levels of monomer. It would appear that the additional amount of tannin slowed the reaction of the anthocyanin with the tannin. This seems counter-intuitive, but the additional tannin could be acting as a preservative, reducing the concentration of reactive oxygen species that may be the anthocyanin-tannin reaction initiators. Seed extract may be preventing losses by precipitation of polymeric pigment, because it also contains flavan-3-ol monomers that can reduce the mean degree of polymerization of tannins (8). Even with the slower reaction, the ultimate result at 8 months was a higher amount of polymeric pigment.

The purpose of the addition of chlorogenic acid was to compare a control wine with wine that had higher levels of copigmentation. In this treatment, there were small increases in incorporation of the label into polymeric pigment from the extraction data, but by chromatography there were no observable differences. Chlorogenic acid is not found in wine (but it is commercially available and is similar to the grape phenolic acid, caftaric acid), and the amount added immediately increased the absorbance at 520 nm by 1.4 units (20% increase versus the control), confirming its effect as a co-pigment. The additions where done post-fermentation, so the initial amount of anthocyanins is already set as opposed to pre-fermentation additions that could have promoted the extraction of more anthocyanins into wine (2). Perhaps as a consequence of that timing, chlorogenic acid did not favor or delay the rate of incorporation into tannins. Like the addition of seed extract, excess of chlorogenic acid may have prevented

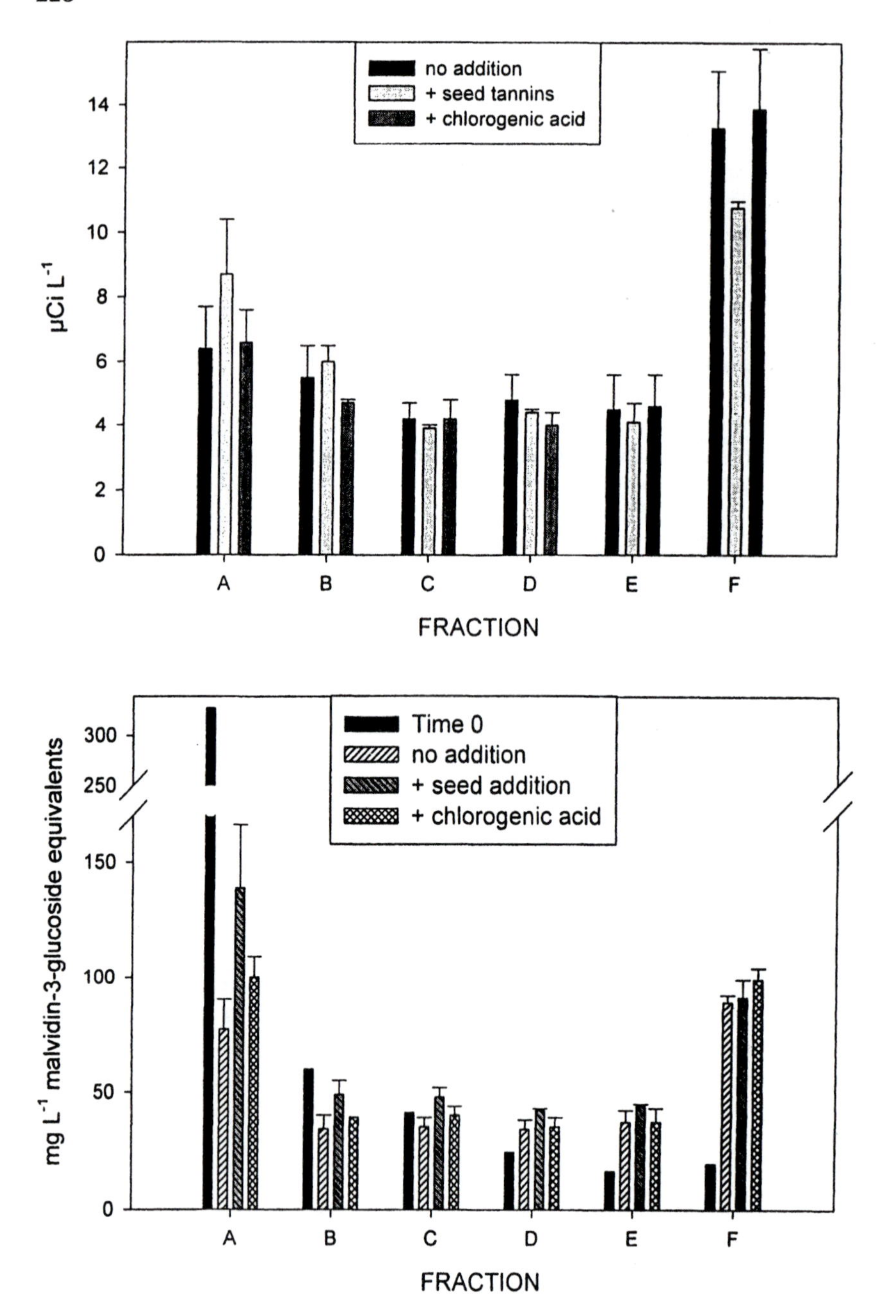

Figure 6. Change in absorbance of aqueous phase fractions from pH 3.6 to 1.5. (Reproduced with permission from Am. J. Enol. *Vitriculture, in press. Copyright 2004 American Society of Enology and Viticulture.)*

oxidation because of the vicinal diphenols which could have slowed down the precipitation of pigmented polymers that were oxidized.

When the samples were prepared, there was adequate material for a single additional set of treatments that were left unopened throughout the 7 months of observation. In the iso-amyl alcohol extraction, all of the treatments had higher concentration of radioactivity in solution than their air-exposed counterparts except for the experiment kept at 5°C that had only 1.4 μCi/L more in solution (4%). The increases of radioactivity in solution for the unexposed experiments compared to the exposed ranged from 16% (seed addition) to 54% (35°C). The increase in the aqueous fraction ranged from 16% (seed addition) to 89% (35°C). Thus, air exposure affects the amount of radioactivity found in solution. Of course, the amount of oxygen exposure that each of these wines experienced is not known, but this result suggests that reactions with oxygen can strongly affect the process. The fact that unexposed wine had more radioactivity in solution indicates that the precipitation of polymeric pigment occurred through an oxidation mechanism. This issue must be directly addressed in future studies.

The amount of anthocyanins polymerized into high molecular weight structures is far larger than the amounts found by Remy et al. where only a small amount of anthocyanins are bound through C-4–C-8 or C-4–C-6 inter-flavonoid bond (3). The authors were able to account for less than 13 mg/L of malvidin-3-glucoside when malvidin-3-glucoside is in the terminal position and only mass spectrometry identification was given for the flavene product. Rather than contradictory, the result here can be considered complementary. As the wine ages, the pigmented polymer appears first to possess flavylium structure which would produce an increase in response at low pH, but on aging, this characteristic is lost. Therefore, in a wine aged for two years (like the wine used by Remy et al.) the low amount of malvidin-3-glucoside recovered by thiolysis would be solely responsible for change in absorbance due to acidification. It also reflects a change in the composition of pigmented tannin.

The rest of the malvidin-3-glucoside in Remy et al. could not be recovered by thiolysis, probably because it is in a flavene-type structure or there are additional linkages to the pigment such as vinyl condensation (9); (10). It has been reported that enzymatic oxidation of caffeic yields an o-quinone that will condense with malvidin-3-glucoside (11) and overall 520 nm absorbance decreases. Similarly, flavan-3-ols could also react with anthocyanins through an oxidized catechol ring (non-enzymatic formation). Such products would have bonds between the tannin and anthocyanins that cannot be cleaved by mild acid.

It has been reported how wines produced from different vineyards produce consistent amounts of polymeric pigment, despite changes in the wine making practices (12). These results corroborate those, as in that experiment there was no treatment at particularly high or low temperatures. Keep in mind that these wines were exposed to higher oxygen concentrations than typical commercial practice, where the loss of polymer would proceed slower. Finally, it is important to remember that color is not the only sensory attribute associated with polymer pigment formation. Throughout aging, the astringency of wine decreases and formation of pigmented polymers and precipitation could be essential in that process.

In conclusion, the aging experiments have revealed several important points. Anthocyanins react almost totally in a short period of time to create polymeric pigments and little is lost to simple degradation to other small molecules. A large fraction of these products are lost to precipitation in the early life of a wine, either due to direct precipitation or indirect co-precipitation. Many of the proposed reaction pathways may not be major contributors to the anthocyanin-tannin reaction because the reaction kinetics of polymeric pigment formation appear to be insensitive to pH. Finally, though observed without proper controls, oxygen has a strong effect on the observed reactions, perhaps more so than in winemaking practice. Future studies should address oxygen exposure directly as well as an analysis of precipitate where it appears a substantial amount of product resides.

References

1. Vasserot, Y.; S. Caillet and A. Maujean American Journal of Enology and Viticulture **1997**, *48*, 433-437.
2. Darias-Martin, J.; M. Carrillo; E. Diaz and R. B. Boulton Food Chemistry **2001**, *73*, 217-220.
3. Remy, S.; H. Fulcrand; B. Labarbe; V. Cheynier and M. Moutounet Journal of the Science of Food and Agriculture **2000**, *80,* 745-751.
4. Mateus, N.; A. M. S. Silva; J. Vercauteren and V. de Freitas Journal of Agricultural and Food Chemistry **2001**, *49*, 4836-4840.
5. Haslam, E. Phytochemistry **1980**, *19*, 2577-2582.
6. Waters, E. J.; Z. K. Peng; K. F. Pocock; G. P. Jones; P. Clarke and P. J. Williams Journal of Agricultural and Food Chemistry **1994**, *42,* 1761-1766.
7. Brouillard, R. and B. Delaporte Journal of the American Chemical Society **1977**, *99*, 8461-8468.
8. Vidal, S.; D. Cartalade; .J.-M. Souquet; H. Fulcrand and V. Cheynier Journal of Agricultural and Food Chemistry **2002**, *50*, 2261-2266.
9. Francia-Aricha, E.; M. T. Guerra; .J. C. Rivas-Gonzalo and C. Santos-Buelga Journal of Agricultural and Food Chemistry **1997**, *45,* 2262-2266.

10. Asenstorfer, R. E.; Y. Hayasaka and G. P. Jones J Agric Food Chem **2001**,*49*, 5957-5963.

11. Sarni-Manchado, P.; V. Cheynier and M. Moutounet Phytochemistry **1997**, *45,* 1365-1369.

12. Zimman, A.; W. S. Joslin; M. L. Lyon; J. Meier and A. L. Waterhouse American Journal of Enology and Viticulture **2002**, *53*, 93-98.

Chapter 14

Matrix-Assisted Laser Desorption–Ionization Time-of-Flight Mass Spectrometry of Anthocyanin-poly-flavan-3-ol Oligomers in Cranberry Fruit (*Vaccinium macrocarpon*, Ait.) and Spray-Dried Cranberry Juice

Christian G. Krueger[1], Martha M. Vestling[2], and Jess D. Reed[1,*]

Departments of [1]Animal Sciences and [2]Chemistry, University of Wisconsin, Madison, WI 53706

Matrix-assisted laser desorption/ionization time-of-flight mass spectrometry (MALDI-TOF MS) was used to characterize the structural diversity of a series of anthocyanin-polyflavan-3-ol oligomers in cranberry fruit and spray dried juice. MALDI-TOF mass spectra provide evidence for a series of compounds corresponding to anthocyanins; cyanidin-hexose, cyanidin-pentose, peonidin-hexose and peonidin-pentose linked to polyflavan-3-ols through a CH_3-CH bridge. Deionizing the isolated anthocyanin-polyflavan-3-ol oligomers with cation exchange resin prior to mass spectral analysis allowed the detection of exclusively $[M]^+$ ions because the singly charged oxonium cation does not require cationization. MALDI-TOF mass spectra of the anthocyanin-polyflavan-3-ols that eluate from Sephadex LH-20 columns with water/ethanol (1:1) contained oligomers with a degree of polymerization (DP) of 1 to 2. MALDI-TOF MS of the Sephadex-ethanol eluate contained a series of masses corresponding to DP1 to DP4 and the Sephadex-ethanol:methanol (1:1) eluate contained a series of masses corresponding to DP3 to DP5. The anthocyanin-polyflavan-3-ol oligomers also show structural variation in the nature of the interflavan bond (A-type vs. B-type). Anthocyanin-polyflavan-3-ol oligomers incorporating a CH_3-CH bridge have been described in red wines and in model reactions as a condensation reaction between anthocyanins and flavan-3-ols via acetaldehyde. Our results indicate that oligomeric anthocyanins may also be present in cranberry fruit and unfermented spray dried cranberry juice.

Introduction

The pigmentation of cranberries (*Vaccinium macrocarpon*, Ait.) is attributed to the presence of six anthocyanins; cyanidin-3-galactoside, cyanidin-3-glucoside, cyanidin-3-arabinoside, peonidin-3-galactoside, peonidin-3-glucoside and peonidin-3-arabinoside (*1*). However, analytic high performance liquid chromatography (HPLC) of spray dried cranberry juice detected a series of polyflavan-3-ols with both a 280 nm and a 520 nm UV absorbance maximum, suggesting that oligomeric pigments are present in the cranberry fruit (*2*). Kennedy *et al* (*3*) reported similar spectral characteristics of proanthocyanidin found in grape skin and concluded that anthocyanins are incorporated in to proanthocyanidins during fruit ripening.

While there are few reports of anthocyanin-polyflavan-3-ol oligomers occurring in fruits and unfermented beverages, there are well documented accounts of complex pigments forming in alcoholic beverages such as red wine (*4,5,6*) or rose cider (*7*). During the aging and storage of red wines, anthocyanins are converted to new pigments through reactions with other phenolics such as polyflavan-3-ols. Condensation of an anthocyanin and a polyflavan-3-ol via an acetaldehyde derived bridge is one mechanism by which anthocyanin-polyflavan-3-ol oligomers may occur (*4,5,8,9,10,11*). Acetaldehyde is found naturally in wine as either a by-product of yeast metabolism or as an oxidation product of ethanol (*4*). Using a model rose cider, Shoji *et al* (*12*) elucidated the structure of such oligomeric pigments by high resolution FAB-MS, ^{1}H and ^{13}C NMR analysis. The dimeric pigments consisted of an anthocyanin linked by a CH_3-CH bridge to a flavan-3-ol.

Recent advances in mass spectrometry allow for the characterization of complex mixtures of anthocyanin-polyflavan-3-ol pigments in wine

(*13,14,15,16*). Both Electrospray Ionization (ESI) and Matrix-Assisted Laser Desorption/Ionization Time-of-Flight mass spectrometry (MALDI-TOF MS) have been applied to the analysis of polyflavans in foods and beverages. While ESI and MALDI-TOF MS are both capable of detecting intact molecular ions with high molar mass (>100,000 Da), ESI is best suited for analysis of monodispersed biopolymers because of complications arising from the formation of multiply charged ions (*17*). Alternatively, MALDI-TOF MS is ideally suited for characterizing polydispersed oligomers (*18*) and is considered the mass spectral method of choice for analysis of polyflavans which exhibit greater structural heterogeneity. MALDI-TOF MS produces only a singly charged molecular ion for each parent molecule and allows detection of high masses with precision (17).

MALDI-TOF MS has been applied to characterize the structural diversity of polyflavans in many foods and beverages. MALDI-TOF mass spectra of grape seed extracts show that polyflavan-3-ol heteropolymers have structural variation in the degree of gallic acid substitutions (*19,20*). Sorghum polyflavan heteropolymers have structural variation in degree of hydroxylation in the B-ring of the glycosylated flavan unit and contain a flavanone (eriodictyol or eriodictyol-5-*O*-β-glucoside) as the terminal unit (*21*). The presence of A-type interflavan bonds in cranberry (*2*), sorghum (*21*), and brown soybeans (*22*) have been described. MALDI-TOF MS has been used to identify anthocyanins from red wine, cranberries, Concord grape juice (*23*) and blueberries (*24*). MALDI-TOF MS is a rapid method for characterizing the mass distribution of oligomeric polyflavans. Polyflavan-3-ols from apples have a mass distribution up to the undecamer (*25*), and the degree of polymerization (DP) of polyflavan-3-ols in brown or black soybean coat was found to be as high as DP30 (*22*). We present results from MALDI-TOF MS that indicate the presence of anthocyanin-polyflavan-3-ol oligomers in cranberry fruit and cranberry juice.

Experimental

Cranberry Fruit and Spray Dried Juice Extraction

Thirty grams of cranberry fruit (Hyred) was frozen in liquid N_2, ground to a fine powder in a blender and extracted with 80% aqueous acetone (v/v, 100ml) in an ultrasonic bath for 10 min. The extract was centrifuged (3500g) for 10 min and the liquid retained. Acetone was removed from the extract under vacuum evaporation at 30° C. The extract was solubilized in 10 ml water. Three grams of 90MX spray-dried concentrate juice powder (Ocean Spray Cranberries® Inc) was dissolved in 10 ml water.

Sephadex LH-20 Separation

Cranberry phenolics were separated by the methods of Porter *et al* (*2*). Sephadex LH-20 (Pharmacia) was equilibrated in water for 2 hours. Cranberry fruit extract and reconstituted spray-dried juice powder were applied to a 15cm x 2.5 cm i.d. semi-preparative column filled with Sephadex slurry. The column was attached to a high-performance liquid chromatograph (HPLC) equipped with two Waters 501 pumps, and the phenolic composition of the eluate was monitored at 280nm (polyflavan-3-ols), 320nm (hydroxycinnamic acids), 360nm (flavonols) and 520nm (anthocyanins) by a Hewlett Packard 1040 diode array detector. Six fractions were obtained by eluting sequentially at a flow rate of 4 ml min^{-1} with water (fraction 1), water/ethanol (1:1; fraction 2), ethanol (fraction 3), ethanol/methanol (1:1; fraction 4), methanol (faction 5) and aqueous acetone (acetone/H_2O, 4:1, v/v; fraction 6). The point at which the solvent was changed was based on the UV absorbance; when the characteristic wavelength for each class of phenolic compounds returned to baseline, the solvent was replaced. Extracts were solubilized in 1mL 80% aq. acetone prior to MALDI analysis.

Matrix-Assisted Laser Desorption/Ionization Time-of-Flight Mass Spectrometry (MALDI-TOF MS)

Mass spectra were collected on a Bruker Reflex II-MALDI-TOF mass spectrometer (Billerica, MA) equipped with delayed extraction and a N_2 laser (337 nm). In the positive reflectron mode, an accelerating voltage of 25.0 kV and a reflectron voltage of 26.5 kV were used. Spectra are the sum of 50-300 shots. Spectra were calibrated with bradykinin (1060.6 MW) and glucagon (3483.8 MW) as external standards.

In accordance with previously published results (*19*) *trans*-3-indoleacrylic acid (*t*-IAA; 5mg/100 µL 80% aq. acetone) was used as a matrix. The polyflavan fractions eluted from the Sephadex column were mixed with the matrix solution at volumetric ratios of 1:2. The polyflavan:matrix mixture was deionized, applied directly (0.2 µL) to a stainless steel target and dried at room temperature. Dowex 50X8-400 cation exchange resin (Supelco), equilibrated in 80% aq.

acetone (v/v) was used to deionize the analyte:matrix solution. Bradykinin, glucagon (Sigma Chemical Co., St. Louis, MO) and *t*-IAA (Aldrich Chemical Co., Milwaukee, WI) were used as received.

Results and Discussion

Rationale for Assigning Structures to Mass

Tentative structures were assigned to cranberry anthocyanin-polyflavan-3-ol oligomers by comparing MALDI-TOF mass spectral distributions to predictive equations. In the case of cranberries, Hong and Wrolstad (*1*) described the anthocyanins which occur in cranberry fruit; cyanidin-3-galactoside, cyanidin-3-glucoside, cyanidin-3-arabnoside, peonidin-3-galactoside, peonidin-3-glucoside and peonidin-3-arabinoside (Figure 1A). Foo *et al* (*26*) elucidated the structure of trimeric polyflavan-3-ols from cranberries, showing that cranberry poly-flavan-3-ols contained both B-type and A-type interflavan bonds (Δ 2 amu). MALDI-TOF MS showed spray dried cranberry juice contained polyflavan-3-ol heteropolymers with structural variation in the degree of hydroxylation of the B-ring (epicatechin or epigallocatechin; Δ 16 amu) and nature of the interflavan bond (A-type and B-type; Δ 2 amu) (*2*). Shoji *et al* (*12*) identified pigments in model rose cider consisting of anthocyanins linked to a flavan-3-ol by a CH_3-CH bridge.

Predictive equations were formulated to describe the heteropolymeric nature of cranberry anthocyanin-polyflavan-3-ol oligomers based on the assumption that the structural diversity seen in polyflavan-3-ols (*2*) and anthocyanins (*1*) can be extrapolated to higher degrees of polymerization (DP) when condensation products involving a CH_3-CH bridge (*12*) are applied. Whereas MALDI-TOF MS has the power to distinguish molecular weight differences due to extent of hydroxylation (Δ 16 amu), nature of interflavan bonds (A-type vs. B-type; Δ 2 amu), degree of polyflavan-3-ol polymerization (Δ288 amu) and substitutions by hexose (Δ 162 amu) or pentose (Δ 132 amu), it lacks the ability to assign specific stereochemistry to the molecule. Thus, glucosides cannot be differentiated from galactosides and are simply referred to as hexosides.

An equation was formulated to predict the series of anthocyanins linked to polyflavan-3-ols via a CH_3-CH bridge. The equation is: *Anthocyanin* + 28 + 288a - 2b, where *anthocyanin* represents the molecular weight of the terminal anthocyanin [cyanidin-3-galactoside (449.4 amu), cyanidin-3-glucoside (449.4 amu), cyanidin-3-arabinoside (419.4 amu), peonidin-3-galactoside (463.4

A

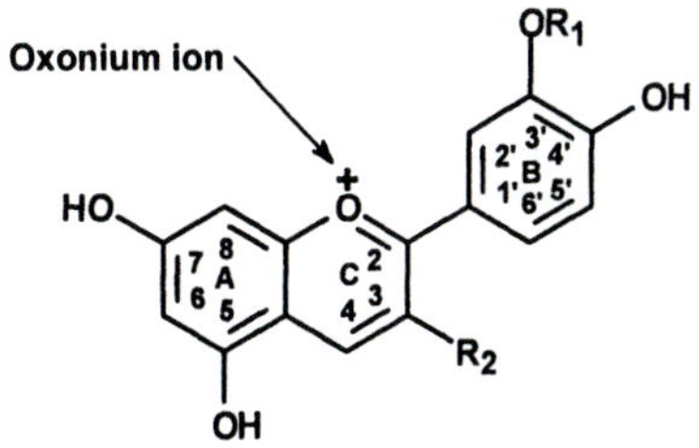

Cyanidin 3-glucoside: R_1 = H , R_2 = glucose (m/z = 449.4)
Cyanidin 3-galactoside: R_1 = H , R_2 = galactose (m/z = 449.4)
Cyanidin 3-arabinoside: R_1 = H , R_2 = arabinose (m/z = 419.4)
Peonidin 3-glucoside: R_1 = CH_3, R_2 = glucose (m/z = 463.4)
Peonidin 3-galactoside: R_1 = CH_3, R_2 = galactose (m/z = 463.4)
Peonidin 3-arabinoside: R_1 = CH_3, R_2 = arabinose (m/z = 433.4)

B

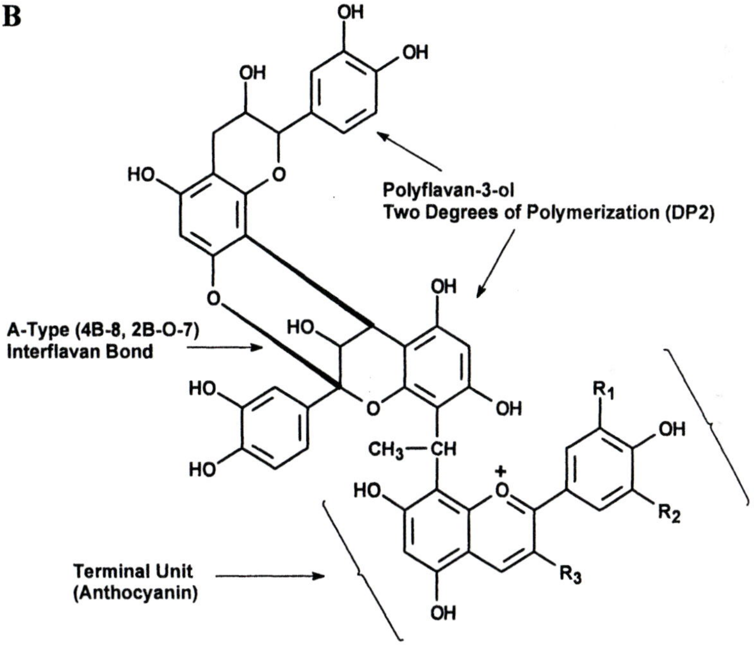

Figure 1. (A) Anthocyanins from Hyred cranberry fruit (Vaccinium macrocarpon, Ait.) and spray dried juice. (B) Tentative structure representing an anthocyanin linked through a CH_3-CH bridge to a polyflavan-3-ol of 2 degrees of polymerization containing an A-type interflavan bond.

amu), peonidin-3-glucoside (463.4 amu) and peonidin-3-arabinoside (433.4 amu)], 28 is the molecular weight of the CH_3-CH bridge, *a* is the DP contributed by the repeating flavan-3-ol unit and *b* is the number of A-type interflavan bonds.

Ionization

Anthocyanins in the presence of acidic matrixes such as *t*-IAA are predominately in the aromatic oxonium ion form (*23*) and are detected as $[M]^+$ ions by MALDI-TOF mass spectral analysis (*23,24*). However, anthocyanins (*23*), in the same manner as polyflavan-3-ols (*21,22,25*), may also associate with naturally abundant sodium $[M + Na]^+$ and potassium $[M + K]^+$ forming alkali metal adducts. To suppress the formation of alkali metal adducts, we deionized the anthocyanin-polyflavan-3-ol/matrix solution prior to deposition on the target. This approach resulted in the detection of anthocyanin-polyflavan-3-ols in the oxonium ion form $[M]^+$.

Anthocyanin-polyflavan-3-ols with B-Type and A-Type Interflavan Bonds

The MALDI-TOF mass spectra of the Sephadex-water/ethanol (1:1) eluate (fraction 2) of both the spray dried cranberry juice and cranberry fruit showed a series of anthocyanins; cyanidin-pentoside (*m/z* 419.3), peonidin-pentoside (*m/z* 433.3), cyanidin-hexoside (*m/z* 449.3) and peonidin-hexoside (*m/z* 463.3) (Figure 2A). MALDI-TOF MS equipped with delayed extraction provide unit mass resolution, allowing for the visualization of isotopic distribution. The reported observed masses (*m/z*) correspond to the monoisotope of the predicted compound. For example; the predicted and observed monoisotope of cyanidin-pentoside is (*m/z* 419.3) representing the contribution of ^{12}C, ^{1}H and ^{16}O to the compound. The mass at (*m/z* 420.3) represents one ^{13}C, or one ^{2}H, or one ^{17}O. The mass at (*m/z* 421.3) represents two ^{13}C, or one ^{13}C and one ^{2}H, or one ^{18}O, or two ^{2}H. Mass calculating programs such as IsoPro 3.0 [Shareware at: http://members.aol.com/msmsoft] can be used to predict the isotopic distribution of compounds and allow for comparison between predicted and observed isotopic distributions.

A series of anthocyanins linked to a single flavan-3-ol (Δ 288 amu) unit via a CH_3-CH (Δ 28 amu) bridge were also seen; cyanidin-pentoside-flavan-3-ol (*m/z* 735.3), peonidin-pentoside-flavan-3-ol (*m/z* 749.3), cyanidin-hexoside-flavan-3-ol (*m/z* 765.4) and peonidin-hexoside-flavan-3-ol (*m/z* 779.3) (Figure 2B). Again, the observed isotopic distribution is in agreement with the predicted distribution.

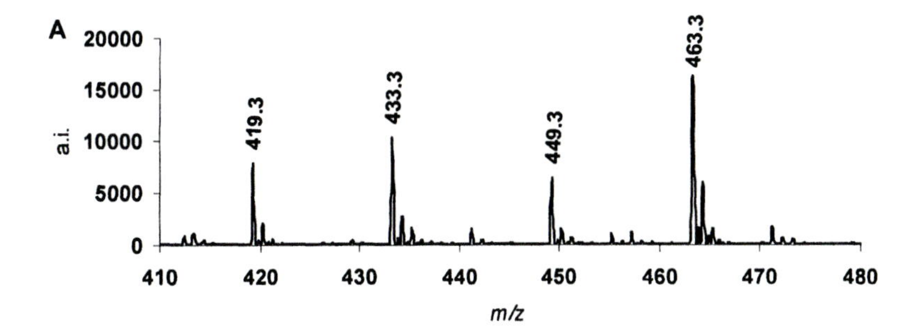

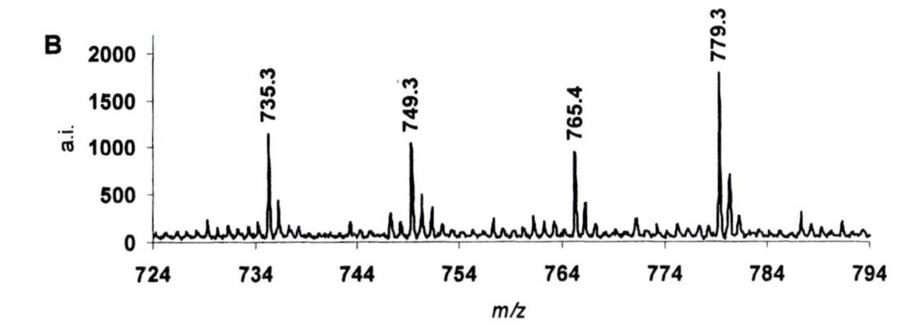

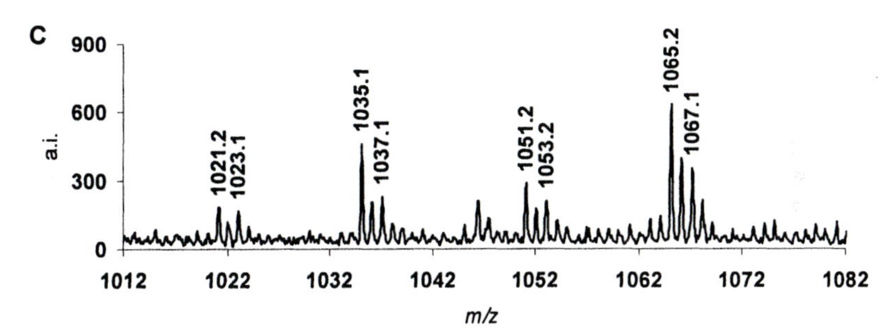

Figure 2. MALDI-TOF positive reflectron mode mass spectra of the anthocyanin-polyflavan-3-ol oligomers of Hyred cranberry fruit and spray dried juice. (A) Anthocyanins $[M]^+$. (B) Anthocyanin linked to a single flavan-3-ol through a CH_3-CH bridge $[M]^+$. (C) Anthocyanin linked to a polyflavan-3-ol of 2 degrees of polymerization through a CH_3-CH bridge, containing either an A-type or a B-type interflavan bond $[M]^+$.

A series of anthocyanins linked to polyflavan-3-ols of 2 degrees of polymerization (DP2) with either an A-type or B-type interflavan bonds (Δ 2 amu) were observed; cyanidin-pentoside-DP2 (A-type = *m/z* 1021.2, B-type = *m/z* 1023.1), peonidin-pentoside-DP2 (A-type = *m/z* 1035.1, B-type = *m/z* 1037.1), cyanidin-hexoside-DP2 (A-type = *m/z* 1051.2, B-type = *m/z* 1053.2) and peonidin-hexoside-DP2 (A-type = *m/z* 1065.2, B-type = *m/z* 1067.1) (Figure 2C). The B-type interflavan bonds consist of flavan-3-ol units linked by a carbon-carbon bond at positions C4-C6 or C4-C8, while A-type interflavan bonds contain both a carbon-carbon bond (4β-8) and an ether linkage (2β-*O*-7) between adjacent flavan-3-ol subunits (Figure 1B). The molecular weight difference between a B-type interflavan bond and an A-type interflavan bond is due to the loss of 2 hydrogen atoms (Δ 2 amu) during the formation of the ether linkage.

The isotopic distribution of the anthocyanin-polyflavan-3-ol containing an A-type interflavan linkage will overlap with the isotopic distribution of the anthocyanin-polyflavan-3-ol, of the same DP, containing a B-type interflavan linkage. The overlapping isotope distributions results in a summation of the observed intensity. For example, the mass observed at (*m/z* 1023.1) is the summation of the monoisotope (^{12}C, ^{1}H and ^{16}O) of cyanidin-pentoside-DP2 with a B-type interflavan bond and the contribution of isotope distribution (two ^{13}C, or one ^{13}C and one ^{2}H, or one ^{18}O, or two ^{2}H) attributed to the cyanidin-pentoside-DP2 with an A-type interflavan bond. The summation of predicted isotopic distributions generated by IsoPro 3.0 is in agreement with the observed overlapping isotopic clusters.

The MALDI-TOF mass spectra of Sephadex-ethanol eluate (fraction 3) contained a series of masses corresponding to anthocyanins linked to polyflavan-3-ols of DP1 to DP4 and the Sephadex-ethanol:methanol (1:1) eluate (fraction 4) contained a series of masses corresponding to anthocyanins linked to polyflavan-3-ols of DP3 to DP5 (Table I). As the degree of polymerization of the extending polyflavan-3-ol units increase the predicted iterations of A-type and B-type interflavan bonds (Δ 2 amu) at each degree of polymerization also increase. As an example; cyanidin-pentoside attached to a polyflavan-3-ol of DP3 is predicted to have 2 B-type interflavan bonds (*m/z* 1311.2), 1 A-type:1 B-type interflavan bond (*m/z* 1309.3) and 2 A-type interflavan bonds (*m/z* 1307.2), all three predicted masses were observed (Table I).

The Sephadex-methanol eluate (fraction 5) and Sephadex-80% aqueous acetone (v/v) eluate (fraction 6) contained a series of heteropolyflavan-3-ols which have been previously described (*2*) and indications of anthocyanin-polyflavan-3-ol oligomers of DP6 and greater. However, due to the large number of individual compounds within the heteropolyflavan-3-ol series and anthocyanin-polyflavan-3-ol series it was difficult to obtain high mass spectral resolution of any individual anthocyanin-polyflavan-3-ol oligomers of higher degrees of polymerization. It is a future goal to further develop methods of chromatographic separation which will provide narrower mass range distributions allowing higher mass spectral resolution of individual compounds within each subfraction.

Conclusion

The combination of liquid chromatographic separation and MALDI-TOF MS indicates that the structural heterogeneity of cranberry pigments is much greater than previously described. Mass spectral data indicates that cranberry fruit and spray dried cranberry powder both contain a series of anthocyanin-polyflavan-3-ol oligomers, linked through an actetaldehyde derived bridge, with both A-type and B-type interflavan bonds. However, MALDI-TOF MS only allows for tentative structural assignments, as this analytic technique suffers from the inability to assign specific stereochemistry to the observed compounds. Conformational NMR analysis of isolated anthocyanin-polyflavan-3-ols of DP2 or DP3, used in concert with MALDI-TOF MS would provide the most solid foundation for the interpretation of novel compounds such as the proposed anthocyanin-polyflavan-3-ols in cranberry fruit.

The incorporation of acetaldehyde derived bridges between anthocyanins and flavan-3-ols via acetaldehyde condensation reactions has been well described in fermented beverages such as red wine. The presence of acetaldehyde in alcoholic solutions is attributed to either oxidatative products of ethanol or microbial byproducts. In the case of cranberry fruit and spray dried juice neither product was subjected to yeast fermentation. It becomes a concern then that the observed anthocyanin-pigments may be an artifact of harvest, storage, juice processing or analytic techniques.

Lu and Foo (*27*) report that anthocyanins may undergo reactions with acetone to give rise to pyranoanthocyanins. This point must be addressed as our extraction of the whole cranberry fruit and subsequent chromatographic isolation employed aqueous acetone solutions. While our mass spectral data does not reveal masses corresponding to pyranoanthocyanins we remain vigilant of the effects organic solvents may have on phenolic compounds.

Ethanol was also used in the elution of phenolics from the Sephadex LH20 column. While acetaldehyde is known to be generated from oxidation of ethanol in the presence of phenolic compounds (*4*), the model solutions used to investigate this reaction are typically carried out under conditions of high temperature (> 50° C) and for long periods of time (> 30 days). Our analysis of anthocyanin-polyflavan-3-ols from cranberry fruit was completed at

Table I. Observed and Calculated Masses of Anthocyanin-polyflavan-3-ols in Spray Dried Cranberry Juice and Hyred Cranberry Fruit by MALDI-TOF MS

Base Anthocyanin	*# Flavan-3-ols units (1 CH-CH3 bridge)*	*# B type bonds*	*# no. A typeb bonds*	*Calculated Mass [M]+*	*90MX Observed Mass [M]+*	*Berry Observed Mass [M]+*
cyanidin pentose	3	0	2	1307.6	1307.2	1307.0
	3	1	1	1309.6	1309.3	1309.0
	3	2	0	1311.6	1311.2	1311.0
peonidin pentose	3	0	2	1321.6	1321.2	1321.0
	3	1	1	1323.6	1323.1	1323.0
	3	2	0	1325.6	1325.3	1325.0
cyanidin hexose	3	0	2	1337.6	1337.3	1337.0
	3	1	1	1339.6	1339.2	1339.0
	3	2	0	1341.6	1341.3	1341.1
peonidin hexose	3	0	2	1351.6	1351.1	1351.0
	3	1	1	1353.6	1353.1	1353.0
	3	2	0	1355.7	1355.2	1355.0
cyanidin pentose	4	0	3	1593.6	*c*	*c*
	4	1	2	1595.6	*c*	*c*
	4	2	1	1597.7	1597.2	*c*
	4	3	0	1599.7	*c*	1599.2
peonidin pentose	4	0	3	1607.6	*c*	*c*
	4	1	2	1609.7	*c*	*c*
	4	2	1	1611.7	1611.2	1611.7
	4	3	0	1613.7	*c*	1613.7
cyanidin hexose	4	0	3	1623.7	*c*	*c*
	4	1	2	1625.7	1625.1	1625.6
	4	2	1	1627.7	1627.0	1627.7
	4	3	0	1629.7	*c*	1629.7

peonidin hexose	4	0	3	1637.7	1637.0	*c*
	4	1	2	1639.7	1639.1	1639.0
	4	2	1	1641.7	1641.1	1641.0
	4	3	0	1643.7	*c*	1643.0
cyanidin pentose	5	0	4	1879.7	*c*	*c*
	5	1	3	1881.7	*c*	*c*
	5	2	2	1883.7	*c*	*c*
	5	3	1	1885.7	*c*	*c*
	5	4	0	1887.7	1887.8	1887.2
peonidin pentose	5	0	4	1893.7	*c*	*c*
	5	1	3	1895.7	*c*	*c*
	5	2	2	1897.7	1897.8	*c*
	5	3	1	1899.7	1899.8	*c*
	5	4	0	1901.8	*c*	*c*
cyanidin hexose	5	0	4	1909.7	*c*	*c*
	5	1	3	1911.7	1911.8	*c*
	5	2	2	1913.7	1913.8	*c*
	5	3	1	1915.7	1915.1	*c*
	5	4	0	1917.8	1917.8	1917.3
peonidin hexose	5	0	4	1923.7	*c*	*c*
	5	1	3	1925.7	*c*	*c*
	5	2	2	1927.8	*c*	*c*
	5	3	1	1929.8	1929.8	1929.4
	5	4	0	1931.8	1931.7	1931.4

[a]Mass calculations were based on the equation: *Anthocyanin* + 28 +288*a* -2*b*, where *anthocyanin* represents the molecular weight (MW) of the terminal anthocyanin, 28 is the MW of the CH-CH_3 bridge, *a* is the DP of the extending flavan-3-ol units, and *b* is the number of A-type interflavan bonds. [b]Formation of each A-type interflavan ether linkage leads to the loss of two hydrogen atoms (Δ 2 amu). [c]Masses were not observed.

temperatures less than 30° C in less than 12 hours, from the time of extraction to the time of mass spectral analysis.

We propose it is more likely that the anthocyanin-polyflavan-3-ol oligomers arise as a result of fruit ripening and processing rather than as an artifact of analytic analysis. Biologically available acetaldehyde in the berry may be available for the condensation reaction between anthocyanins and polyflavan-3-ols. Fermentative metabolism in cranberry gives rise to acetaldehyde as a result of fruit ripening or during harvest and storage. In grapes and other fruits that accumulate malic acid, such as cranberries (*28*), induction of ethanol production may be associated with increased levels of pyruvate within the fruit (*29*). Induction of malate decarboxylation increases level of pyruvate at the onset of fruit ripening (*30*). In the process of ethanol synthesis from pyruvate, the conversion of pyruvate to acetaldehyde by pyruvate decarboxylase precedes the reduction of acetaldehyde to ethanol by alcohol dehydrogenase (*29, 30*). Pyruvate is located at the branching point between two alternative energy producing processes: conversion to acetyl CoA by pyruvate dehydrogenase leads to respiration via the TCA cycle, whereas conversion to acetaldehyde by pyruvate decarboxylase leads to fermentation (*29*).

Alternatively, acetaldehyde may arise from fermentative metabolism as a result of harvest and storage. The initial step in cranberry harvesting often involves the flooding of the cranberry bog. Submersion of the plants in water decreases the external oxygen concentration (*31*). In response to this environmental challenge the plant shifts from respiration to fermentation as a means of generating ATP, resulting in the production of acetaldehyde via pyruvate decarboxylase (*31, 32*). Cranberries are known to accumulate both acetaldehyde and ethanol as a result of storage (*33*).

Acknowledgements

Financial support was provided through USDA-CSREES-NRI Award # 2002-355030-12297. 90MX spray-dried cranberry juice powder was provided by Ocean Spray Cranberries, Inc.

References

1. Hong, V.; Wrolstad, R. *J. Agric Food Chem.* **1990**, *38*, 708-715.
2. Porter, M. L.; Krueger, C. G.; Wiebe, D. A.; Cunningham, D. G.; Reed, J. D. *J. Sci. Food Agric.* **2001**, *81*, 1306-1313.
3. Kennedy, J. A.; Matthews, M. A.; Waterhouse, A. L. *Am. J. Enol. Vitic.* **2002**, *53*, 268-274.
4. Wildenradt, H. L; Singleton, V. L. *Am. J. Enol. Vitic.* **1974**, *25*, 119-126.
5. Timberlake, C. F.; Bridle, P. *Am. J. Enol. Vitic.* **1976**, *27*, 97-105.

6. Remy, S.; Fulcrand, H.; Labarbe, B.; Cheynier, V.; Moutounet, M. *J. Sci. Food Agric.* **2000**, *80*, 745-751.
7. Shoji, T.; Yanagida, A.; Kanda, T. *J. Agric. Food Chem.* **1999**, *47*, 2885-2890.
8. Rivas-Gonzalo, J. C.; Bravo-Haro, S.; Santos-Buelga, C. *J. Agric. Food Chem.* **1995**, *43*, 1444-1449.
9. Dallas, C.; Ricardo-da-Silva, J. M.; Laureno, O. *J. Agric. Food Chem.* **1996**, *44*, 2402-2407.
10. Saucier, C.; Guerra, C; Pianet, I.; Laguerre, M.; Glories, Y. *Phytochemistry.* **1997**, *46*, 229-234.
11. Es-Safi, N.; Fulcrand, H.; Cheynier, V.; Moutounet, M. *J. Agric. Food Chem.* **1999**, *47*, 2096-2102.
12. Shoji, T.; Goda, Y.; Toyoda, M.; Yanagida, A.; Kanda, T. *Phytochemistry* **2002**, *59*, 183-189.
13. Asenstorfer, R. E.; Hayaska, Y.; Jones, G. P. *J. Agric. Food Chem.* **2001**, *49*, 5957-5963.
14. Hayaska, Y.; Asenstorfer, R. E. *J. Agric. Food Chem.* **2002**, *50*, 756-761.
15. Mateus, N.; Silva, A. M.; Santos-Buelga, C.; Rivas-Gonzalo, J. C.; De Freitas, V. *J. Agric. Food Chem.* **2002**, *50*, 2110-2116.
16. Reme-Tanneau, S.; Le Guerneve, C.; Meudec, E.; Cheynier, V. *J. Agric. Food Chem.* **2003**, *51*, 3592 -3597.
17. Montaudo, G.; Montaudo, M. S.; Samperi, F. In *Mass Spectrometry of Oligomers*; Montaudo G., Lattimer R. P. Eds.; CRC Press: Boca Raton, FL 2002; pp 419-521.
18. Hanton, S. D. *Chem Rev.* **2001**, *101*, 527-569.
19. Krueger, C. G.; Dopke, N.; Treichel, P. M.; Folts, J.; Reed, J. D. *J. Agric. Food Chem.* **2000**, *48*, 1663-1667.
20. Yang, Y.; Chien, M. *J. Agric Food Chem.* **2000**, *48*, 3990-3996.
21. Krueger, C.G.; Vestling, M. M.; Reed, J. D. *J. Agric. Food Chem.* **2003**, *51*, 538-543.
22. Takahata, Y.; Ohnishi-Kameyama, M.; Furuta, S.; Takahashi, M.; Suda, I. *J. Agric. Food Chem.* **2001**, *49*, 5843-5847.
23. Wang, J.; Sporns, P. *J. Agric. Food Chem.* **1999**, *47*, 2009-2015.
24. Wang, J.; Wilhelmina, K.; Sporns, P. *J. Agric. Food Chem.* **2000**, *48*, 3330-3355.
25. Ohnishi-Kameyama, M.; Yanagida, A.; Kanda, T.; Nagata, T. *Rapid Commun. Mass Spectrom.* **1997**, *11*, 31-36.
26. Foo, L.Y.; Lu, Y.; Howell, A.B.; Verosa, N. *J. Nat. Prod.* **2000**, *63*, 1225-1228.
27. Lu, Y.; Foo, Y. *Tetrahedron Lett.* **2001**, *42*, 1371-1373.
28. Jensen, H.D.; Krogfelt, K. A.; Cornett, C.; Hansen, S. H.; Christensen, S. B. *J. Agric. Food Chem.* **2002**, 50, 6871-6874.
29. Or, E.; Baybik, J.; Sadka, A.; Ogrodovitch, A. *Plant Sci.* **2000**, 156, 151-158.

30. Ruffner, H.P. *Vitis.* **1982**, 21, 346-358.
31. Kursteiner, O.; Dupuis, I.; Kuhlemeier, C. *Plant Physiol.* **2003**, 132, 968-978.
32. Geigenberger, P. *Curr. Opin. Plant Biol.* **2003**, 6, 247-256.
33. Gunes, G.; Liu, R. H.; Watkins, C. B. *J. Agric. Food Chem.* **2002**, 50, 5932-5938.

Chapter 15

Compositional Investigation of Pigmented Tannin

James A. Kennedy[1] and Yoji Hayasaka[2]

[1]Department of Food Science and Technology, Oregon State University, Corvallis, Oregon 97331 (telephone 1–541–737–9150; fax: 1–541–737–1877; email: James.Kennedy@oregonstate.edu)
[2]The Australian Wine Research Institute, P.O. Box 197, Glen Osmond, South Australia 5064, Australia and the Cooperative Research Centre for Viticulture, P.O. Box 154, Glen Osmond, South Australia 5064, Australia

The composition of tannin isolated from a red wine (cv. Pinot noir) was investigated in order to improve our understanding of the associated pigmentation. Utilizing a variety of analytical techniques including high performance gel permeation chromatography, acid catalyzed cleavage, nuclear magnetic resonance spectroscopy, mass spectrometry and UV/Vis spectrophotometry, evidence for the direct condensation between anthocyanins and proanthocyanidins was found. By mass spectrometry, polymeric direct condensation products consistent with an anthocyanin linked to a proanthocyanidin containing 7 subunits were observed.

INTRODUCTION

The color of red wine is an important aspect of red wine quality, and is a major factor by which wine is initially assessed, often affecting purchasing decisions. Because of this, improving and stabilizing red wine color is a perennial concern in the wine industry.

Anthocyanins are grape-derived flavonoid compounds and are initially responsible for the color of red wine. The red appearance of anthocyanins in wine is dependent upon many factors including wine pH, ethanol concentration, copigmentation and anthocyanin substitution pattern.

Although the anthocyanin substitution pattern in grape-derived material can vary, for most of the world's important wine varietals, malvidin-3-glucoside is the most significant anthocyanin. Yet, while the substitution pattern of extracted anthocyanins is initially somewhat limited, the complexity of anthocyanin structure increases tremendously once extracted into wine, and during subsequent aging. By the time red wine is bottled, the grape-derived anthocyanin structures have undergone significant transformation (*1*). Investigations into this structural transformation have been a perennial topic of research for many decades. Out of this research, new structures have been isolated and characterized (*2-9*).

Despite these advances, it is generally accepted that the vast majority of the observed color in aged red wine remains to be characterized. The obvious factor that separates characterized compounds from uncharacterized compounds appears to be molecular weight, with much of the high molecular weight material being unknown in structure. What is apparent is that as wine ages, the observed red color increases in size, apparently due to interaction between anthocyanins and proanthocyanidins (tannin) (*10*).

Several products and plausible mechanisms have been described for the incorporation of anthocyanins into the tannin pool (*11-13*). Confirming the structure of these species is likely to be a formidable challenge due to the complexity of the likely products combined with limitations in the chromatographic and characterization sciences. Nevertheless progress continues to be made (*14,15*).

The purpose of this chapter is to summarize the findings of a research project conducted on the characterization of pigmented tannin isolated from red wine. In this work, a deliberate attempt was made to exclude lower molecular weight material. The wine selected for study was a 3-year-old Pinot noir. This wine was selected because it was at an age that is significant from a market standpoint, and also because cv. Pinot noir has a very high proportion of malvidin-3-glucoside and lacks acylation. Because of this, the complexity of the pigmented tannin should be simplified.

MATERIALS AND METHODS

With the exception of the NMR work, the materials and methods have been previously described (*16*).

^{13}C NMR Spectroscopy

Purified pigmented tannin (100 mg/mL, 1:1 acetone-d_6:D_2O) was characterized by ^{13}C-NMR (100 MHz, Bruker DPX400), with chemical shifts (δ) in ppm referenced internally with acetone-d_6. The proton-decoupled, inverse-gated sequence, with 30° pulse length, 25,000 Hz spectral width, 64 K data points, 1.3 s acquisition time, relaxation delay of 3 ms, 50 K scans and with 3 Hz line broadening was carried out at a temperature of 300 K.

RESULTS AND DISCUSSION

The long-term practical goal of grape and wine tannin research is to improve our ability to manage wine texture and color. Understanding the structural modifications that tannins undergo during wine aging is an important part of achieving this goal. With an understanding of structural modifications, it is hoped that process modifications can be adopted which will enhance the quality of tannin perception and improve color stability. Beginning with the knowledge that red wine color becomes increasingly associated with polymeric material as wine ages, the first part of the research discussed here was designed to isolate and purify pigmented tannin from red wine (*16*).

The wine selected for isolation was a 3-year old Pinot noir. When analyzed by reversed-phase HPLC, this wine contained a combination of grape-derived anthocyanins (Figure 1), with malvidin-3-glucoside (Mv-3-glc) being the most prominent resolved peak. In addition, when magnified, a significant amount of polymeric material (Figure 1 inset) was present (41% of pigment in Mv-3-glc equivalents). This broad unresolved material was the target for isolation. Previous studies have shown that this material increases as wine ages (*7*).

The pigmented wine tannin was isolated using a low-pressure adsorption chromatography technique used for tannin isolation (*17*). After this treatment, a deep red, pigmented isolate was obtained. The isolated material made up a significant amount of material in the wine (0.47 g/L).

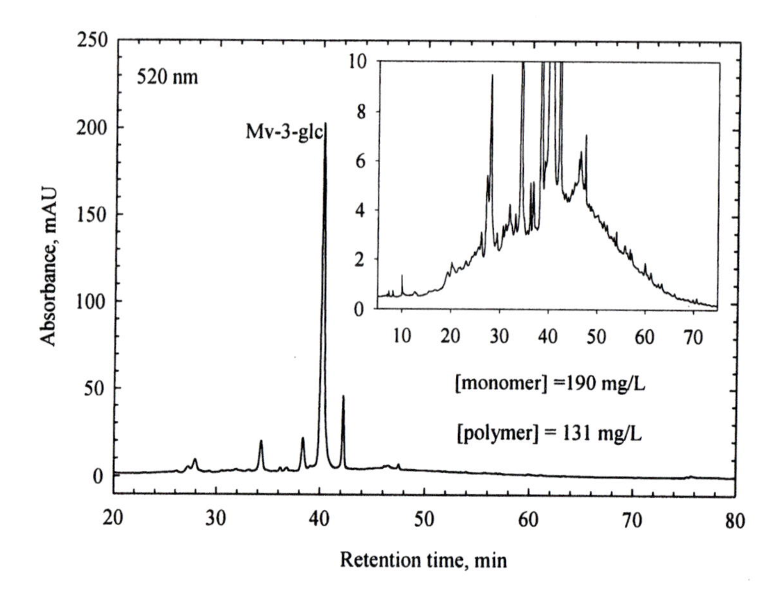

Figure 1. Reversed-phase HPLC chromatogram with monitoring at 520 nm of wine prior to wine tannin isolation. Inset shows detail of the baseline. (Reproduced from reference 16, with permission from the Australian Society of Viticulture and Oenology)

The pigmented tannin was analyzed using a combination of chromatography techniques including acid-catalysis in the presence of phloroglucinol (*17,18*) and high performance gel permeation chromatography (*18*). Phloroglucinol analysis (Table 1) indicated that a large portion of the pigmented tannin was composed of proanthocyanidin (45% w/w). Comparison of the pigmented tannin with grape seed and skin proanthocyanidins isolated from cv. Pinot noir, suggested that the tannin had undergone a significant amount of modification given that the conversion yield was much lower than grape derived proanthocyanidins. Comparing the subunit composition of the three isolates suggested that a significant amount of seed and skin derived material was present in the pigmented tannin.

Based upon analysis by high performance gel permeation chromatography (Figure 2), the isolation and purification strategy was successful in that very little monomeric material was present in the isolate. In comparing the retention properties of the pigmented tannin with that of purified grape seed and skin proanthocyanidin isolates, and using calculated molecular weight estimates based upon phloroglucinol analysis, the estimated molecular mass of the pigmented tannin as estimated by GPC and phloroglucinol analysis, was similar. So, despite the low conversion yield as determined by phloroglucinol analysis, the overall hydrodynamic volume of the pigmented tannin was similar to expected results (based upon phloroglucinol analysis).

Interestingly, the elution pattern of the pigmented tannin was bimodal. Given that the molecular weight distribution of seed and skin tannins is very different, one possible explanation for the bimodal distribution of the tannin is that it represents seed (later eluting material) and skin (earlier eluting material) extraction.

From these results it could be confirmed that the tannin isolated from wine (in a manner consistent with techniques used in grape tissues), was deeply pigmented. It was confirmed that the isolate contained very little low molecular weight material indicating that most of the pigment was polymeric in nature. Finally, the pigmented tannin contained a significant amount of proanthocyanidin material.

^{13}C NMR Spectroscopy

Although a considerable amount of the isolate was characterized as containing proanthocyanidin material, approximately 55% w/w of the material was unknown. The most obvious non-proanthocyanidin material contained by the isolate was the pigment. To acquire additional information on this pigment within the isolate, ^{13}C NMR spectroscopy was employed.

Given that a major portion of the pigmented tannin was unknown (~55% w/w), it was somewhat surprising that the spectrum was fairly straightforward in its interpretation (Figure 3). When compared to the ^{13}C NMR spectrum for grape skin proanthocyanidins (*19*), the pigmented tannin spectrum was very similar, indicating that much of the spectrum was consistent with a proanthocyanidin.

Some signals however were inconsistent with proanthocyanidin structure (noted in the figure). The pigmented tannin spectrum was clearly more complex in the regions δ 80-50 ppm. In this region, there were several signals that could be assigned to an anthocyanidin (specifically Mv-3-glc) and glucose (*20,21*). Specifically, δ 57 ppm is consistent with the methoxy group of malvidin and δ 62 ppm is consistent with the glucose C-6.

Additional signals consistent with a glucosyl moiety were also present although signal assignment was more ambiguous. Several downfield signals

Table 1. Results from acid-catalysis in the presence of excess phloroglucinol for the pigmented polymer (Wine) and grape seed and skin proanthocyanidins isolated from cv. Pinot noir. (Reproduced from reference 16, with permission from the Australian Society of Viticulture and Oenology)

Sample	*EGC-P*[a]	*C-P*	*EC-P*	*C*	*ECG-P*	*EC*	*ECG*	*Yield*[b]	*mDP*[c]	*est. MW*[d]	*log MW*
Wine	0.143	0.050	0.657	0.088	0.029	0.031	-	0.45	8.3	2459	3.39
Skin	0.294	0.023	0.610	0.053	0.012	0.008	-	0.66	16.4	4841	3.68
Seed	-	0.123	0.637	0.101	0.089	0.050	0.066	0.79	4.8	1597	3.20

[a] Proportional composition of proanthocyanidins (in moles), and with the following subunit abbreviations:

EGC-P: (-)-epigallocatechin extension subunit

C-P: (+)-catechin extension subunit

EC-P: (-)-epicatechin extension subunit

C: (+)-catechin terminal subunit

ECG-P: (-)-epicatechin-3-*O*-gallate extension subunit

EC: (-)-epicatechin terminal subunit

ECG: (-)-epicatechin-3-*O*-gallate terminal subunit

[b] Conversion yield (by mass) in the conversion of proanthocyanidin fraction to known subunits

[c] Mean degree of polymerization

[d] Estimated average molecular mass based upon proportional composition and mDP

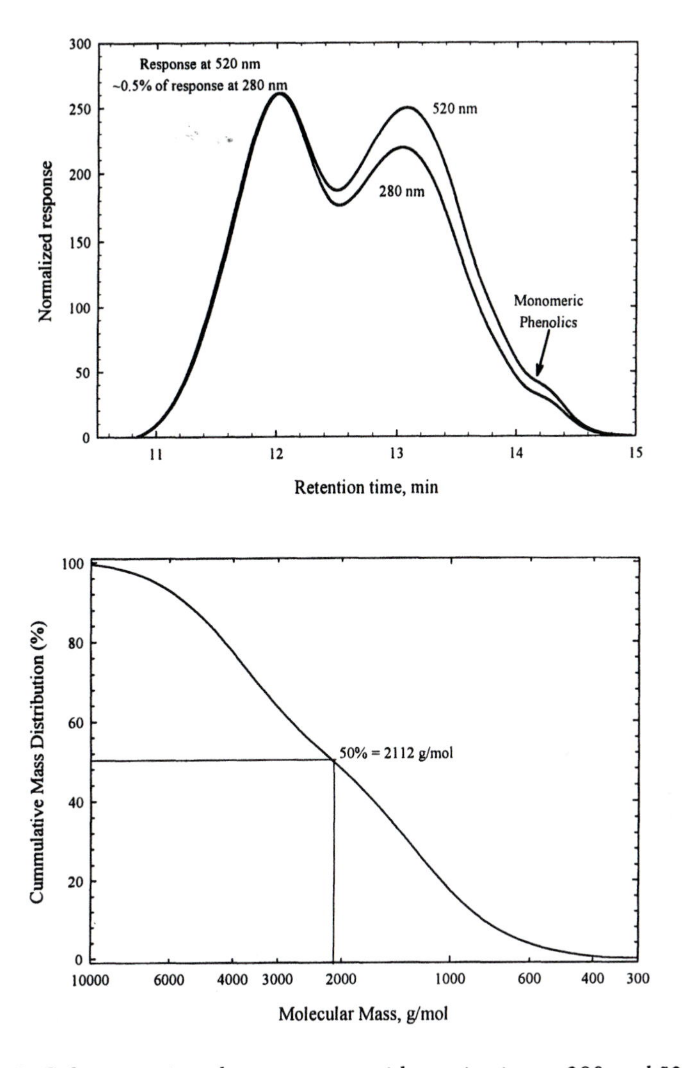

Figure 2. Gel permeation chromatogram with monitoring at 280 and 520 nm of wine tannin. (a) Normalized mass elution profile and (b) cumulative mass elution versus molecular mass. (Reproduced from reference 16, with permission from the Australian Society of Viticulture and Oenology)

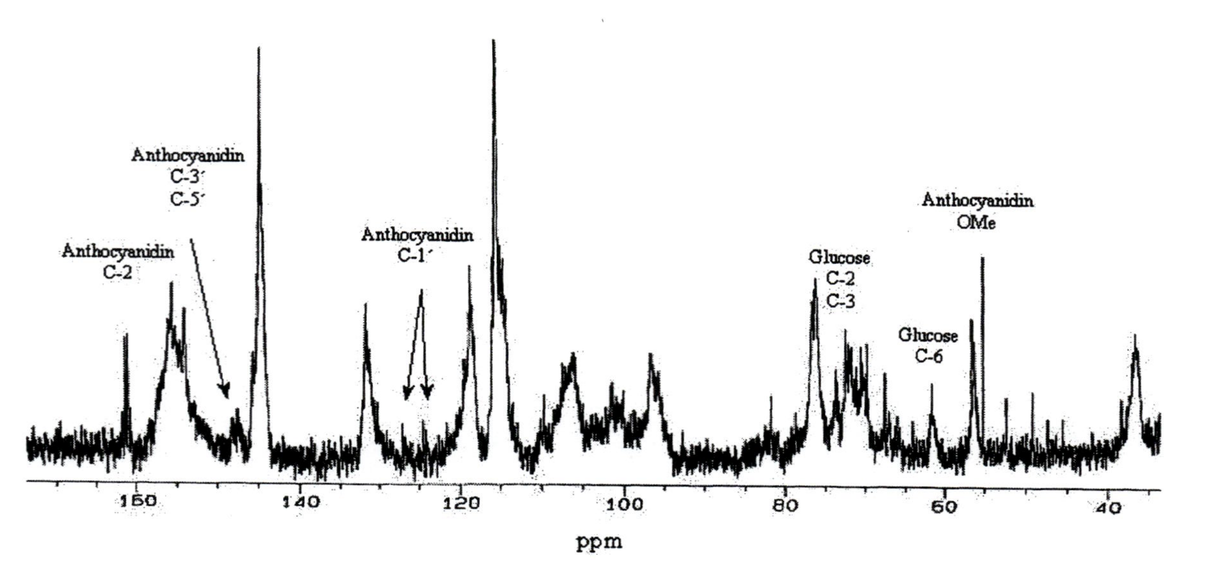

Figure 3. ^{13}C *NMR spectrum of pigmented tannin isolated from red wine, with anthocyanin signals indicated.*

inconsistent with proanthocyanidins (δ 161, 147, 125 ppm) were also present. Again, these signals were consistent with anthocyanidin substitution. Given the broadness of the signals, combined with the results of GPC, the presumed anthocyanin pigmentation was consistent with being polymeric.

Mass Spectrometry

Mass spectrometry has become increasingly important in the identification of new wine pigments. Its use has historically been restricted to lower molecular weight compounds, although recent research suggests that it has the potential for providing information for higher molecular weight flavonoid material (*22*).

Analysis of the pigmented tannin by LC-MS indicated that the material eluted as an unresolved broad peak at 280 and 520 nm, although several minor resolved pigment peaks were present (*16*). The mass spectrum of one of the resolved peaks agreed with malvidin-3-glucoside. Overall, LC-MS analysis of the pigmented tannin confirmed that much of the material was polymeric, consistent with results obtained by high performance gel permeation chromatography.

After a progressive decline in anthocyanin concentration, the presence of a broad unresolved peak is typically observed from aged wine when analyzed by reversed-phase HPLC (*23*) or gel permeation chromatography (*7*). Historically this hump was considered to be composed primarily of pigmented polymers, although until recently (*16*) evidence for this was based upon chromatographic evidence.

The pigmented tannin was largely composed of proanthocyanidin material when analyzed by infusion mass spectrometry, consistent with expectations. Specifically, there was a series of dominant ions separated by a mass of 288 and which covered a range from *m/z* 579 to 2307 (nominal mass). This series corresponds to the respective molecular ions of procyanidins from dimers to octamers (Figure 4). In addition, a series of ions derived from a broad range of proanthocyanidins containing epigallocatechins or epicatechin-3-*O*-gallates was also observed.

Evidence for the presence of covalent adducts involving proanthocyanidins and anthocyanidins was also observed by infusion mass spectrometry. These minor ions appeared in two series, one series starting at *m/z* 781 and one at *m/z* 783. Each of these series is observable up to *m/z* 2509/2511, with ions separated by a mass of 288 (Figure 4). Based upon previous studies and speculation, these ions were consistent with two types of anthocyanin-proanthocyanidin adducts. Specifically, these ions agree with the direct condensation products of

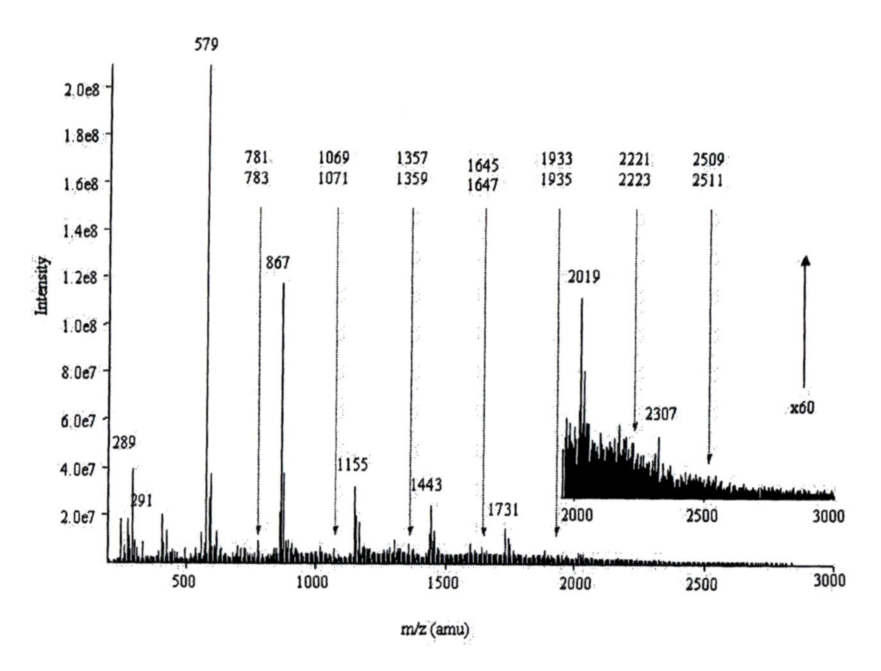

Figure 4. Electrospray mass spectrum of infused pigmented tannin. (Reproduced from reference 16, with permission from the Australian Society of Viticulture and Oenology)

malvidin-3-glucoside and procyanidin, with the anthocyanin bound to the flavan-3-ol via the A ring of the anthocyanin (T-A type) or the C-ring of the anthocyanin (A-T type) or an A-type linkage (indistinguishable from the A-T type) as reported previously (*14*, Figure 5).

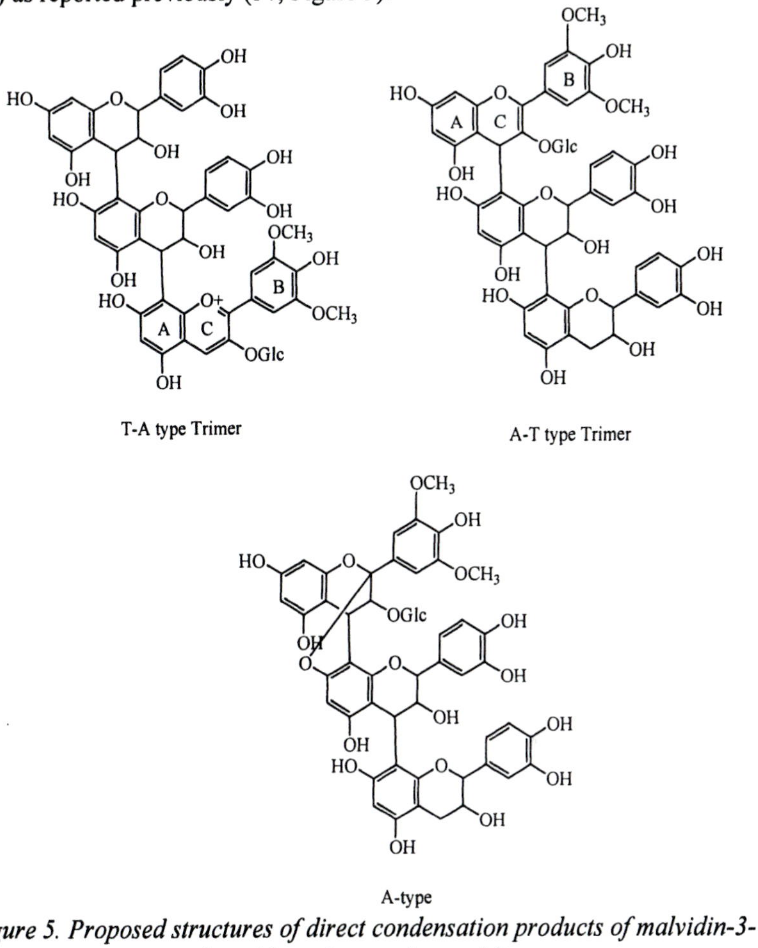

T-A type Trimer

A-T type Trimer

A-type

Figure 5. Proposed structures of direct condensation products of malvidin-3-glucoside and proanthocyanidins.

Ions corresponding to dimers but containing other anthocyanins were also found at *m/z* 751/753 for peonidin-3-glucoside, *m/z* 753/755 for delphinidin-3-glucoside and *m/z* 767/769 for petunidin-3-glucoside (Figure 6a). In addition, ions consistent with a dimer composed of malvidin-3-glucoside and epigallocatechin (*m/z* 797/799) were also observed. This general pattern increased through both series although the signal intensity declined with each additional flavan-3-ol subunit.

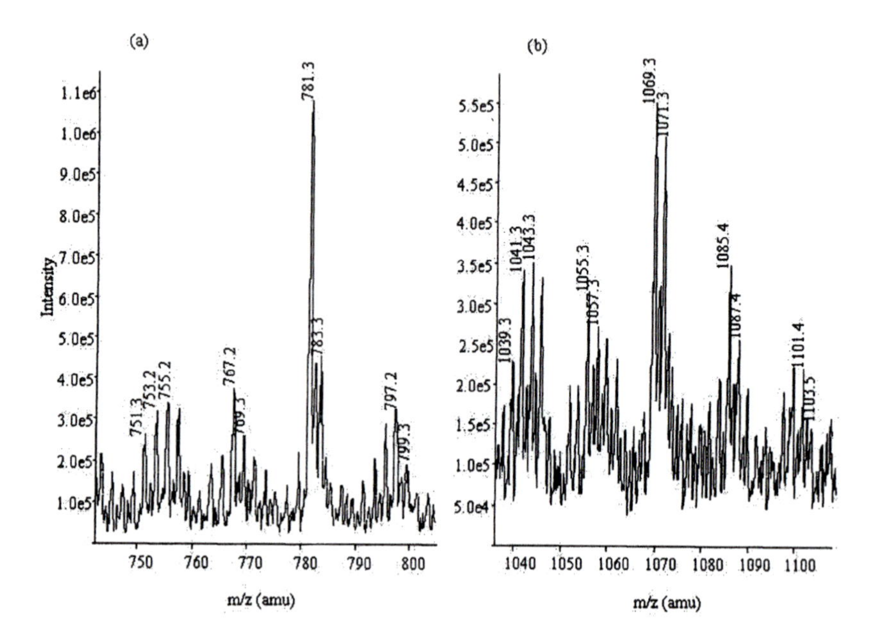

Figure 6. Partial mass spectra representing (a) dimers, (b) trimers, obtained from pigmented tannin and analyzed by infusion ES-MS. (Reproduced from reference 16, with permission from the Australian Society of Viticulture and Oenology)

To further characterize the pigmented tannin, the product ion spectra were obtained by infusion ES-MS/MS. Procyanidins were predominantly fragmented by cleavage at the interflavonoid bonds to produce sequence ions separated by a mass of 288 (*24*). Retro-Diels-Alder (RDA) fission of the flavan-3-ol subunit was also evident based upon the production of a characteristic ion (*m/z* 152, *24*). Anthocyanins were fragmented into anthocyanidins as a result of the elimination of the dehydrated sugar moiety with a mass of 162. Together, these elimination masses of 152, 162 and 288 were observed in the product ion spectra derived from the respective ions of T-A and A-T type polymers (Table 2), providing good evidence that the respective ions were composed of malvidin-3-glucoside and (epi)catechin units.

The A-T type polymers gave a more complicated fragmentation pattern and thus appeared to have an additional fragmentation pathway. For instance, A-T type polymers had a tendency to eliminate a mass of 290 instead of 288 for the T-A type polymers. In the case of procyanidins, cleavage of the interflavonoid bond (C4-C8 or C4-C6) gives rise to two types of fragment ions. One is the carbonium ion at the C4 position of the upper procyanidin unit and another is a protonated lower procyanidin unit (*25*). These fragment pathways probably occurred from anthocyanin-procyanidin polymers that apparently contained malvidin-3-glucoside as a top unit of the polymer (Figure 5). On the other hand, the T-A type polymers gave ions resulting from the loss of 288, suggesting the lower units of polymers contained malvidin-3-glucoside as a bottom unit.

In the formation of anthocyanin-proanthocyanidin polymers, it is hypothesized that the anthocyanin position of the polymer depends upon whether an electrophilic position (C-4) or nucleophilic position (C-8 or C-6) of the anthocyanin becomes involved in the interflavonoid linkage. The former and latter reactions would involve a nucleophilic position (C-8 or C-6) and electrophilic position (C4: a carbonium ion derived by acidic cleavage) of proanthocyanidins leading to the formation of the A-T (*14,26-27*) and T-A type (*12,14*) polymers, respectively.

The linkage of anthocyanin and proanthocyanidins in the A-T type polymer has been proposed to constitute a flavene and/or A-type (*14,28*) and both forms of the A-T type polymers would lack pigmentation based upon their structures (Figure 5).

Because the observed T-A and A-T type polymers could have been derived either by an ion-molecule reaction of anthocyanin and proanthocyanidin molecule (adduct ions) during the electrospray ionization process or from the same compound as fragment ions, the individual polymers were separated by LC and monitored by MS in the step scan mode, to ensure that the pigmented polymers were present in the sample. The results indicated that both polymer types were separated from each other, consistent with the T-A and A-T type polymers being different polymer series (*16*). The T-A type polymers eluted earlier than the A-T type polymers, indicating that the T-A type interacted less with the reverse-phase column. One possible explanation for this would be that the acidic conditions used in the mobile phase and the subsequent presence of the flavylium form would result in a more hydrophilic species. Additional evidence that the observed T-A and A-T type polymers are not artifacts is that the retention times of peaks were obviously different from that of malvidin-3-glucoside, suggesting that the ions were not adduct ions of malvidin-3-glucoside and proanthocyanidins formed during the electrospray process.

An interesting observation made in this study is that ethyl bridged anthocyanin-proanthocyanidin products were not observed. These adducts are believed to play a significant role in the color of aged red wine. The absence of evidence in this study suggests that ethyl bridged compounds are do not play a significant role in red wine color, or are unstable and are rapidly converted to other pigmented material.

Table 2. Characteristic elimination masses observed in product ion spectra. (Reproduced from reference 16, with permission from the Australian Society of Viticulture and Oenology)

T-A type polymers				*A-T type polymers*			
Precursor ions	*Masses eliminated by fragmentation*			*Precursor ions*	*Masses eliminated by fragmentation*		
	162	*288*	*152*		*162*	*288*	*152*
Dimer 781	781→619 493→331	781→493		*Dimer 783*	783→621 493→331	783→495	783→631 621→469
Trimer 1069	1069→907 781→619 493→331	1069→781 781→493	907→755	*Trimer 1071*	1071→909 783→621 781→619 495→333 493→331	1071→783 783→495 781→493	1071→919 783→631 621→469
Tetramer 1357	1357→1195 781→619 493→331	1357→1069 1069→781 781→493		*Tetramer 1359*	1359→1197 1071→909 783→621 781→619 493→331	1359→1071 1071→783 1069→781 783→495 781→493	783→631

Increase in anthocyanin upon acid-catalysis

UV/Vis spectra of the wine tannin in methanol obtained before and immediately after acidification indicated that the proportion of anthocyanin in the flavylium form could be increased rapidly (*16*). This increase in absorbance at 540 nm immediately following acidification can be based simply upon anthocyanin equilibria. Anthocyanin in the pigmented polymers (T-A type) should be shifted towards the flavylium form with a decrease in pH.

Interestingly, the red color increase observed upon acidification was accompanied by an increase in free malvidin 3-glucoside when analyzed by reversed-phase HPLC. The malvidin-3-glucoside concentration increased as the reaction time increased (22% and 32% increase after 5 and 80 minutes respectively) with a concomitant increase in red color, demonstrating that the malvidin-3-glucoside was released from the non-pigmented polymers.

The flavene and bicyclic forms (Figure 5) of the non-pigmented polymers (A-T type) are analogous to the B-type and A-type linkage of proanthocyanidin, respectively. In the case of proanthocyanidins, the B-type polymer is susceptible to acid-catalyzed cleavage while the A-type (bicyclic form) is resistant to cleavage (*29*). Remy-Tanneau et al (*15*) have provided structural confirmation for the presence of the bicyclic form of the non-pigmented (A-T type) dimers in wine. The increase in the total observed color in the present study can be explained by the release of the anthocyanin from the flavene type of the non-pigmented polymers (Figure 7), and therefore provides some evidence for this type of A-T type polymer. Definitive proof however is still needed.

Acknowledgements

We thank Dr M. J. Herderich, Dr E. J. Waters and Prof. P. B. Høj, of the Australian Wine Research Institute for their enthusiastic support. We also thank Rodger Kohnert of Oregon State University for NMR support. This project was supported financially by the Oregon Wine Advisory Board and Australia's grapegrowers and winemakers through their investment body the Grape and Wine Research and Development Corporation, with matching funds from Federal Government, and by the Commonwealth Cooperative Research Centre Program of Australia.

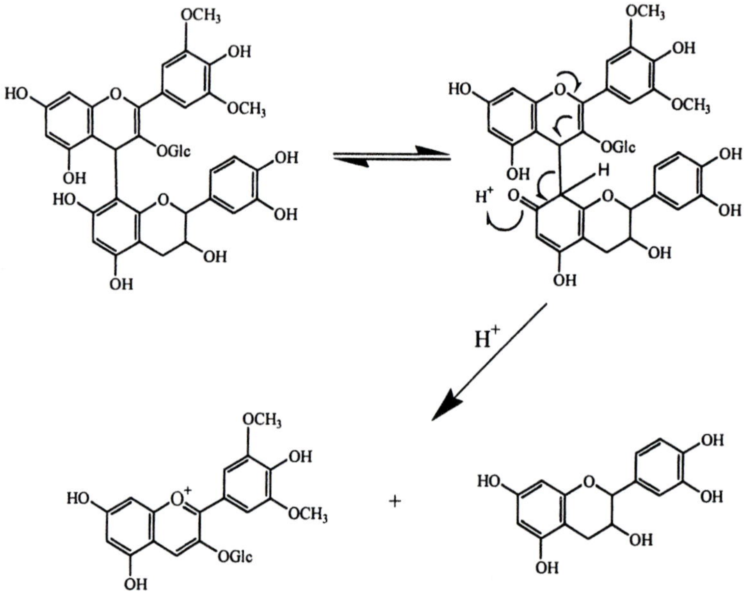

Figure 7. Proposed reaction mechanism explaining the increase in malvidin-3-glucoside under acid-catalyzed cleavage conditions.

References

1. Nagel, C.W.; Wulf, L.W. *Am. J. Enol. .Vitic.* **1979**, 30, 111-116.
2. Bakker, J.; Timberlake, C.F. *J. Agric. Food Chem.* **1997**, 45, 35-43.
3. Benabdeljalil, C.; Cheynier, V.; Fulcrand, H.; Hakiki, A.; Mosaddak, M.; Moutounet, M. *Sci. Aliment.* **2000**, 20, 203-220.
4. Fulcrand, H.; Benabdeljalil, C.; Rigaud, J.; Cheynier, V.; Moutounet, M. *Phytochemistry* **1998**, 47, 1401-1407.
5. Hayasaka, H.; Asenstorfer, R.E. *J. Agric. Food Chem.* **2002**, 50, 756-761.
6. Mateus, N.; Carvalho, E.; Carvalho, A.R.F.; Melo, A.; González-Paramás, A.M.; Santos-Buelga, C.; Silva, A.M.S.; de Frietas, V. *J. Agric. Food Chem.* **2003**, 51, 277-282.
7. Es-Safi, N.E.; Fulcrand, H.; Cheynier, V.; Moutounet, M. *J. Agric. Food Chem.* **1999**, 47, 2096-2102.
8. Dallas, C.; Ricardo-da-Silva, J. M.; Laureano, O. *J. Agric. Food Chem.* **1996,** 44, 2402-2407.
9. Fulcrand, H.; dos Santos, P.; Sarni-Manchado, P.; Cheynier, V.; Favre-Bonvin, J. *J. Chem. Soc., P1* **1996**, 735-739.
10. Somers T.C. *Phytochemistry* **1971**, 10, 2175-2186.
11. Jurd, L. *Am. J. Enol. Vitic.* **1969**, 20, 191-195.
12. Haslam, E. *Phytochemistry* **1980**, 19, 2577-2582.
13. Timberlake, C.F.; Bridle, P. *Am. J. Enol. Vitic.* **1976**, 27, 97-105.
14. Remy, S.; Fulcrand, H.; Labarbe, B.; Cheynier, V.; Moutounet, M. *J. Sci. Food Agric.* **2000**, 80, 745-751.
15. Remy-Tanneau, S.; Le Guerneve, C.; Meudec, E.; Cheynier, V. *J. Agric. Food Chem.* **2003**, 51, 3592-2597.
16. Hayasaka, Y.; Kennedy, J.A. *Austral. J. Grape Wine Res.* **2003**, 9, 210-220.
17. Kennedy, J.A.; Jones, G.P. *J. Agric. Food Chem.* **2001**, 49, 1740-1746.
18. Kennedy, J.A.; Taylor, A.W. *J. Chromatogr. A* **2003**, 995, 99-107.
19. Kennedy, J.A.; Hayasaka, Y.; Vidal, S.; Waters, E.J.; Jones, G.P. *J. Agric. Food Chem.* **2001**, 49, 5348-5355.
20. Mateus, N.; Silva, A.M.S.; Santos-Buelga, C.; Rivas-Gonzalo, J.C.; de Frietas, V. *J. Agric. Food Chem.* **2002**, 50, 2110-2116.
21. Berké, B.; Chèze, C.; Vercauteren, J.; Deffieux, G. *Tetrahedron Lett.* **1998**, 39, 5771-5774.
22. Hayasaka, Y.; Waters, E.J.; Cheynier, V.; Herderich, M.J.; Vidal, S. *Rapid Commun. in Mass Spectrom.* **2003**, 17, 9-16.
23. Revilla, I.; Pėrez-Magariňp, S.; González-SanJosė, M. L.; Beltrán, S. *J. Chromatogr. A* **1999**, 847, 83-90.
24. Karchesy, J.J.; Hemingway, R.W.; Foo, Y.; Barofsky, E.; Barofsky, D.F. *Anal. Chem.* **1986**, 58, 2563-2567.

25. de Pascual-Teresa, S.; Rivas-Gonzalo, J.C.; Santos-Buelga, C. *Int. J. Food Sci. Tech.* **2000**, 35, 33-40.
26. Liao, H.; Cai, Y.; Haslam, E. *J. Sci. Food Agric.* **1992**, 59, 299-305.
27. Santos-Buelga, C.; Bravo-Haro, S.; Rivas-Gonzalo, J.C. *Z. Lebensm. Unters. F. A.* **1995**, 201, 269-274.
28. Bishop, P.D.; Nagel, C.W. *J. Agric. Food Chem.* **1984**, 32, 1022-1026.
29. Foo, L.Y.; Lu, Y.; Howell, A.B.; Vorsa, N. *J. Nat. Prod.* **2000**, 63, 1225-1228.

Chapter 16

Tannin–Anthocyanin Interactions: Influence on Wine Color

Cédric Saucier, Paulo Lopes, Marie Mirabel, Célito Guerra, and Yves Glories

Laboratory of Applied Chemistry, faculty of Enology, University Victor Segalen Bordeaux 2, 351 Cours de la Liberation, 33405 Talence Cedex, France

Tannin and anthocyanin interaction were studied by UV-Visible spectrometry to understand both copigmentation phenomenon and polymerization reactions. The aim of the work was to conduct studies in wine like model solution (pH 3.2 and 0 or 12% Ethanol) to understand the role some parameters on color. The parameter studied were: Tannin size, presence of ethanol, acetaldehyde. The molecules studied were catechin, epicatechin and grape seed proanthocyanidins which were fractionated in monomeric, oligomeric, and polymeric fractions. Concerning copigmentation, tannins can a.: as copigments regardless of their mean molecular weight although monomers more efficient. Even if better copigments exist in wine, because of their relative high content, they are likely to play a major role in wine color. Concerning reactivity with anthocyanins, the relative influence of molecular weight and acetaldehyde presence were studied. Results show that the molecular weight has little influence on the kinetic of anthocyanin disappearance. In contrast the presence of acetaldehyde greatly increase the reaction kinetic, especially when both anthocyanin and tannin are present.

1. INTRODUCTION

Anthocyanins, extracted from grape skins, are mainly responsible for the color of young wines (Ribereau-Gayon 1964). In wines made from Vitis vinifera grapes, their initial concentrations vary from about 500 to 1200 mg/l and are monoglucosides (Figure 1a). Their chemical structures in aqueous media is influenced a lot by pH conditions (Brouillard and Dubois 1977, Brouillard and Delaporte 1977), and few anthocyanins are in the flavylium colored form at wine pH (Glories 1984). Hopefully other uncolored phenolic compounds present in the grape, such as proanthocyanidins (Figure 1b), are present and can enhance the anthocyanin colors and protect them from oxidation:

First, some non covalent interactions take place which modify the visible absorbance and modify the maximum wavelength of absorption. These phenomenon are grouped in the term copigmentation and exist in many colored fruit or flowers (Robinson and Robinson 1934, Asen et al. 1972). Copigmentation could be a first step in wine ageing before coupling reactions occur (Brouillard and Dangles 1994). In the case of red wine, many phenolic compound could act as effective copigment but there is still some debate about the possibility for condensed tannins (i.e proanthocyanidins) to be efficient (Mirabel *et al.* 1999, Boulton 2001).

Second, free anthocyanins are gradually disappearing to form new polymeric pigments (Somers 1971). Acetaldehyde, produced by yeasts or coupling oxidation of ethanol, is able to bridge the flavanols (catechins or tannins) together (Fulcrand *et. al.* 1996, Saucier *et al.* 1997) or with anthocyanins (Timberlake and Bridle 1976, Guerra *et al.* 1997a) to create new pigments. Some reactions involving anthocyanins and ketone or vinyl compounds have more recently focused the attention of researchers. These reactions lead to new pigments with a cyclisation reaction involving the carbon 4 of the anthocyanin and are called vitisins (Bakker and Timberlake 1997, Fulcrand *et al.* 1998, Mateus *et al.* 2003). Unfortunately, quantification of these compounds and reaction kinetics have not yet been studied so their relevance to wine color remains uncertain.

In this paper, we focus our attention to the effect of tannin on wine color by the two types of phenomenon: Copigmentation and reactions with anthocyanins in relation to the initial color of wine and its evolution.

2. Effect of tannins on anthocyanin copigmentation

The role of tannin in copigmentation in general, and in wine in particular, has been pointed out by researcher in early studies (Robinson and Robinson 1934, Ribereau-Gayon 1973). We report here some results obtained in experiments obtained in model solutions involving malvidin-3-glucoside, the

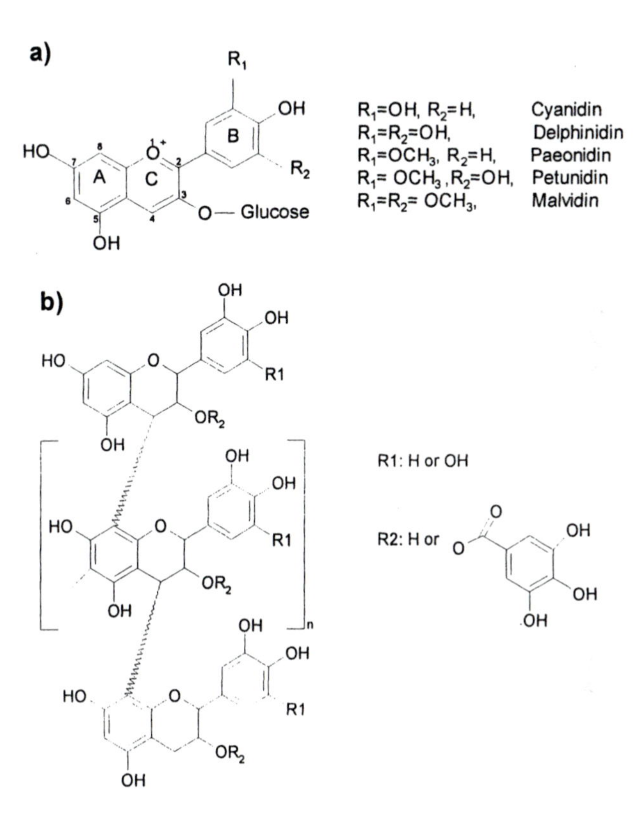

***Figure1.** General structure of anthocyanins (a) and proanthocyanidins (b) found in grapes*

major anthocyanin found in grape, and various tannins extracts. In a first series of experiments (Mirabel *et al.* 1999), an oligomeric pine bark extract, containing mainly oligomers but also monomers was used to study the copigment effect at various tannin/anthocyanin ratios and the effect of ethanol. The experiences were also conducted with epicatechin alone. The results obtained are shown in Table 1. The effect of ethanol is similar for the two type of compounds: The hyperchromic (augmentation of absorbance) is increased when the ethanol content or copigment/anthocyanin ratio (r) is increased. In both cases, the presence of ethanol reduce the bathocromic shift (augmentation of λmax).

These experiments shows that tannin could be at least as efficient as monomers in copigmentation. The hyperchromic effect seems to be even more pronounced for the tannins in presence of ethanol. In order to study the influence of polymerization degree of tannin on the hyperchromic effect, another experiment was more recently made in our lab. A grape seed extracts was fractionated into three classes (Saucier et al. 2001): Monomers, Oligomers, Polymers. These extracts were then placed in solution in the presence of malvidin and the hyperchromic effect was measured. The corresponding results are shown on table 2. They show that tannin are indeed able to enhance efficiently the color of anthocyanins. This effect is more pronounced when increasing molecular weight. However the control solutions containing the tannins alone are also colored at 523 nm so the copigmentation effect *stricto sensu* is much more effective with the monomeric fraction.

Table 1. Copigment effect on Malvidin-3-Glucoside in tartrate buffer (pH 3.2) solutions containg various amount of copigment (r is the copigment/malvidin ratio). A and λ are the maximum absorbance and corresponding wavelength in the presence of copigment (Ao and λo without).

		0% Ethanol		*12% Ethanol*	
Copigment	r	(A-Ao)/ Ao **X100**	λ-λo (nm)	(A-Ao) /Ao **X100**	λ-λo (nm)
Epicatechin	1	11	3	15	0
	3	42	6	41	3
	5	63	7	86	5
	10	82	11	143	8
Oligomeric tannin extract	1	33	2	31	2
	3	53	4	80	3
	5	61	7	127	4
	10	75	12	215	5

Table 2. Hyperchromic effect (measured at 523 nm) of grape seed tannin fraction on Malvidin-3-Glucoside in tartrate buffer (pH 3.5). mDP is the mean degree of polymerization measured by thioacidolysis. r the molar ratio between tannin and anthocyanin (a fixed concentration of 1 g/l of tannin was used). T is the absorbance of tannin solution (control) at 523nm.

Fraction	mDP	r	(A-Ao)/Ao x100	T/Ao x100
Monomers	1.1	11.3	62.5	25.2
Oligomers	4.0	2.8	77.0	71.9
Polymers	10.0	1.1	92.8	74.9

In fact all these phenomena (copigmentation, color addition) are certainly taking place in wine and the human eyes will detect the sum of these absorbances. In order to have a easy method for measuring a global copigmentation effect in wines, we propose the following index:

$$Icopig = C.I / o.d\ 280$$

Where C.I, the color intensity is the sum of absorbance at 420, 520, 620 nm, and o.d 280 is the absorbance measured at 280nm (in a 1cm quartz cell) after a one hundred fold dilution. This index is in fact measuring the coloring efficiency of the phenolic compounds present in red wines.

3. Effect of tannins in new pigments formation

During ageing, the anthocyanins can react with other compounds produced from the yeast like pyruvic acid (Fulcrand *et al.* 1998), acetaldehyde (Es-Safi *et al.* 1999) or other aldehydic compounds such as furfural and hydroxymethyl furfural coming from oak wood (Es-Safi *et al.* 2002). In order to assess the possibility for acetaldehyde to be produced by oxidation of ethanol (Wildenradt and Singleton 1974) and to study the structures of compounds formed, the model solution previously used for copigmentation studies (Mirabel *et al.* 1999) were stored for ten weeks in the dark at ambient temperature. Two model solutions (pH3.75) containing 0 or 12% Ethanol were analyzed by LC/MS after 10 weeks of incubation.

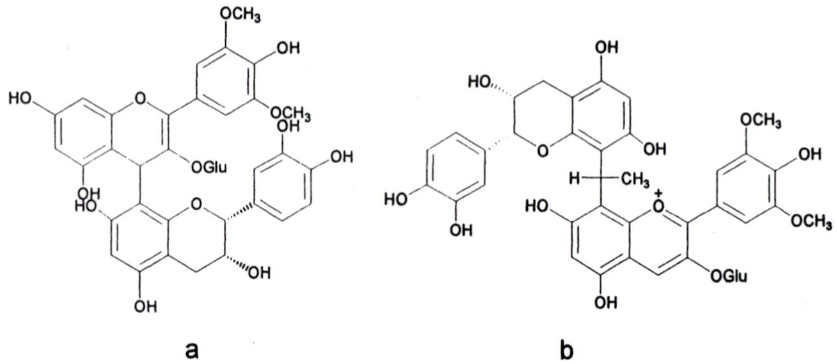

a

b

Figure 2. Possible structures for compounds formed in model solution containing (-)-epicatechin and mavlvidin. Compound A was formed in solutions containing 0 or 12% ethanol whereas compound B was formed only in the ethanol containing solution.

In both solution, a compound with m/z=783 was detected. This correspond to a mass of 782 (positive mode was used) consistent to an Anthocyianin-Tannin adduct with a direct carbon-carbon linkage (Figure 2a) and the anthocyanin in the uncolored flavene form. Such compounds have been detected in a two year old red wine (Remy *et al.* 2000). In the solution containing ethanol a compound with m/z=809 was detected. This is consistent with an epicatechin connected to the anthocyanin *via* an ethyl bridge. The proposed structure (Figure 2b) involve both carbon 8 from the two units. This has been demonstrated in our lab for catechin and malvidin-3-Glucoside (1997b) in accordance with similar reactions (Escribano-bailon *et al.* 1996).

In order to study the influence of tannin length in the acetaldehyde on the kinetics of bridging reaction, we used the different grape seed tannin fraction (Table 2 and incubated them with malvidin 3-glucoside in presence or absence of a relative excess of acetaldehyde. We followed then the disappearance of anthocyanin by HPLC. The kinetics observed were assimilated to first order kinetics. The corresponding rates were then measured and are reported in table 3.

The kinetics parameters obtained show that the presence of tannin increases the kinetic of reaction and so does the presence of acetaldehyde. The molecular weight of the tannin has very little influence on the kinetics compared to the presence of acetaldehyde. These preliminary results should be completed in the future by varying the different factors independently to determine the key factors on the kinetics.

Table 3. Reaction kinetics obtained in the reaction of malvidin 3-glucoside (150 mg/l) in presence or absence of arelative excess of acetaldehyde. (100 mg/l) and tannins (1g/l). All reaction were performed at pH 3.5 with 12% Ethanol at 20°C in the dark. Coding: MV3G:malvidin 3-glucoside, Mon: Monomers, Olig: Oligomers, Poly:Polymers, acet: Acetaldehyde.

Compounds	k ($days^{-1}$)	r^2 ($C=Coe^{-kt}+B$)
Mv3G	0,016	0,655
Mv3G-Mon	0,029	0,939
Mv3G-Olig	0,030	0,976
Mv3G-Poly	0,027	0,957
Mv3G-acet-Mon	0,125	0,993
Mv3G-acet-Olig	0,122	0,945
Mv3G-acet-Poly	0,078	0,927

4. Conclusion

Tannin are key compounds involved in red wine color. Even if better copigment exist in wine, they are still effective and their abundance compared to other phenolic compounds is likely to play in their favor. Their oxidized forms clearly also have a direct impact on red wine color and there is un urgent need to understand their chemical structures and physical properties.

Concerning the reactions between tannin and anthocyanins, several pathways involving direct coupling or bridging via ketone compounds have been elucidated by the researchers. However, few kinetics studies have been undertaken to assess which are the key reactions. The influence of these compounds on wine color has yet to be fully understood and compared to the compounds involving anthocyanins and not tannins such as vitisins. There is hopefully a lot of mysteries that remains in the glass of red wine for the pleasure of researchers and oenophiles.

Literature Cited

Asen S., Stewart R.N., Norris K.H. 1972. Co-pigmentation of anthocyanins in plant tissues and ist effect on color. *Phytochemistry*. 11, 1139-1144.

Bakker and Timberlake 1997. Isolation, identification, and characterisation of new color-stable anthocyanins occurring in some red wine. *J. Agric. Food Chem.*, 45, 35-43.

Boulton R. 2001. The copigmentation of anthocyanins and its role in the color of red wine: a critical review. *Am. J. Enol. Vitic.* 52, 67-87.

Brouillard R., Dubois J.E. 1977. Mechanism of the structural transformations of anthocyanins in aqueous media. *J. Am. Chem. Soc.* 99, 1359-1366.

Brouillard R., Delaporte B. 1977. Chemistry of anthocyanin pigments. 2: Kinetic and thermodynamic study of proton transfer, hydratation and tautomeric reactions of malvidin-3-O-glucoside. *J. Am. Chem. Soc.* 99, 8461-8467.

Brouillard R., Dangles O. 1994. Anthocyanin molecular interactions : The first step in the formation of new pigments during ageing ? *Food chem.* 51, 365-371.

Es-Safi N.E, Fulcrand H., Cheynier V., Moutounet M. 1999. On the acetaldehyde-induced condensation of (-)-epicatechin and malvidin 3-O-glucoside in a model solution system. *J Agric Food Chem.* 47, 2096-102.

Es-Safi NE, Cheynier V, Moutounet M. 2002. Role of aldehydic derivatives in the condensation of phenolic compounds with emphasis on the sensorial properties of fruit-derived foods. *J Agric Food Chem.* 50, 5571-5585.

Escribano-Bailon T., Dangles O., Brouillard R. 1996. Coupling reaction between flavylium ions and catechin. *Phytochemistry.* 41, 1583-1592.

Fulcrand H., Doco T., Es Safi N. et Cheynier V. 1996. Study of the acetaldehyde induced polymerisation of flavan-3-ol by liquid chromatography-Ion spray mass spectrometry. *J. Chromatogr.* 752, 85-91.

Fulcrand H., Benabdeljalil C., Rigaud J., Cheynier V., Moutounet M. 1998. A new class of wine pigments generated by reation between pyruvic acid and grape anthocyanins. *Phytochemistry.* 47, 1401-1407.

Glories Y. 1984. La couleur des vins rouges. I – Les équilibres des anthocyanes et des tanins. *Conn. Vigne Vin.* 18, 195-217.

Guerra C., Saucier C., Bourgeois G., Vitry C., Busto O., Glories Y. 1997a. Partial characterization of coloured polymers of flavan-3-ols-anthocyanins by mass spectrometry. 1 st Symposium "In vino analytica scientia". Soc. Fr. Chim (Ed). 124-127

Guerra C. 1997b. Recherches sur les interactions anthocyanes-Flavanols : Application à l'interpretation chimique de la couleur des vins rouges. *PhD Thesis.* University Victor Segalen Bordeaux2. n°502.

Mateus N., Silva M.S.A., Rivas gonzalo C.J., Santos-Buelga C., de Freitas V. 2003. A new class of blue anthocyanin-derived pigments isolated from res wines. *J. Agric. Food Chem.*, 51, 1919-1923.

Mirabel M., Saucier C., Guerra C., Glories Y. 1999. Copigmentation in model wine solution: occurrence and relation to wine ageing. *Am. J. Enol. Vitic.* 50, 211-218.

Remy S., Fulcrand H., Labarbe B., Cheynier V., Moutounet M. 2000. First confirmation in red wine of products resulting from direct anthocyanin-tannin reactions. *J. Sci Food Agr.* 80, 6745-751.

Ribereau-Gayon P. 1964. Les composés phénoliques du raisins et du vin. II. Les flavonosides et les anthocyanosides. *Ann. Physio. Veg.* 6, 211-242.

Ribereau-Gayon P. 1973. Interpretation chimique de la couleur des vins rouges. *Vitis.* 12, 119-142.

Robinson G.M., Robinson R. 1931. A survey of anthocyanins. *Biochem. J.* 25, 1687-1705.

Saucier C. Little D. et Glories Y. 1997. First evidence of acetaldehyde-flavanol condensation products in red wine. *Am. J. Enol. Vitic.48,3,369-373.*

Saucier C., Mirabel M., Daviaud F., Longieras A., Glories Y. 2001. Rapid fractionation of grape seed proanthocyanidins. *J. Agric. Food Chem.* 49, 5732-5735.

Somers T.C. 1971. The polymeric nature of wine pigments. *Phytochemistry,* 10, 2175-2186.

Timberlake C.F Briddle.P. 1976. Interactions between anthocyanins phenolic compounds and acetaldehyde , and their significance in red wines. *Am.J.Enol.Vitic.* 27, 97-105.

Wildenradt H.L., Singleton V.L. 1974. The productions of aldehydes as a result of oxidation of polyphenolic compounds and its relation to wine aging. *Am.J.Enol.Vitic.* 25, 119-126.

Chapter 17

Fractionation of Red Wine Polymeric Pigments by Protein Precipitation and Bisulfite Bleaching

Douglas O. Adams, James F. Harbertson, and Edward A. Picciotto

Department of Viticulture and Enology, One Shields Avenue, University of California, Davis, CA 95616

Bovine serum albumin (BSA) precipitates tannins from red wine and also removes some of the red pigments. The pigments that bind to BSA are not released from the precipitate by washing and they are stable in the presence of bisulfite. Together these observations suggest that the pigments removed from wine by BSA precipitation are polymeric pigments. The pigments removed from wines by BSA do not account for all of the polymeric pigments in the wine. After removal of the precipitated pigments by centrifugation the supernatant fraction still contains pigments that are stable to bisulfite bleaching. Thus, protein precipitation fractionates the polymeric pigments into two distinguishable classes; large polymeric pigments (LPP) that precipitate along with the tannins, and small polymeric pigments (SPP) that do not. The number that best expresses the relative amounts of the two classes of polymeric pigment distinguished by protein precipitation is the LPP/SPP ratio. This ratio was found to be highly variable in 454 commercial red wines and could vary by more than a factor of 20 even in wines from a single variety (Cabernet Sauvignon). Composition of must and conditions during fermentation favor formation of LPP compared to SPP, and during barrel aging LPP is also preferentially formed compared to SPP.

Introduction

For the past few years we have been using a protein precipitation assay developed by Hagerman and Butler to study tannin development in skins and seeds of grape berries during ripening (*1,2*). The procedure is shown in the polygon in Figure 1 and uses bovine serum albumin (BSA) as the protein in the precipitation step. We scaled the analysis down and adapted it for use with grape extracts, and also used it to measure tannin in wines. In doing so we always encountered a 510 nm absorbance when the protein/tannin pellet was resuspended in TEA/SDS buffer prior to the addition of ferric chloride. The background absorbance was observed in wine samples and skin extracts where anthocyanins are present along with tannins, but was absent from seed extracts. This background absorbance had to be subtracted from the final absorbance obtained after ferric chloride addition, and we soon recognized that the background absorbance was generally larger in wines than in grape extracts. Because of the well known phenomena of polymeric pigment formation during winemaking (*3*) we set out to determine if this background absorbance could be used as a direct measure of polymeric pigments in wine and grape extracts. In the course of investigating this possibility we found that protein precipitation fractionates the polymeric pigments into two classes, those that precipitate with BSA and those that do not. Because our previous studies indicated that procyanidin dimers and trimers do not precipitate with BSA whereas higher oligomeric forms do (*4*), we have designated the polymeric pigments that precipitate with protein as large polymeric pigments (LPP) and those that do not as small polymeric pigments (SPP).

Materials and Methods

Bovine serum albumin (BSA, Fraction V powder), sodium dodecyl sulfate (SDS; lauryl sulfate, sodium salt), triethanolamine (TEA), ferric chloride hexahydrate, potassium metabisulfite and (+)-catechin were purchased from Sigma, St. Louis MO, as were all of the reagents used for preparing buffers.

Analysis of Polymeric Pigments

The Hagerman and Butler method for tannin analysis was combined with bisulfite bleaching of monomeric anthocyanins to give estimates of polymeric pigments in grape-skin extracts and wines (Figure *1*). A model wine consisting of 12% aqueous ethanol (v/v) containing 5 g/L potassium bitartrate (pH 3.3) was used to dilute wines or aqueous extracts prior to analysis. Assay of polymeric pigments by bisulfite bleaching, and precipitation of tannins and polymeric pigments were both conducted in a buffer containing 200 mM acetic acid and 170 mM NaCl (pH 4.9). The tannin precipitation reaction was carried out in this buffer by including bovine serum albumin (BSA) to give a final protein concentration of 1 mg/mL.

Analysis of polymeric pigments in parallel with tannin in wines or grape extracts required two 1.5 mL microfuge tubes for each sample. The first tube was made up by adding 1 mL of the acetic acid/NaCl buffer to the tube and then adding 500 μL of the diluted skin extract or wine. One mL of the mixture was transferred to a cuvette and the absorbance at 520 nm was determined (reading A). Then 80 μL of 0.36M potassium metabisulfite was added, and the absorbance at 520 nm was re-determined after a 10 minute incubation (reading B). From this tube the absorbance due to monomeric anthocyanin could be determined (A-B) where reading B represents the total amount of polymeric pigment (SPP+LPP).

The second tube contained 1mL of the acetic acid/NaCl buffer along with BSA (1mg/mL) into which 500 μL of the diluted skin extract or wine was added. The mixture was allowed to stand at room temperature for 15 minutes with slow agitation, after which the sample was centrifuged for 5 minutes at 13,500g to pellet the tannin-protein precipitate. One mL of the supernatant was transferred to a cuvette, then 80 μL of 0.36M potassium metabisulfite was added. After a 10 minute incubation the absorbance was determined at 520 nm (reading C). This absorbance represents polymeric pigment that did not precipitate along with the tannin and protein (SPP), and this value was used to calculate the amount of polymeric pigment that precipitated with the tannin and protein (B-C).

Background Absorbance in the Analysis of Tannin

The tannin-protein pellet from the second tube described above was washed with 250 μL of the acetic acid /NaCl buffer to remove residual monomeric anthocyanins. The precipitate was re-centrifuged for 1 minute at 13,500g and the wash solution was discarded. Then 875 μL of a buffer containing 5% TEA (v/v) and 5% SDS (w/v) was added and the tube was allowed to stand at room temperature for 10 minutes. This buffer dissolves the precipitate containing

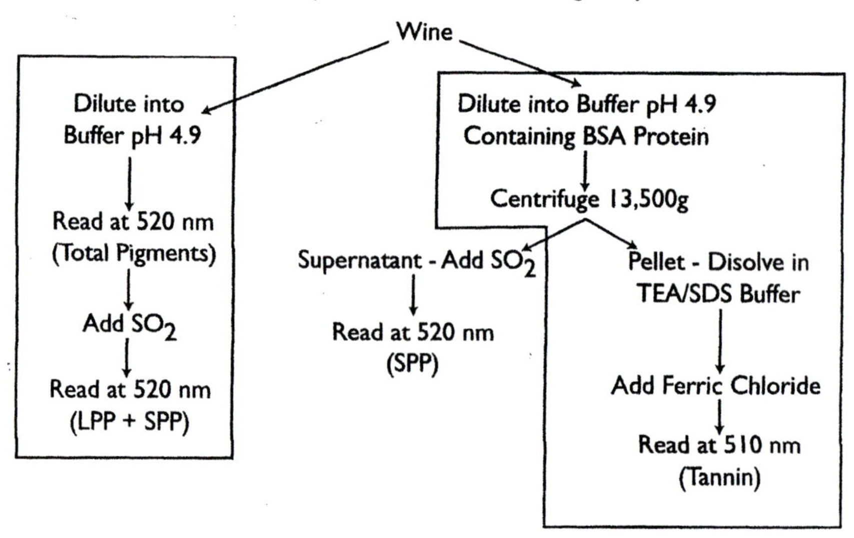

Figure 1. Parallel Determination of Tannins and Polymeric Pigments. The procedure shown in the rectangle (left) is a Somers assay performed at pH 4.9 and provides a measure of total pigments and total polymeric pigments (LPP + SPP) at that pH. The procedure shown in the polygon (right) is the Hagerman Butler assay for tannin. © 2003 American Society for Enology and Viticulture. AJEV 54:301-306

tannin, protein and any polymeric pigment that precipitated with the tannin and protein. After incubation the tube was vortexed to dissolve any of the remaining precipitate and the absorbance at 510 nm was measured after allowing the solution to stand at room temperature for 10 minutes. This is the background absorbance in the tannin assay we described previously (*1*). For tannin analysis 125 µL of 10 mM ferric chloride was added and the absorbance at 510 was re-determined after 10 minutes.

Berry Extraction

Cabernet Sauvignon berries were collected near Oakville California from eight-year-old vines planted on 110 Richter rootstock. Pinot noir fruit was obtained from the Carneros region of Southern Napa County from two six-year-old vineyards (Dijon 115) planted on 3309 rootstock. Syrah berries used to determine how polymeric pigments changed as fruit ripened were gathered from a five-year-old vineyard near Esparto California planted on 110 Richter rootstock.

Sample collection was conducted as described previously (*1*). Briefly, three twenty-berry samples were collected, put into plastic bags and transported to the laboratory on ice. The samples were weighed, and the skins were removed and extracted for polymeric pigment analysis. Berries were sliced in half with a razor blade and skin was carefully collected from each berry-half using a small metal spatula. Skins from the twenty-berry sample were put into a 125 mL Erlenmeyer flask containing 20 mL of 70% aqueous acetone (v/v). The flasks were sealed with a rubber serum cap and extracted overnight with gentle shaking (100 rpm).

After overnight incubation the extraction solution was filtered and the acetone was removed at 38° C using a rotary evaporator at reduced pressure. The residual aqueous extract was adjusted to 10 mL with deionized water and frozen at -20° C until used for analysis.

Wines for Total Polymeric Pigment Comparison

Twelve commercial 1998 red wines that were part of a different study were used to study the correlation between polymeric pigments as determined by the method of Somers and Evans, and the polymeric pigments (background absorbance) in the protein-tannin precipitate. The twelve commercial wines were from nine different wineries; nine of the wines were Cabernet Sauvignon, two were Merlot and one was Petite Sirah.

For comparing polymeric pigments in fruit and the resulting wine we collected three 20-berry samples from picking bins and extracted skins in 70 % aqueous acetone as described above. We measured polymeric pigments in the grape skin extracts and then analyzed the resulting wines 90 days after pressing.

Results

In order to determine if the background absorbance in our tannin assay could be used as a direct measurement of polymeric pigment, we compared the background absorbance to the polymeric pigment values measured by the Somers method in a set of wines. Figure 2 shows the correlation between the background absorbance and polymeric pigments as determined by the Somers assay, and clearly demonstrates that the measurements are poorly correlated.

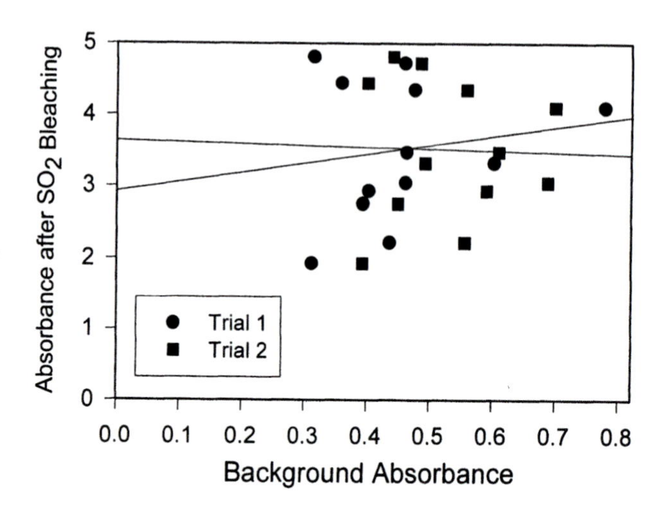

Figure 2. Absence of a correlation between the background absorbance in the tannin assay and the amount of polymeric pigment in 12 wines as determined in the standard Somers assay at wine pH. The analysis was performed twice about eight months apart. © 2003 American Society for Enology and Viticulture. AJEV 54:301-306

Absorbance measurements indicated that the reason for the poor correlation was that some of the polymeric pigments were not precipitated by protein and we set out to measure this fraction of polymeric pigments directly. We did this by first precipitating tannin and the accompanying polymeric pigments with BSA under the conditions of the Hagerman Butler tannin assay (pH 4.9, acetate buffer) so as to obtain maximum precipitation of protein along with tannin and polymeric pigment. After removal of the precipitate by centrifugation, we used bisulfite to bleach any monomeric anthocyanins remaining in the supernatant. The residual absorbance at 520nm represents polymeric pigment in the sense that it does not bleach with bisulfite. The absorbance due to the monomeric anthocyanins was taken to be the amount of 520 nm absorbance that was bleached with bisulfite. The results of one experiment to directly observe SPP is shown in Figure 3.

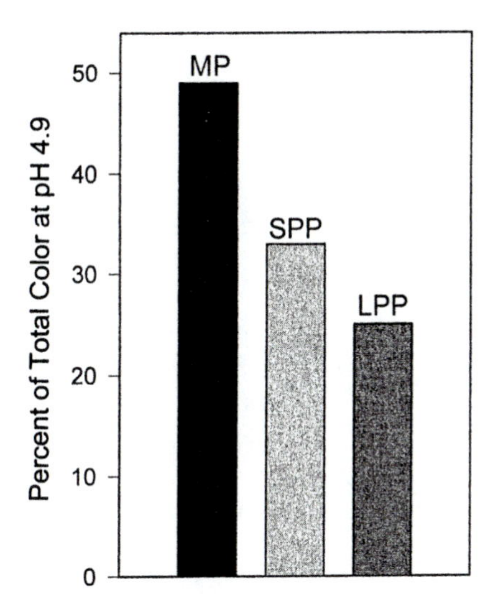

Figure 3. Percentage of the total color contributed by the three classes of pigments found in a Syrah wine shortly after pressing. MP, monomeric pigments (anthocyanins); SPP, small polymeric pigments; LPP, large polymeric pigments.

The wine was a Syrah obtained soon after pressing and we could attribute over 30 percent of the total color at pH 4.9 to polymeric pigments that did not precipitate with protein. When the wines shown in Figure 2 were re-examined and the amount of LPP and SPP were added together, the correlation with polymeric pigments by the Somers assay was found to be quite good (Figure 4).

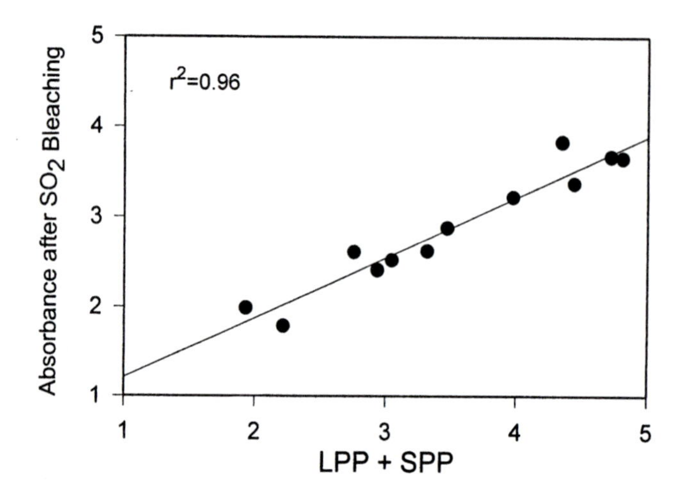

Figure 4. Correlation between the amount of polymeric pigment as determined in the Somers assay at wine pH and the sum of LPP and SPP as determined by combined protein precipitation and bisulfite bleaching. © 2003 American Society for Enology and Viticulture. AJEV 54:301-306

This is not surprising since the way in which total pigments and total polymeric pigments (LPP +SPP) are determined is nothing more than a Somers assay at pH 4.9 (Figure 1). From these results (Figures 3,4) we concluded that protein precipitation fractionates polymeric pigments into two classes, those that precipitate with protein and those that do not.

Our next objective was to determine if polymeric pigments were present in fruit during ripening. The results obtained with Syrah are shown in Figure 5. Since analysis of the grape extracts was conducted at pH 4.9 the contribution of monomeric anthocyanins to the total color is minimized and thus is greatly underestimated in Figure 5. However, at this pH the polymeric pigments are most easily observed because they change very little in absorbance with pH, unlike the monomeric anthocyanins (*3,5*). The data indicate that fruit at harvest may contain small amounts of polymeric pigment and that most of it is SPP.

After finding that protein precipitation fractionated polymeric pigments, we recognized that the ratio of large to small polymeric pigments must be different in the wines used in Figure 2 and Figure 4, otherwise a good correlation would have been observed in both cases. Recognizing that the ratio of LPP/SPP was variable in wines prompted us to measure the ratio in fruit at harvest and compare that with the ratio in the resulting wine.

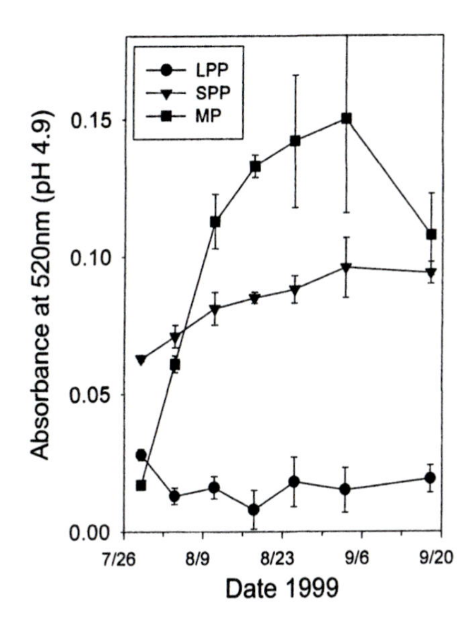

Figure 5. Pigments in Syrah Berries During Ripening Determined at pH 4.9 The color contributed by monomers is underrepresented because at this pH very little of the monomeric pigments are in the flavillium form. © 2003 American Society for Enology and Viticulture. AJEV 54:301-306

The results for four wines made from three different grape varieties are shown in Table I. In these four examples the absolute amount of LPP and SPP both increased (data not shown). However, since LPP increased more than SPP the ratio showed an increase of nearly two fold from fruit to wine in the case of Cabernet Sauvignon, to over seven fold in one of the Pinot noir experiments.

Since we found that even a limited set of wines could exhibit a large range of LPP/SPP values we wished to characterize this range in a larger set of finished wines. Table II shows the range and average of LPP/SPP values for 454 commercial bottled wines of Zinfandel, Pinot noir, Cabernet Sauvignon and Syrah, all of various ages and origins.

Table I. LPP/SPP Ratio in Fruit of Four Grape Varieties Compared to Wine Made from the Fruit

Grape Variety	*LPP/SPP in Fruit*	*LPP/SPP in Wine*	*Fold Increase*
Zinfandel	0.37	0.62	1.8
Pinot noir	0.18	0.64	3.6
Cabernet Sauvignon	0.24	1.73	7.2
Syrah	0.20	1.14	5.7

NOTE: The ratio was measured in fruit at harvest and in the resulting wine 90 days after pressing.

Table II. Analysis of the LPP/SPP Ratio in Wines of Four Grape Varieties.

Variety	*N*	*Min.*	*Max.*	*Average*	*Std. Dev.*
Zinfandel	200	0.08	7.19	1.10	0.93
Pinot Noir	134	0.14	2.20	0.79	0.39
Cabernet Sauvignon	87	0.18	3.93	1.16	0.63
Syrah	33	0.28	1.71	0.85	0.30

NOTE: All wines were commercial products. N, number of wines analyzed; Min., minimum ratio; Max, maximum ratio; Std. Dev., standard deviation of the Average.

Figure 6 shows the frequency distribution of LPP/SPP values in 85 of the commercial Cabernet Sauvignon wines. The distribution of the Cabernet Sauvignon wines about the mean is fairly normal, but the range of values was surprising, over 20 fold in the set of Cabernet Sauvignon wines (Table II).

The large increase in the LPP/SPP ratio we observed between fruit and the resulting wine 90 days after pressing (Table I) prompted us to study the effects of winemaking practices and aging on LPP and SPP levels. Thus far we have found that fermentation temperature is one of the most important factors that influence LPP and SPP formation during winemaking. Our first indication of this came from interrogating a fermentation temperature experiment performed in David Block's lab (Department of Viticulture & Enology, University of California, Davis). In this experiment Cabernet Sauvignon from a common must-pool was fermented at different temperatures. We measured the amount of LPP, SPP and monomeric pigment (at pH 4.9) in wines from a 17° C and a 27° C fermentation at bottling and again after three years of aging. The results are shown in Figure 7. From these data we can see the effects of both fermentation temperature and bottle aging on LPP and SPP formation. In the year the wines were made (1999) the one from the 27° C fermentation had more

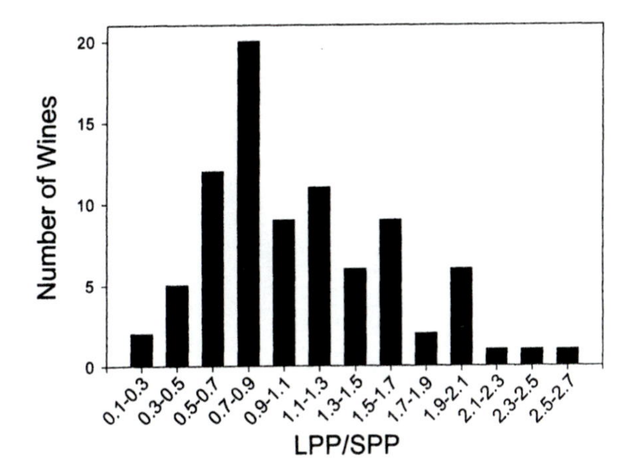

Figure 6. Frequency distribution of LPP/SPP ratios in 85 commercial Cabernet Sauvignon wines.

than twice as much LPP as the wine fermented at 17° C, whereas SPP was only about 60% greater in the wine from the 27° C fermentation. This observation is consistent with commercial scale experiments that we have monitored during the past two years (data not shown). The effect of bottle aging on LPP and SPP formation in these two wines can also be seen from the results in Figure 7 by comparing the pigment composition in the wine at bottling with the composition three years later. The data show that LPP was preferentially formed compared to SPP during the three years of bottle aging, and that the amount of LPP increased about three-fold in both wines during this period.

Discussion

The routine analysis of polymeric pigments in wine is based on the work of Somers who showed that monomeric anthocyanins are bleached with bisulfite whereas polymeric pigments are not (*3,5*). Thus, the difference in absorbance readings at 520nm before and after bleaching with bisulfite is widely used as a measure of the amount of polymeric pigment present in red wine.

Figure 1 shows the scheme we currently use for the parallel determination of tannin and polymeric pigments in wines and berry extracts. The boxed branch on the left is a Somers assay at pH 4.9. This pH was chosen so that the bisulfite bleaching reaction would take place in the same buffer system used in the Hagerman Butler assay for tannin.

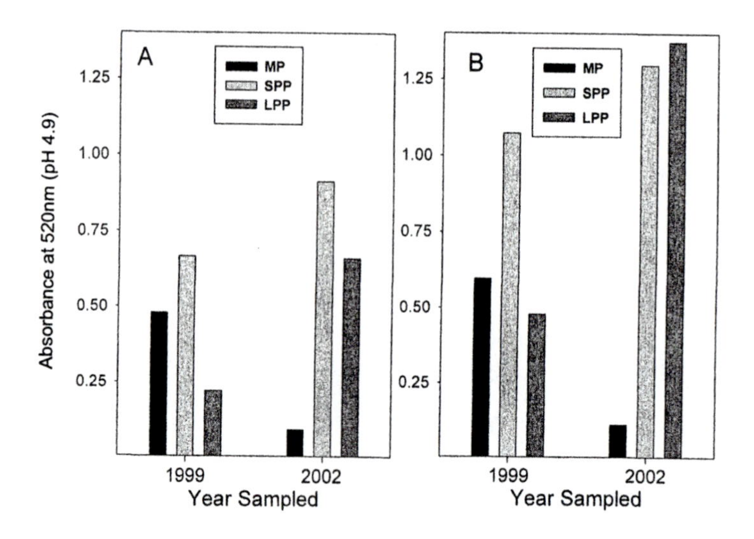

Figure 7. Pigment composition of Cabernet Sauvignon wines fermented at 17° C (7A) and 27° C (7B). The composition of each wine was determined at bottling and after three years of bottle aging. MP, monomeric pigment; SPP, small polymeric pigment; LPP, large polymeric pigment. Monomeric pigment declined during aging whereas SPP and LPP increased.

The acetate buffer is set at pH 4.9 because 4.9 is the pI of BSA and thus affords maximum precipitation of the tannin/protein complexes formed during the precipitation reaction. Coincidentally pH 4.9 is particularly good for measuring absorbance due to polymeric pigments because the anthocyanins have their minimum absorbance at that pH (*6*). Since any remaining monomeric anthocyanins are bleached with bisulfite, this procedure assures that all of the remaining 520 nm absorbance is due to polymeric pigments (LPP + SPP).

The procedure in the polygon in Figure 1 is the Hagerman Butler assay for tannin. The only modification is that we retain the supernatant after centrifugation of the precipitation reaction mixture and bleach the monomeric anthocyanins with bisulfite. The residual absorbance represents pigments that do not precipitate with protein and do not bleach in the presence of bisulfite. We have designated this material "small polymeric pigment" solely to denote these two characteristics. The LPP removed from the sample by protein precipitation is responsible for the background absorbance in the tannin assay. However, since LPP is only a variable fraction of the total polymeric pigment, the background absorbance is poorly correlated with the total polymeric pigment as determined by the Somers assay (Figure 2). When both large and small polymeric pigments are taken into account the correlation is seen to be very good (Figure 4).

We recognize that the small polymeric pigment fraction is likely to be a heterogeneous mixture containing anthocyanins as acetaldehyde cross linked reaction products, direct reaction products, and cycloaddition products, in various proportions. The LPP fraction probably contains acetaldehyde cross linked reaction products and direct reaction products, where anthocyanin has reacted to produce a pigmented molecule "large" enough to precipitate with protein; thus the designation "large polymeric pigment".

The number that best expresses the relative amounts of the two classes of polymeric pigment distinguished by protein precipitation is LPP/SPP. We chose to express the relative amounts of LPP and SPP as the LPP/SPP ratio rather than SPP/LPP ratio, because we found that LPP usually increases relative to SPP during winemaking and aging (e.g. Figure 7). In the varieties we have studied thus far this ratio is typically very low in acetone extracts of berry skins at harvest, but shows a consistent increase during winemaking and early barrel aging (Table I). Syrah fruit at harvest had little polymeric pigment and most of what was present was SPP (Figure 5). Thus the increase in the LPP/SPP ratio during winemaking and aging is a result of LPP formation rather than a decline in the amount of SPP. This is most easily seen in the bottle aging experiment shown in Figure 7, where the amount of LPP in the wine increased nearly 3 fold in 3 years while the amount of SPP increased by less than 25 percent.

We found that commercial wines of four varieties exhibited an extraordinary range of LPP/SPP values (Figure 6 and Table II). This suggests that polymeric pigment populations are quite different even in wines made from the same variety. That is to say, if protein precipitation did nothing more than fractionate similar populations of polymeric pigments, then the ratio of LPP/SPP would be similar among red wines. In fact the LPP/SPP ratio shows remarkable variation even among wines of the same variety (Table II). Winemaking conditions and the age of the wine when assayed clearly are factors in explaining the large range of values. Nevertheless, as we study more examples of polymeric pigment evolution in wines during aging (such as shown in Figure 7) we should be able to identify other variables that contribute to the wide range of LPP/SPP values seen in finished wines.

References

1. Harbertson, J. F.; Kennedy, J. A.; Adams, D.O. Tannin in skins and seeds of Cabernet Sauvignon, Syrah, and Pinot noir berries during ripening Am. J. Enol. Vitic. **2002**, 53, 54-59.

2. Hagerman, A. E.; Butler L. G. Protein precipitation method for the quantitative determination of tannins. J. Agric. Food Chem. **1978**, 26, 809-812.

3. Somers, T. C. Polymeric nature of wine pigments. Phytochemistry. **1971**, 10, 2175-2186.

4. Adams, D. O.; Harbertson J. F. Use of alkaline phosphatase for analysis of tannins in grapes and red wines. Am. J. Enol. Vitic. **1999**, 50, 247-252.

5. Somers, T. C.; Evans M. E. Spectral evaluation of young red wines: anthocyanin equilibria, total phenolics, free and molecular SO_2, "chemical age". J. Sci. Food Agric. **1977**, 28, 279-287.

6. Cabrita, L.; Fossen T.; Andersen O. M. Colour stability of the six common anthocyanidin 3- glucosides in aqueous solutions. Food Chem. **2000**, 68,101-107.

Indexes

Author Index

Subject Index

B

C

D

G

N

O

P

S

V

W

X

Y

Z